QUELEA QUELEA
AFRICA'S BIRD PEST

Frontispiece The Red-billed Quelea *Quelea quelea* is one of the most destructive agricultural pests in Africa (photo: J. Jackson).

Quelea quelea

Africa's Bird Pest

EDITED BY

RICHARD L. BRUGGERS
International Programs Research Section
US Department of Agriculture
Denver Wildlife Research Center

AND

CLIVE C. H. ELLIOTT
Food and Agriculture Organization
of the United Nations

OXFORD NEW YORK TOKYO
OXFORD UNIVERSITY PRESS
1989

Oxford University Press, Walton Street, Oxford OX2 6DP
Oxford New York Toronto
Delhi Bombay Calcutta Madras Karachi
Petaling Jaya Singapore Hong Kong Tokyo
Nairobi Dar es Salaam Cape Town
Melbourne Auckland
and associated companies in
Berlin Ibadan

Oxford is a trade mark of Oxford University Press

Published in the United States
by Oxford University Press, New York

British Library Cataloguing in Publication Data
Quelea quelea
1. Africa, crops, pests, quelea
1. Bruggers, Richard L. II. Elliott,
Clive C.H.
632.6'8
ISBN 0–19–857607–2

Library of Congress Cataloging in Publication Data
Quelea quelea.
Bibliography: p.
Includes index.
1. Quelea quelea—Africa. 2. Quelea quelea—
Control—Africa. I. Bruggers, Richard L. II. Elliott,
Clive C.H.
SB996.Q44Q44 1989 632'.6873 88–33022
ISBN 0–19–857607–2

Set by Latimer Trend & Company Ltd

Printed in Great Britain
by Butler & Tanner Ltd, Frome, Somerset

This book is dedicated to the memory of the late Dr Peter Ward. His initial research and insight into quelea biology and management formed the basis for so much of the work on quelea described in this book. His infectious but highly critical enthusiasm for quelea studies personally influenced almost all authors; it was a tragedy that his life was prematurely cut short.

WAINA CHENG WARD
Ali Meleri.
NIGERIA. 1973

Foreword

Since time immemorial, cereal crops in Africa have been attacked by grain-eating birds. The damage, varying as it does according to ecological and environmental factors, can be a cruel hardship for the traditional farming population. Dispersed subsistence agriculture is the most severely affected by quelea, especially when it is near plentiful water supplies which are available as drinking places for the birds. Sometimes bird pressure on crops can force out local farming populations. Often fertile land has to be abandoned in favour of less fertile areas with less bird pressure.

It was after World War II that Africa's bird pest problem, particularly the quelea, took on a new dimension. At that time, the human population explosion, accelerating urbanization, and a degree of economic progress raised the question of how to feed people who had fled from the world of agriculture. The immediate solution was to increase food importation, including rice from Asia, but at a cost that could be scarcely supported, least of all by those countries suffering from plagues of birds. The long-term solution in most of the continent was to increase cereal cultivation, and this has now become a top priority. While the exigencies of development may vary from one country to another, food self-sufficiency is the overall objective.

Great expanses of cereals were created. Enormous irrigation schemes were opened in arid areas. But it is exactly in such places that bird damage can have its greatest impact, quickly reaching catastrophic proportions and causing losses of up to 80 per cent of production. Traditional control methods that can work for small, scattered cultivation are useless for large-scale mechanized agriculture with limited manpower.

The failure of some of these extensive cereal production schemes, as a result of bird attacks, has had many repercussions, including mobilizing the appropriate authorities into action. Initially, the solution favoured by economic and political leaders was the destruction of all bird pests. Bird control was, to some extent, absorbed into locust control that had begun to achieve some success. However, the problems posed by birds proved to be different, and the military-type operations geared to eliminate them usually did not produce the desired results. Only in certain limited areas, where local quelea populations had no possibility of escape, were satisfactory results achieved.

It was then decided that if success was to be realized, some parallel work was needed better to identify and locate the bird pest populations attacking the crops. At the same time, efforts were made to improve the bird control techniques. Again, the results did not satisfy the high expectations. Insufficient knowledge of quelea population dynamics and the real impact of control on local populations prevented proper orientation of control campaigns and objective analysis of results.

It was in these circumstances that the first UNDP-funded project under FAO implementation was created towards the end of the 1960s. Between 1970 and 1979, FAO projects employed about 20 researchers covering all the Sahelian countries from Senegal to Somalia. Bilateral projects were begun by the British, French, Germans, and Americans. From the start, the projects attempted to investigate all aspects of the problem. Co-operative investigations were initiated between researchers of different organizations in areas of mutual interest.

A number of important accomplishments that resulted from these research efforts related to: (i) delineating local and migratory movements; (ii) identifying particular populations; (iii) estimating losses; (iv) evaluating non-lethal control methods; (v) improving spray operations; (vi) monitoring environmental hazards; and (vii) using quelea as a resource. Despite the efforts made over several years, this vast programme did not produce an exhaustive knowledge of the quelea and did not lead to a 'miraculous' elimination of the pest, as many had hoped. Officials of the countries concerned and of UNDP began in 1975 to realize the difficulties and to realize that after all the inputs of funds and years of work, it was likely that the quelea problem could never permanently be solved in the sense of its elimination, but needed to be treated as a regular agricultural pest requiring control when necessary. It was likely, as Peter Ward prophesied, that success would not be achieved by eliminating the species, but by controlling quelea populations that were damaging or threatening cereal crops. In the 1980s, budgetary restrictions have prevented donor organizations from continuing the programmes in full, and the work has been continued by small teams directing their efforts at practical problems likely to produce immediate results. It is certain that many of the expected results were never produced because of this premature interruption of the in-depth studies which were under way.

This book brings together the results of much of the work that has been carried out during the past 20 years by the many different investigators and research teams. However, the 26 authors representing 9 nationalities are only a small portion of those who have contributed to this advancement of knowledge of quelea. The large international effort was recently underscored in 1985, when an 'International Conference on the Quelea: Ecology, Management, and Policy' was held in Kenya and attended by 80 participants representing 25 countries and 18 international organizations. This was the first international conference since 1957, when the quelea problem gained international attention, and was designed to bring together the world's foremost experts in quelea biology and control. They and others who have contributed over the years through their research and their efforts to prevent quelea damaging crops—and there are many—will hopefully find in this book a permanent recognition of the multiplicity of their contributions. It is also fitting that the book should be dedicated to Peter Ward, whose untimely death was a heavy loss so keenly felt by us all.

Jean Roy

Acknowledgements

Research into the biology and management of quelea in Africa began in the 1950s. Since that time, many individuals from numerous organizations have been involved in defining the pest status of this species, evaluating lethal and non-lethal methods to reduce the damage it causes, training individuals in quelea management techniques, and implementing national and regional control strategies.

The donor organizations that have assisted in this effort and a brief summary of their contributions are mentioned in Chapter 2; none of this work would have been possible without their support. However, many other organizations in Africa have been specifically involved in the results generated from all studies which are presented in each chapter. In particular, we wish to acknowledge the support and assistance of the Ministries of Agriculture of each country where research and control efforts were undertaken, the three regional pest control organizations, OCLALAV, Organisation International de Criquet Migrateur Africain (OICMA), and DLCO-EA, the agronomic research centres where much of the initial crop protection efforts were tested, and the many farmers who assisted us in crop protection evaluations and operations. We would also like to express our appreciation to our many colleagues, who have moved on to other work but who either assisted us in these efforts or offered constructive criticism of study designs or reported results. Many of these individuals are specifically acknowledged in the numerous publications on quelea generated over the years. We also owe a great debt to our many African colleagues with whom we have shared companionship, discussions, hardships, laughter, and excitement in numerous far-flung corners of the quelea habitat.

We wish to express our appreciation to the individuals who kindly reviewed the chapters in this book. In particular, we acknowledge with thanks R. G. Allan, R. A. Dolbeer, C. J. Feare, C. Halvorson, M. M. Jaeger, P. J. Jones, and G. J. Morel, all of whom reviewed several chapters and made major contributions. We also thank the following other reviewers:

D. Anderson
J. S. Ash
J. F. Besser
P. Berthold
L. Bortoli
S. A. Clark
N. E. Collias
K. J. Cook
A. J. Craig
J. L. Cummings
R. D. Curnow
R. M. Engeman
M. W. Fall
L. A. Fiedler
H. C. Fry
P. Gramet
L. House
D. C. Houston

W. E. Howard
M. P. S. Irwin
W. B. Jackson
C. Jones
F. L. Knopf
M. LaGrange
P. W. Lefebvre
R. Luder
R. E. Marsh
L. Martin
L. McDonald
D. F. Mott
P. J. Mundy
J. Nagy
I. Newton
P. A. Opler
G. Orians
A. Peterle
R. L. Phillips
J. Pinowski
J. G. Rogers, Jr.
J. Roy
P. J. Savarie
E. W. Schafer, Jr.
R. Skaf
E. Urban
R. E. Williams
E. N. Wright

We are most grateful to Dolores Steffen and Jacqueline Bruggers for assistance in preparing the illustrations, all individuals who graciously provided photographs in the text, Nigel Harding for writing the model of quelea population dynamics used in Chapter 15, and in particular Annaliese Valvano, Editorial Assistant, International Programs Research Section, Denver Wildlife Research Center, who assisted in many ways, including typing all chapters and preparing this time-consuming document in its final form for publication.

We also are grateful for the funds provided by USAID, FAO, Bayer, Micronair, and Bowling Green State University which helped defray certain preparation and printing costs.

The book is a joint effort of 26 authors from 9 different nations. We thank our co-authors most warmly for their contributions and for the hard work that was involved. Their interest and enthusiasm have greatly helped us to see the project through to its conclusion.

Throughout the book, reference to trade names does not imply endorsement by the US Government and the Food and Agriculture Organization of the United Nations.

Contents

Plates fall between pages 138 and 139, and between pages 234 and 235

Contributors

Richard G. Allan Ornithologist/Project Manager with the Food and Agriculture Organization/United Nations Development Programme (FAO/UNDP) regional and national quelea projects in Somalia, Sudan, and Kenya, 1971–1986. Present address: FAO/UNDP Vertebrate Pest Project, Box 521, Kampala, Uganda.

El Sadig A. Bashir Government of Sudan, Ministry of Agriculture, Plant Protection Department staff, 1963–1968, and Head, 1971–1978, working on the research and management of quelea and rodent pests. FAO Ornithologist in Pakistan, 1978–1980; FAO Crop Protection Ornithologist in eastern Africa, 1980–1983, and western Africa, 1983-1986; FAO Migrant Pest Specialist for locusts and quelea in Botswana, 1986 to present. Present address: Migrant Pest Specialist, FAO/UNDP, PO Box 54, Gaborone, Botswana.

Louis Bortoli Ornithologist with FAO/UNDP regional quelea projects in Chad and Senegal, 1970–1978. Present address: CILSS, BP 7049, Ouagadougou, Burkina-Faso.

Brhane Gebrekidan Plant Breeder and Leader, Ethiopian Sorghum Improvement Project, Ethiopia, 1974–1982; Coordinator for Sorghum and Millet in eastern and southern Africa OAU/SAFGRAD/ICRISAT, 1982–1985. Present address: Regional Maize Breeder, Centro Internacional de Mejoramiento de Maiz y Trigo (CIMMYT) International Centre for Maize and Wheat Improvement, PO Box 25171, Nairobi, Kenya.

Richard L. Bruggers Crop Protection Specialist with the FAO/UNDP regional quelea projects in Senegal and Somalia, 1974–1979. Frequent consultant on quelea research in Africa from 1980 to present as Wildlife Biologist, International Programs Research Section (IPRS), Denver Wildlife Research Center (DWRC), Colorado, USA. Present address: Chief, IPRS/DWRC, USDA/APHIS/S & T, Denver, CO 80225–0266, USA.

Roger W. Bullard Research Chemist investigating avian repellents and attractants in Section of Bird Damage Control, DWRC, 1963–1988. Present address: Chemical Development/Registration Section, DWRC, USDA/APHIS/S & T, Denver, CO 80225–0266, USA.

John W. De Grazio† Wildlife Biologist and Chief, IPRS/DWRC, 1974–1986. Consultant to FAO quelea projects, 1972–1985; Project Manager of DWRC vertebrate pest projects in Tanzania and Sudan, 1976–1981.

Clive C.H. Elliott Ornithologist/Project Manager with the FAO/UNDP regional and national quelea projects in Chad, Tanzania, and Kenya, 1975–1989. Present address: FAO, Via delle Terme di Caracalla, 00100 Rome, Italy.

William A. Erickson Survey Ornithologist with the FAO/UNDP national project in Ethiopia, 1976–1981. Present address: Department Wildlife and Fisheries Biology, University of California at Davis, Davis, CA 95616, USA.

Jan-Uwe Heckel Plant Protection Specialist/Project Manager with the Deutsche Gesellschaft für Technische Zusammenarbeit (GTZ) specializing on bird best problems in Niger and Somalia, 1979 to present. Present address: GTZ Bird Damage Prevention Project, PO Box 3487, Mogadishu, Somalia.

Jeffrey J. Jackson Ornithologist/Team Leader with the FAO/UNDP regional quelea project in Chad and Sudan, 1971–1976. Present address: Department of Wildlife Extension, College of Agriculture, University of Georgia, Athens, GA 30602, USA.

William B. Jackson Retired Professor of Biology, Bowling Green State University (BGSU), Bowling Green, Ohio. Consultant to FAO/UNDP quelea projects in eastern Africa. Graduate Adviser to 60 international students including 13 Africans who were in the past or are currently involved in quelea research and control activities. Currently Distinguished University Professor Emeritus of Biological Sciences, BGSU, and President of BioCenotics, Michigan. Present address: Life Sciences, BGSU, Bowling Green, OH 43403–0212, USA.

Margaret E. Jaeger Volunteer assistant to FAO/UNDP national and regional quelea projects in Ethiopia, 1975–1982, and eastern Africa, 1984–1985. Present address: USAID/Dhaka, Air Pouch, Agency for International Development, Washington, DC 20520, USA.

Michael M. Jaeger Ornithologist/Project Manager with the FAO/UNDP national quelea project in Ethiopia, 1975–1982. Ornithologist/Liaison Officer of FAO/UNDP regional quelea project in eastern Africa, 1984. Present address: Project Leader, DWRC/Vertebrate Pest Control Project, USAID/Dhaka, Air Pouch, Agency for International Development, Washington, DC 20520, USA.

†Deceased.

Brad E. Johns Research Physiologist at DWRC, 1964 to present; specializing since 1981 in developing wildlife techniques in relation to improving management techniques. Present address: Research Physiologist, Bird Control Research Section, DWRC, USDA/APHIS/S & T, Denver, CO 80225–0266, USA.

Peter J. Jones Ornithologist on quelea projects with the Botswana Government, 1969–1972, and the Centre for Overseas Pest Research in Nigeria, 1973–1975. Lecturer in Animal Ecology, University of Edinburgh, 1979 to present. Editor of *Ibis*. Present address: University of Edinburgh, Department of Forestry and Natural Resources, Edinburgh, EH9 3JU, UK.

James O. Keith Wildlife Biologist at DWRC and the Patuxent Wildlife Research Center, 1961 to present; specializing in investigating the impact of insecticides on wildlife. Consultant in quelea survey and pesticide research, 1981, 1985, and 1986. Present address: Wildlife Biologist, IPRS, DWRC, USDA/APHIS/S & T, Denver, CO 80225–0266 USA.

Graham M. Lenton Ornithologist with the FAO/UNDP regional quelea project in Sudan, 1980–1982. Present address: Department of Educational Studies, University of Oxford, 15 Norham Gardens, Oxford, UK.

Stanislaw Manikowski Ornithologist with the FAO/UNDP regional quelea projects in Chad, 1974–1977, and Mali, 1980–1982. Migratory Pests Officer at FAO, Rome, 1984–1986. Present address: 4680 Mariette, Montreal, Quebec H4B 2G4, Canada.

Wolfgang M. Meinzingen Bird Control Officer with the GTZ quelea project in Sudan, 1965–1967, and Nigeria, 1968–1979. Spray Application Specialist with the FAO/UNDP regional project in Ethiopia and Kenya, 1982–1985, and then with Technical Cooperation Programme (TCP) as coordinator for migrant pests in eastern, central, and southern Africa. Present address: c/o FAO, Box 30470, Nairobi, Kenya.

Alioune N'diaye Bird Control Officer with the Organisation Commune de Lutte Antiacridienne et de Lutte Antiaviaire (OCLALAV), 1970–1975. With the FAO/UNDP regional quelea project in western Africa, specializing on training, 1975–1979. Project Manager of a new phase of this project that covered bird pests in general and extended its area outside the Sahelian countries into Liberia, 1979–1986. Continuing with FAO on short-term programmes and consultancies on bird pests training. Present address: c/o OCLALAV, Box 1066, Dakar, Senegal.

Iwao Okuno Project Leader of Chemical Research and Analytical Services Unit, US Fish and Wildlife Service, DWRC, 1976–1983. Present address: Research Chemist, Analytical Chemistry Section, DWRC, USDA/APHIS/S & T, Denver, CO 80225–0266, USA.

David L. Otis Biometrician in the Section of Bird Damage Control, DWRC, 1977–1986. Consultant in quelea research in eastern Africa in 1984. Present address: Chief, Bird Control Research Section, DWRC, USDA/APHIS/S & T, Denver, CO 80225–0266, USA.

John D. Parker Lecturer in the Bio-aeronautics Section at the Cranfield Institute of Technology, 1979–1985; Director of the International Centre for the Application of Pesticides (ICAP), at Cranfield, 1985 to present. Frequent consultant on aspects of quelea spraying, 1983 to present. Present address: ICAP, Cranfield Institute of Technology, Cranfield, Bedford MK43 OAL, UK.

Jean-Marc Thiollay Ornithologist/Researcher of the National Centre of Scientific Research at the Ecole Normale Superieure in Paris, 1967 to present. Specialist on birds of prey, especially in West Africa. Consultant to the FAO/UNDP regional project in Mali, 1973, Chad, 1975, and Benin, Ivory Coast and Senegal, 1980. Present address: Ecole Normale Supérieure, Laboratoire d'Ecologie, 46 Rue d'Ulm, 75230 Paris Cedex 05, France.

Plates

Plate 25 After spraying, quelea are sometimes collected by the sackful (5000 birds per sack) to be used as food by the local people.

Plate 26 In-service training in the field provides a most effective method of learning the routines of quelea work.

1

Pest birds—an international perspective

JOHN W. DE GRAZIO

Introduction

Birds have affected man, both positively and negatively, for thousands of years. When birds conflict with man's endeavours, they become a pest problem that can cause serious economic losses or threaten human health and safety. The most serious economic losses are caused by birds feeding on agricultural crops. Birds also can damage stored grains; consume high-protein commercial feed at livestock feedlots and mink and poultry farms; damage plastic irrigation pipes, cables, wooden utility poles, and houses; damage trees, structures, and equipment; cause a potential health hazard by forming large roosting concentrations at urban and rural areas; and cause hazards to aircraft.

Bird damage to both sprouting and ripening grains occurs worldwide. Most pest species are in the order Passeriformes. Where economies are largely based on agriculture, particularly in developing countries, losses caused by birds can result in inadequate food supplies for millions of people in scores of nations. Countries with highly developed agricultural technology spend considerable time and effort in attempting to improve methods for prevention of bird damage. In developing countries, however, resources are limited and insufficient scientific or technological effort has been devoted to resolving pest bird problems. Under these conditions, pest birds can severely limit agricultural production on a local scale, particularly in areas where farmers are dependent for sustenance and income on the harvest from their small fields.

The crop and geographic region determine the species that are implicated in damage situations. Most pest species are wide-ranging and many are migratory. Many species have learned to take advantage of man's agricultural activities and feed upon newly introduced crops within their range. In some cases, these crops replace wild foods present before the land was cultivated, although more often they are an important supplement taken when native foods are scarce. A variety of crops are eaten or damaged by birds including cereal grains, oil crops, fruit and nut crops, and vegetables.

Damage to cereal crops (source: De Grazio 1978, unless otherwise indicated)

Africa

The pest status of the Red-billed Quelea is fully discussed in the next chapter. It suffices here to state that both in global terms and of course in Africa, the available evidence strongly suggests that the quelea is the most serious of all agricultural pests.

Apart from the quelea, other examples of bird pest problems in Africa are rice damage by House Sparrows *Passer domesticus* in Egypt, damage to grain sorghum by Laughing Doves *Streptopelia senegalensis* and Cape Turtle Doves *S. capicola* in Botswana, loss of rice to the Red Fody *Foudia madagascariensis* in Madagascar (Fahlund 1965), loss of wheat to House and Spanish Sparrows *Passer hispaniolensis* in Morocco, an instance of US $2.8 million lost in ripening cereal in one Province in Nigeria (Fahlund 1965), and loss of 100 000 to 200 000 t of cereal to birds in Senegal (Mallamaire 1959*a*). Black-headed Weavers *Ploceus cucullatus* were reported as a serious pest in millet in Chad and sparrows in wheat and barley in Libya (Jackson and Jackson 1977). In Somalia, quelea, bishops, and *Ploceus* weavers damage rice and grain sorghum, and *Ploceus* weavers damage maize (Bruggers 1980).

Tree-ducks *Dendrocygna* spp., Ruffs *Philomachus pugnax*, and Sparrow Larks *Eremopterix* spp. damage newly sown and emerging rice and wheat in Cameroon, Senegal, and Sudan (Bruggers 1979*a*; Hamza *et al.* 1982; J. De Grazio *et al.*, unpubl. data). In the Senegal River Valley, Senegal, a 9000 ha Government rice scheme experienced losses by wading species of >15 per cent in 1975; Treca (1976) confirmed the occurrence of considerable localized aquatic bird damage to newly sown seeds in this valley.

Other species that can cause economic losses to African agriculture are Golden Sparrows *Passer luteus*, starlings *Lamprotornis* spp., doves *Streptopelia* spp., Bimaculated Larks *Melanocorypha bimaculata*, Crowned Cranes *Balearica pavonina*, European Starlings *Sturnus vulgaris*, White-vented Bulbuls *Pycnonotus barbatus*, and Rose-ringed Parakeets *Psittacula krameri*.

Asia and the Middle East

Rice is the grain most often damaged in Asian and Middle Eastern countries. The main pest species throughout most of Asia are mannikins and munias *Lonchura* spp., sparrows, larks, weavers, mynahs *Acridotheres* spp., crows, bulbuls (Pycnonotidae), finches (Fringillidae), and parrots (Psittacidae). In

Borneo, the Pin-tailed Parrot-finch *Erythrura prasina* is a serious rice pest (Grist and Lever 1969). Up to 50 per cent loss of rice has been caused by weavers in Thailand. Other important cereal damage caused by a variety of species occurs in rice, grain sorghum, and millet in India (Prakash 1982; Shivanarayan 1980), barley, millet, and emerging cereal grains in Korea (Howard *et al.* 1975), and wheat and corn in Nepal. The White-rumped Munia *Lonchura striata* damages rice in Malaysia (Avery 1979). Emerging grains are damaged by House Sparrows, mynahs, crows, and doves in Bangladesh (Poché *et al.* 1980). In Pakistan, sparrows, parakeets, doves, crows, and weavers have been reported to cause an estimated annual loss of $31 million to ripening cereals (Roberts 1981), and Bashir (1978) reported that Rose-ringed Parakeets regularly destroy in excess of 50 per cent of the maize crop. In the Near East, House and Tree Sparrows cause major problems in wheat in Turkey. In Israel, three species of larks (Alaudidae) cause problems in wheat and barley. Despite this extensive list of damage situations, losses in Asia are probably considerably less than those in Latin America and Africa.

Oceania

Rice and wheat are the cereals most frequently damaged by House Sparrows in New Zealand (Dawson 1970), and by parrots, crows, and waterfowl in Australia (Grist and Lever 1969). In New Guinea, the Spotted Tree-duck *Dendrocygna guttata*, Whistling Tree-duck *D. arcuata*, Spotbill Duck *Anas superciliosa*, and Pied Goose *Anseranas semipalmata* damage emerging and ripening rice (Grist and Lever 1969). The Galah *Eolophus roseicapillus* is also an agricultural pest in Oceania.

Latin America

Grain sorghum, corn, rice, and wheat are the cereals most frequently damaged. Examples include up to 80 per cent loss of rice in Argentina to the Brown Pintail *Anas georgica* and Chestnut-capped Blackbird *Agelaius ruficapillus*; $7.0 million loss to rice by Black-headed Weavers in 1971 in the Dominican Republic (Peña 1977); and $0.25 million loss to emerging wheat primarily by Eared Doves *Zenaida auriculata* in 1974 in Uruguay (J. De Grazio and J. Besser, unpubl. data). Two widespread pest bird problems in Latin America are damage by White-faced Tree-ducks *D. viduata* to emerging rice (it occurs in Argentina, Colombia, Costa Rica, Honduras, Nicaragua, Surinam, Uruguay, and Venezuela) and parrot damage to corn and grain sorghum. The Dickcissel *Spiza americana* is an important but sporadic pest of rice in Colombia and Mexico (Elias 1977) and to rice and grain sorghum in Costa Rica, Trinidad, and Venezuela (De Grazio and Besser

1970). Other agricultural pest birds in Latin America include the White-winged Dove *Zenaida asiatica*, Monk Parakeet *Myiopsitta monachus*, parakeets of the genus *Aratinga*, Great-tailed Grackle *Cassidix mexicanus*, and cowbirds *Molothrus* spp. In the Caribbean, the Hispaniolian Woodpecker *Melanerpes striatus* and the Black-headed Weaver are the key pests.

Europe

In Great Britain, the classic bird pests of cereals are the Wood Pigeon *Columba palumbus* (Murton 1965), the Rook *Corvus frugilegus* (Dunnet and Patterson 1968), and the European Starling *Sturnus vulgaris* (Feare 1984). The Wood Pigeon also causes serious problems to cereals, especially wheat, in The Netherlands. In France, the Rook and the Carrion Crow *C. corone* are important pests to cereals. House Sparrows and Tree Sparrows are a problem in Germany, The Netherlands, and other parts of Europe. The Mallard *Anas platyrhynchos* damages rice in Bulgaria. In Poland, the Rook and the Common Jackdaw *C. monedula* are serious pests to both spring-planted and ripening cereals; House Sparrows damage ripening wheat and barley; and Rock Doves *Columba livia* damage newly planted wheat and maize (Pinowski 1973). Newton (1968) has summarized the importance of Bullfinches *Pyrrhula pyrrhula* as pests to fruit buds in Europe. Other potential European pest species include geese (Anserinae).

Canada

Losses to agricultural crops, primarily wheat, were estimated to amount to $86 million to seed-eating birds in 10 provinces and $33 million to waterfowl in 4 provinces. Losses were greatest in Ontario ($40 million), Quebec ($24 million), and Manitoba ($21 million) (Alsager 1976). The Red-winged Blackbird *Agelaius phoeniceus*, Canada Goose *Branta canadensis*, and several species of surface-feeding ducks (Anatinae) were the most serious pests.

United States

Red-winged Blackbirds, Common Grackles *Quiscalus quiscula*, and European Starlings are the most serious agricultural pest birds in the United States. Surveys of blackbird damage to ripening corn in 1970, 1971, and 1981 showed nation-wide losses of $15, $20, and $35 million, respectively. Damage to emerging corn by Common Grackles and the introduced Ring-necked Pheasants *Phasianus colchicus* may be equally important; losses in 1971 may have ranged from $6 to $49 million (Stone and Mott 1973*a*). J. Besser (unpubl. data) evaluated current bird damage estimates in the United States to ripening and sprouting cereals at about $66 million; corn was most often damaged, with losses amounting to about $48 million. Regional examples of bird damage to cereal grains in the United States include blackbird damage to rice in 22 counties in Arkansas in 1963 ($4.2 million; Meanley 1971) and blackbird and sparrow damage to grain sorghum in 23 states in 1974 ($5.8 million; Knittle and Guarino 1976).

Damage to some non-cereal crops

Important examples of bird damage to non-cereal crops worldwide include damage to oilseeds, particularly sunflowers. In India, sunflower field damage may reach 100 per cent in unprotected fields (Shivanarayan 1980). In Pakistan, parakeets may even prevent farmers from growing promising oilseed crops or cause them to abort their production attempts after only a single trial (Anon. undated; Roberts 1974). Agriculturalists in Pakistan are trying to close the gap between oilseed production and consumption (Muhammed and Khan 1982), which is currently costing the country $239 million, but finding methods of reducing bird damage will be imperative to their success (J. Besser, unpubl. data). In Latin America, parrot damage to sunflowers occurs in Argentina, Bolivia, Brazil, Paraguay, and Uruguay. It is most acute in Uruguay where, in 1974, $600 000 was lost to the Monk Parakeet. Damage to sunflowers is also caused by Tree Sparrows in Korea, House and Tree Sparrows in Morocco (De Grazio 1978), and House Sparrows in Poland (Pinowski 1973). J. Besser (unpubl. data) estimated the annual sunflower loss in the United States to blackbirds at $7 million.

In Latin America, Eared Doves are a major pest to emerging soybeans in Argentina, Colombia, and Uruguay; White-winged Doves damage the same crop in Honduras, Mexico, and Nicaragua. The Rufous Turtle Dove *Streptopelia orientalis* hampers soybean cultivation in Japan by feeding on newly planted seeds and emerging cotyledons (Nakamura and Matsuoka 1983).

Bird damage to fruit is primarily a problem in the United States, Canada, and Europe. In the United States, bird damage to fruit amounts to about $19 million annually, with cherries ($11 million) being the fruit crop most often damaged (J. Besser, unpubl. data). Bird damage to cherries is also a problem in Ontario, Canada. In Europe, the Bullfinch is a serious problem to the fruit industry. The European Starling is another major pest of fruit in various parts of Europe.

Many species of birds also attack fruit crops in developing countries. In Latin America, Monk Parakeets damage a variety of fruit in Argentina, Bolivia, Brazil, Paraguay, and Uruguay. The Brown Jay *Cyanocorax morio* is implicated in Costa Rica. Parrots attack mangoes in Honduras and Mexico. In North-east Africa, the European Starling and a thrush damage olives, and the White-vented Bulbul is a problem on other fruit. In South Africa, the European Starling, Red-winged Starling *Onychognathus morio*, and Cape Sparrow *Passer melanurus* cause problems. In Tunisia, the European Starling damages olives to the extent of 15 000 t loss in yield per year (De Grazio 1978). In Pakistan, bulbuls and parakeets are the chief pests; Red-vented Bulbuls *Pycnonotus cafer* are problems on pears, plums, and persimmons, and the Black Bulbul *Hypsipetes madagascariensis* on apricots and peaches in Northwest Frontier Districts (Anon. undated); parakeets caused problems on a variety of fruit in these districts (J. Besser, unpubl. data), including apples and pears (Ali and Ripley 1969). Losses to fruit are caused by parrots, finches, and crows in Bangladesh, and by crows in Korea.

Similarities and differences

It is clear that bird pest species are present in almost every habitat on every continent and that the damage involves many of the same crops and genera of birds. Species causing damage may comprise a flock of only a few individuals such as bulbuls feeding on ripening mangoes in an Ethiopian village, or swarms of thousands such as quelea devastating ripening grain sorghum in Botswana. Damage to crops at particular plant growth stages is quite similar regardless of locality, and species causing problems exhibit many similar destructive habits. Because more similarities than dissimilarities exist, bird damage research conducted in one part of the world sometimes has much value to farmers in another area. For instance, if an effective method were found to prevent parakeets from damaging corn in Pakistan, it might need only slight modifications to prevent blackbirds from damaging corn in the United States. If an effective method were found for dealing with dickcissel damage to rice in Nicaragua, it would have value as a basis for dealing with blackbird depredations in rice in the United States and quelea damage to cereals in Africa (De Grazio and Besser 1970). Likewise, in northern, temperate regions, planting of grains is largely restricted to the months of April and May, and ripening grains are vulnerable to birds mostly in the months of August and September. These seasons give researchers only short periods each year to carry out field tests of damage control procedures. In tropical regions, many of the same crops ripen almost every month of the year. Whereas research in the United States now takes 5 years and usually longer to work out an effective procedure to control a specific kind of damage, the availability of similar problems in tropical and southern regions could substantially shorten the time needed for a solution.

Developing a management strategy

No single method of reducing losses to birds is applicable to all damage situations. Numerous vertebrate pest management materials, methods, and techniques have been used or suggested for particular situations. These include frightening devices, traps, protective netting, nest destruction, electric shock, shooting, cultivation of less susceptible crop varieties, modified planting and harvest schedules, alternative crops or foods, decoy crops, varying seeding depths, habitat alteration, lethal baits, lethal sprays, contact toxicants, repellent sprays, stressing agents, glues, soporifics, chemosterilants, hunting, predators, diseases, and parasites (Boudreau 1975). Methods must not only be effective in reducing damage, but also should be evaluated for safety to humans and non-target animals, cost, practicality, environmental effects, acceptability to farmers, and availability of materials. Unfortunately, even the effectiveness of most potential crop protection methods has not been adequately evaluated in many situations, and even under ideal conditions, effectiveness varies.

Managing pest birds is a very challenging endeavour. Successfully developing an effective vertebrate pest control technology involves problem identification, species identification and biology, materials research and laboratory and field evaluation, and training and technology transfer. Before undertaking crop protection experiments, it is important to confirm that the damage is actually being caused by birds. Birds frequently are blamed for lower yields that may have been attributable, wholly or in part, to plant diseases, insects, nocturnal mammals, or poor management. In some cases, birds may actually have been feeding on insects in the field. It also is important to identify correctly the pest species, because species vary in their susceptibility to different control methods.

To gauge the amount of research a pest bird situation warrants, it is necessary to obtain a general estimate of the amount of loss being sustained per unit area (a group of farms, a village, a province). Loss information both on a national and local scale entails the greatest coordinated effort. Because increased yield and subsequently the cost:benefit ratio is the basis for determining the value of control methods, loss data influence the type and intensity of control efforts and are the measure of success or failure.

Damage control for vertebrates such as birds is perhaps even more complex than that for invertebrates. Although most agricultural organizations are staffed with entomologists, few have personnel trained in animal ecology or wildlife management with experience in managing vertebrate pests. Because of their association with pests and pesticides, staff entomologists often inherit problems of vertebrate pest management. However, a wildlife biologist with experience in managing vertebrate pest situations is no more qualified to provide professional leadership in entomology than the entomologist is in the management of vertebrate populations (Besser 1971).

Differences among countries in economics and cultures also contribute to the challenge of managing pest birds. A device or technique applicable in a developed country may not be applicable under conditions prevailing in a developing country. For example, expensive devices that produce bird distress calls may provide economic protection to a high-value crop such as lettuce in California, but would be impractical for farmers in Sudan who are trying to protect sorghum fields and whose earnings amount to only a few hundred dollars annually. In many developing countries, some cultures strongly oppose any manner of killing even though birds may be causing real hardship to the family, village, or community. Present-day research in bird damage is becoming more sophisticated, and long-range benefits and safeguards are now being considered at the planning stage.

It is imperative that research findings in one part of the world should be shared with individuals and agencies confronted with similar damage situations elsewhere. Efforts should be made to optimize resources and expertise through co-operative planning and complementary research. Since vertebrate pests play a major role in limiting agricultural production and often have the greatest impact on the poorest farmers in marginal production areas, training of counterpart personnel and extension of research recommendations should be an integral part of a developing country's vertebrate pest programme. Transferring information can take many forms and will include exchange of letters between interested parties, workshops, demonstrations, seminars, training manuals, brochures, preliminary reports, formal publications, and media materials (De Grazio 1984).

Techniques to manage pest birds must be simple, safe, available, cost-effective, and adaptive to local farming practices; products and devices must be locally available, and farmers must be trained in their proper use. In addition, the correct use of recommended techniques should be reinforced by extensionists to ensure that practical and useful methods are maintained and do not fall into disrepute.

2

Historical overview of quelea research and control

JEFFREY J. JACKSON and
RICHARD G. ALLAN

The quelea is the most destructive and perhaps the most numerous species of grain-eating bird in the world. Anyone who knows quelea birds never forgets them. They are intensely social at all times. During the dry season, flocks feeding on open savannahs give the impression of a single, monstrous creature as they rise and settle with a roar of tiny wings. In the evening, flocks pour into roosts in millions, with bending and sometimes breaking of branches. Nesting colonies may exceed 100 ha of acacia bush land. Each tree in the colony may be crowded with hundreds of nests. Breeding is synchronized: egg-laying and then hatching begins on the same day over large areas. At hatching time the females remove and drop the eggshells. The falling eggshells resemble the sound of light rain. In the following days, the screeching of young becomes almost deafening.

After nesting, birds return to their migratory, sometimes nomadic ways. They may feed on grassheads, fallen seeds, or insects, but sometimes swarms descend on maturing cereal crops. Crop loss may be total. Quelea have always threatened the livelihood of traditional African farmers, but as recently as 40 years ago these birds were little known to African governments and professional ornithologists.

Today, papers on quelea and their control are an occasional feature at international meetings on ornithology and vertebrate pest control in the United States, Europe, and Africa. Literature on biology, control, and administration of quelea control organizations and operations is scattered, and much of the information is unpublished. This introduction attempts to summarize the efforts and progress of knowledge that brought work on quelea to where it is today. Details are provided in the following chapters.

History of quelea control organizations

Early days

African folklore dates the granivorous bird problem to antiquity. In Senegal, for example, local tradition traces the quelea problem back for centuries (Grosmaire 1955). African farmers developed bird-scaring methods centuries ago to protect their crops as best as they could. (One of these, the cracking sling, which is still used today throughout the Sahel, is pictured on pharaonic tombs.) African traditions also included cutting down quelea nesting trees near crops to try to force the birds elsewhere, and sometimes collecting nestlings for food. Early ornithologists in Africa collected specimens, named species, and made some general observations on the biology of quelea and their destructive habits. Haylock (1959) refers to the 1881 famine caused by quelea depredations in the Ugogo district of what is now Tanzania.

We found less than a dozen accounts of quelea publications prior to 1940. Other than destruction of nesting colonies organized by district officers and traditional chiefs, no 'official' attempt was made to control the quelea problem during that time, and farmers had to cope with quelea on their own, much as some traditional farmers still do today.

During World War II this situation changed. Governments, especially in eastern Africa, became interested in producing grain to feed the Allied Armies fighting in North Africa and the Middle East. Quelea were then recognized as an important obstacle to efficient grain production. In western Africa there was concern for poor areas that were not self-sufficient in food. Importation of food was impractical, and 1942 and 1943 were drought years. Since then, departments of agriculture, ornithologists, and pest control workers have paid increasing attention to quelea, as profit margins and efficiency became more important to commercial agriculture.

Post-war period

Government-supported quelea control began in western Africa between the 1940s and 1950s. A direct attack on the birds to reduce numbers seemed like a logical approach. Military-type weapons were the first choice. Flame throwers, poison gases, mortar bombs, and other explosives were used to kill quelea in Senegal, Sudan, Tanganyika (as it was then called), and South Africa. C. E. Wilson, an Inspector of Agriculture in Sudan, reported in 1945 that he tried bird lime, poison gases, flame throwers, poisoned water, and bird-resistant varieties of sorghum. Wilson correctly forecast that 'the dropping of death-dealing substances from the air onto the roosts' would develop as a means for controlling quelea.

Crop production experts began to correspond with one another on the quelea problem, and they soon felt the need for an international meeting to share information. In 1955, the Commission for Technical Cooperation in Africa south of the Sahara and its Council (CCTA/CSA) organized the first meeting of quelea specialists in Dakar, Senegal. Representatives came from southern, western, and eastern Africa. Participants proposed that regional committees be formed for co-operation on research in three regions within the quelea's range: West, North-east and East, and southern Africa. The participants proposed to meet again at the first Pan-African Ornithological Conference scheduled for Livingstone in 1957. There, a questionnaire was circulated to obtain and share information on distribution, crop damage, movements, and other data. A promising development was aerial spraying with parathion, first tried in South Africa in 1955.

Independence

As colonial nations transferred governance to African nations, they also passed on the responsibility for resolving the quelea problem. The next CCTA/CSA meeting was co-sponsored by FAO. Soon after this meeting, the CCTA/CSA was replaced by the Technical and Scientific Research Committee (CSTR) of the Organisation for African Unity (OAU). The CSTR had many concerns to deal with during the early years of independence and gave little or no attention to the quelea problem. Unfortunately, representatives from southern Africa have been, for the most part, absent from regional meetings since the independence period. An exception was the presence of representatives from Botswana in 1985 at the International Conference on the Quelea: Ecology, Management, and Policy, at Taita Hills Lodge, Kenya. This loss of contact with quelea control workers in southern Africa may have slowed the progress of quelea research and control efforts.

During the 1960s, most African governments controlled quelea birds on their own. Some countries, Nigeria and Sudan, for example, had their own control organizations. In 1955, French-speaking countries in western Africa formed first the Organisme de Lutte Antiaviaire (OLA), which in 1958 became the Organisation Commune de la Lutte Antiaviaire (OCLA) and later it joined the locust control organization to become the Organisation Commune de Lutte Antiacridienne et de Lutte Antiaviaire (OCLALAV). The new OCLALAV had responsibility for controlling both locusts and birds. Many countries, however, remained without organized quelea control. Notable ecological studies on the quelea were carried out at this time by Morel and Bourlière who worked for the Office de la Recherche Scientifique et Technique Outre-Mer (ORSTOM). During this period there was no international conference to air new research or exchange ideas on quelea control. Knowledge of quelea was advanced by individuals. Some researchers were sponsored by universities. Outstanding among these was Peter Ward, a British doctoral student, who worked in north-eastern Nigeria. In the early 1960s he produced three landmark papers on quelea biology.

In other papers, Ward proposed that quelea control should focus only on birds threatening damage, usually by their proximity to cereal crops, rather than on exterminating the population as a whole. Ward reasoned that extermination would be extremely expensive, probably impractical and ecologically unsound. This idea became widely quoted and repeated. Ward produced numerous other papers on various aspects of African ornithology including one on the function of roosts as communication centres for birds. This short article became a 'citation classic' being quoted a record number of times. Ward later worked with the Centre for Overseas Pest Research (COPR) and FAO. His productivity came to an untimely end with his tragic death in 1979.

Technical assistance begins

African crop protection authorities continued to find quelea control an intransigent problem and sought technical assistance first from Europe and later from the United States. France assisted the OCLA countries beginning in 1954 (Senegal and Mauritania) and then in the 1960s extended this support to other French-speaking countries. German technical assistance began in the Sudan in 1962. After the FAO Conference on Quelea and Water Hyacinth Control in Douala, Cameroon, in 1965, 11 western African nations jointly requested a regional quelea project. This project became the first of a series of FAO regional projects.

About this time, the COPR in England became interested in funding quelea research. There was brief competition between COPR and FAO to be the contractor for the UNDP regional project. FAO prevailed, and after some delay, the FAO/UNDP project became operational in 1970, with headquarters in Fort Lamy, Chad, serving 11 member nations stretching between Chad and Senegal. COPR began its quelea work by providing technical assistance in northern Tanzania.

The United States Agency for International Development (USAID) also became interested in research on quelea control. Denver Wildlife Research Center (DWRC) staff (under USAID funding) visited the FAO project headquarters in Chad in 1971. The three agencies shared information, and by the mid-1970s, the approach to the quelea problem and its management gradually became more sophisticated.

The FAO regional project in western Africa emphasized the standardization of scouting techniques and prepared forms for recording of data to monitor populations and for evaluating control operations. These standardized techniques made it possible to learn more precisely the distribution of the races of quelea originally described by Peter Ward. Various techniques for assessing crop damage were also tested and used. Regular training courses helped to develop a cadre of experienced African staff capable of collecting field data. With precise scouting of bird targets under way, more effective spray operations were possible. A picture of the north–south movements of the quelea population was gradually built up, and in Chad and

Cameroon it was possible to advise rice growers of specific planting and harvest dates which might avoid, to a large extent, the crop being at a susceptible stage when the quelea passed on their north–south migrations.

At the same time, a West German technical assistance team from the Deutsche Gesellschaft für Technische Zusammenarbeit (GTZ) in Nigeria was field testing low-volume applications with helicopters of the insecticide fenthion. This research led to improved aerial spray techniques. In Chad, there was parallel research with fixed-wing aircraft. FAO/UNDP, in addition to their role of regional coordination, began to support national technical assistance projects in eastern African countries. These projects were aimed at encouraging national teams to adopt the scouting and control techniques developed in western Africa. In western Africa, similar developments were taking place with a technical assistance project to the West African Rice Development Association (WARDA) in Liberia and assistance by FAO/UNDP and Belgium to the regional locust and bird control organization OCLALAV.

Recent developments

Interest in new developments in bird damage control has recently spread to other parts of Africa. In 1975, FAO visited 33 countries with known bird damage problems and interviewed agricultural personnel. Based on responses to the survey, a new regional project was established in Nairobi in 1979 for Sudan, Somalia, Ethiopia, Kenya, and Tanzania. DWRC extended its activities in eastern Africa with staff moving from Tanzania to Sudan, and later collaborated with the FAO project with numerous consultant visits and technical assistance. New interest is now developing in southern Africa where various governments are currently requesting technical assistance in the field of migratory pest control including quelea.

In the early 1980s, research efforts included the development of ground sprayers, phytoecological investigations of food resources, assessment of environmental contamination by spray, use of radio-telemetry to locate nesting colonies, mass-marking quelea with fluorescent particles, and analysis of trace elements in feathers to help clarify movement patterns and identify populations of quelea. However, in the late 1980s this momentum for innovative research has been, to some degree, lost. UNDP-funded assistance has swung away from support of projects directed solely at quelea, probably on the grounds that quelea damage should be seen in the context of general crop protection. By 1985, country projects in West Africa continued only in Niger with GTZ assistance. The FAO Regional Project in eastern Africa was taken over in July 1985 by the Desert Locust Control Organization for Eastern Africa (DLCO-EA), which was given the responsibility for coordinating quelea control and research. DLCO-EA received some continuing small-scale assistance from FAO and has the possibility of future support from UNDP in which quelea work would be only one component. Projects being supported since 1986 in Kenya and Botswana are on general crop protection in which quelea control is only a part. As a result of these changes, the opportunities for further long-term research on the quelea have decreased dramatically. Only one project remains in 1988 in which such research is a strong component—the major GTZ project on bird damage prevention in Somalia.

Quelea control research and development

We found 496 titles of articles and reports in bibliographies, project documents, and other summary sources. These are primarily unpublished papers about quelea and their control. More than 95 per cent of the articles were written in the last 35 years. There was a rapid rate of publication in the 1950s. Then a decline followed in the 1960s, which might be called a 'research recession'.

One cause of this slowing down was the development of aerial spray technology, first with parathion in South Africa and shortly thereafter with fenthion. This breakthrough no doubt convinced some people that effective control of quelea was imminent. Another factor causing a decline of research was that most African nations received their independence during the 1960s. These new governments faced a myriad of problems and quelea control received little attention. By the mid-1960s, however, it was clear that the quelea problem was as severe as ever. Biologists were beginning to question whether simple destruction of all quelea swarms, particularly those in remote areas, was cost-effective, so there was pressure not to control quelea in many areas.

Research efforts diversified rapidly once the technical assistance period entered the 1970s. Distribution and migration patterns were studied. The first nesting colonies in Ethiopia and Somalia, and later in the 1980s in Uganda, were found and new nesting locations in many other countries were discovered. Formerly, many species of birds had been lumped under the heading of 'quelea'. Researchers now found that the granivorous bird problem was more complex than had been imagined. Several species were identified as important crop pests. The effects of colours, sounds, baits, and resistant crops were explored with poor to moderate success. Logistics and reporting were improved. New technology, including dyes, fluorescent particles, and radio transmitters was applied to migration studies. Repellents to protect crops were tried; some showed promise.

Quelea continue to damage cereals despite the efforts of control organizations. Details of the extent of quelea damage have always been a bit sketchy. (Good damage assessments did not become a priority until the 1970s.) For example, one estimate went like this: 'there are upwards of 2000 million quelea in West Africa. Each bird is capable of eating 2–3 g of seed per day and spoiling an additional 20 g. Therefore, the population consumes 3000 t of grain per day for a total of 1.1 million tonnes per year. Of this, perhaps 44 000–88 000 t might be cereal grains—equal to 140 000–170 000 ha of millet.' This reasoning led to the conclusion that the annual losses in the mid-1960s amounted to a cash value of US $8 million. Such estimates may be accurate, but the possible margin of error in these figures is incalculable. There was demand for estimates of losses, and these were the best that could be made without expensive and time-consuming field studies. The statements did stimulate international interest in the problem. Damage reporting remains a persistent problem. Crops in remote areas may be severely damaged without the knowledge of quelea workers or agricultural officials.

The 1970s produced well over 300 mainly unpublished papers on quelea. Only a portion of this writing is new research on quelea, much of it being in the form of progress reports.

The future of quelea work

In 1985, after 35 years of research and control, quelea are still abundant and still damaging cereal crops. Why can quelea damage not be prevented? There are many reasons. A primary weakness of many wildlife research and management efforts is the difficulty of censusing quelea populations. Quelea are especially difficult to find in the remote bush where they nest. Mention of obstacles and difficulties frequently appear in reports and research papers. Anyone who has ever faced a few kilometres of flooded black cotton soil knows it is impossible to traverse in any land vehicle until the rain ends, and that may take weeks or months. Poor support systems including roads, power, and communications are a fact of life in much of Africa. Water shortages, lack of certain supplies, many languages, low rates of literacy, political unrest, and wars make almost every aspect of life difficult including quelea research and control.

The main reasons why quelea remain a major pest are the same ones that allow rats and mosquitoes to continue their depredations. A discovery of an exploitable weak link in the biology of a generalized animal is much less likely than when dealing with a specialized one. Therefore, a 'breakthrough' is unlikely, and quelea are likely to remain as a serious pest for the foreseeable future, along with many other crop-destroying organisms both in Africa and developed countries.

There is much talk these days of failed technical assistance projects in

African countries, failed factories, or failed crop production schemes. This should not be an excuse to turn away from development efforts. Quelea research has made consistent progress—control methods have been improved, biological knowledge increased, and African citizens are taking an increasing prominence in every aspect of quelea work. At the same time, it must be stated that at the time of writing, the pendulum is swinging away from continuing research into the biology and migrations of quelea. Present plans concentrate mainly on purely practical attention to control techniques. Studies should continue, however, on both aspects, with the dimension of economics and cost-effectiveness badly needing attention. Otherwise progress may slow down.

As more becomes known about quelea, and as more people become experienced in bird control, there is increasing opportunity for debates and disagreements. These are desirable as they may lead to new ideas. One current controversy stems from the opinion that perhaps the net influence of quelea on cereal production in Africa as a whole is minor and that in most circumstances part of the quelea problem stems from farmers' attitudes toward the pest rather than from the actual weight of grain lost. Other controversies deal with how to organize quelea work regionally, nationally, or both. What are the landowner's responsibilities? What should be research priorities? Which control efforts should be deleted when funds are short? Should control efforts be abandoned in certain areas? Is reporting sufficiently accurate? Are the best people being selected for training? Will there be opportunities for their advancement?

There will continue to be no easy answers, only difficult decisions. We hope the following chapters will serve as a landmark summarizing knowledge of quelea up to 1989 and that they will be a useful source of information for those who continue to work on granivorous bird problems in Africa.

3

The pest status of the quelea

CLIVE C.H. ELLIOTT

Introduction

The Red-billed Quelea *Quelea quelea* is a pest because it attacks sorghum, bullrush millet, finger millet, italian millet, rice, wheat, barley, oats, triticale, and teff (an Ethiopian cereal) in cultivated fields, throughout much of Africa (Fig. 3.1; Plate 1) before harvest. Its status as a pest becomes significant if sufficient losses of these cereals occur to justify some level of intervention. Yet, as Wiens and Dyer (1977) stated, an assessment of the potential impact of a granivorous bird in any ecosystem including agriculture should cover all its positive and negative aspects.

The purpose of this chapter is to evaluate the available evidence of the impact of quelea on the environment and on man. Such an evaluation has to be made in the light of a curious ambivalence in the literature. Quelea are regularly described as 'the most destructive and possibly the most numerous bird in the world' (Ward 1965*a*), 'one of the most serious agricultural pests in Africa' (Crook and Ward 1968), 'the most serious avian pest in the world' (Magor 1974), 'a superabundant species ... truly comparable with locusts, which they have replaced as the main plague of seed-growing farmers throughout Africa', and even 'the only bird known to us which has necessitated the calling of an International Conference' (Mackworth-Praed and Grant 1973).

Supporting evidence for this notoriety seems to be scanty. For example, Ward (1973*a*) said that 'even after many years of field research devoted to *Q. quelea*, the author is quite unable to make any quantitative assessment of the importance of *Q. quelea* beyond stating that in some years they do serious damage and occasionally causing the total loss of the cereal crop of certain villages'.

The ambivalence on the quelea's status is rooted partly in the fact that flocks of quelea in cultivated fields are conspicuous and impressive, with the result that farmers often exaggerate the bird's importance. Also, quelea damage is usually so unevenly distributed that collecting statistically valid assessments is difficult and arduous. In addition, its pest status seems to be related to its population status.

(*a*)

(*b*)

Fig. 3.1. Large flocks of quelea: (*a*) leaving a roost on the Logone River at dawn; (*b*) attacking irrigated rice at Yagoua, Cameroon (photos: M-T. Elliott).

Population status

Crook and Ward (1968) have estimated the quelea population in Africa as 1×10^9 or 10^{11} based on the reported number of birds killed by control teams. Ward (1972*a*) pointed out that the insistence of officialdom on bird control units to report their results in terms of numbers of birds killed is 'to invite fictitious returns'. As a result, I have tried to estimate quelea populations using data on the number of hectares of breeding colonies in different regions. I recognize that these are only estimates and not statistically reliable population numbers.

Probably the best known quelea populations in Africa are those of eastern Africa, incorporating Tanzania, Kenya, Somalia, and Ethiopia. In these countries, ornithologists have studied quelea for a number of years and have had at least the occasional use of the most efficient means of survey, the helicopter. The past 6 years of surveying have added only 14 degree squares to the total distribution of quelea breeding colonies in East Africa, first compiled by Magor and Ward (1972). It is unlikely that this distribution will be expanded by additional surveys except perhaps in some of the presently inaccessible parts of its range such as the Ogaden in Ethiopia.

I have estimated the population of breeding adults in several areas for an average year. These estimates were obtained as follows: firstly, the number of degree squares regularly surveyed was expressed as a percentage of the total potential breeding distribution. This shows that regular surveys covered only about 45–62 per cent of the potential area. Secondly, recently reported breeding colony data (see Minutes and Proceedings of the Annual Technical Meetings of the FAO/UNDP Regional Quelea Projects RAF/77/042 and RAF/81/023, 1979–1984) were assembled, and the average total area of colonies found in the areas regularly surveyed was estimated. This figure was proportionally increased to account for colonies that probably existed in unsurveyed areas within the total potential breeding distribution. Finally, the total adult breeding population was calculated on the basis of about 60 000 adults/ha per colony.

These estimates involve making several assumptions. I assume that within the degree squares regularly surveyed all colonies have been found. If some were missed, the population would be underestimated. I also assume that the average density of colonies in the surveyed and unsurveyed areas is comparable. Table 3.1 presents more than 20 years of accumulated data. In any one year, it is most unlikely that with uncertain rainfall more than half the degree squares would support colonies. Thus, this assumption will overestimate the population. The figure of 60 000 adults/ha is based on counts made by C. Elliott and S. Manikowski (unpubl. data) near Lake Chad. Erickson *et al.* (Chapter 14, Table 14.4) show that nest densities in eastern Africa can vary enormously; average density for 12 colonies was 50 000 adults/ha, a figure

Table 3.1. Estimated populations of breeding quelea in regions of Africa.

	Regions					
	Tanzania	Kenya	Somalia	Ethiopia	Lake Chad Basin	Niger Delta
Potential breeding distribution[a]	29	14	11	15	9	6
Area regularly surveyed[a]	18	7	5	8	7	4
Breeding survey coverage (%)	62	50	45	53	78	67
Mean area of breeding colonies (ha)	640	155	175	195	1218	260
Total estimated area of colonies (ha)	1032	310	481	366	1566	390
Average annual adult breeding population (millions)	62	19	29	22	94	23

[a] Number of degree squares.

close to the one used here. A third assumption is that each colony found in any one year contains a separate quelea population, but birds breeding in early colonies may breed again later in another colony either within the same area or a neighbouring country. This will result in both national and regional overestimates of the total population.

Colony area data are available from the Lake Chad Basin and from the Niger River Delta in Mali. In the Lake Chad Basin, colonies totalling 1002 ha and 1434 ha were found in 1976 and 1977, respectively (C. Elliott and S. Manikowski, unpubl. data), and the population was estimated at 94 million. However, Manikowski (1980) estimated only 62 million for this population using the 1976 hectarage and assuming 100 per cent coverage. In Mali, the breeding area was estimated at 260 ha in 1980 (Manikowski 1981). Assuming 67 per cent coverage, the population estimate is 23 million adults.

No reliable survey data are available for three quelea breeding areas in Africa: the Senegal River Valley, the Sudan, and southern Africa. Therefore, I have assumed that the breeding distribution for these areas follows that given by Magor and Ward (1972) and that the density is similar to the nearest quelea population where density has been measured. In Senegal, since Magor and Ward's (1972) survey, the breeding range has moved south; now no breeding occurs north of 17°N. With 12 degree squares presently covering the breeding distribution, and assuming that the density is similar to that of the Niger River Delta, i.e. 65 ha/degree square, the population estimate is 47 million. Because of drought-related habitat changes, this is likely to be an overestimate.

For the Sudan, confirmed breeding has been recorded in 45 degree squares. The breeding density is assumed to be the average of the two adjacent countries; the Lake Chad Basin with a density of 174 ha/degree square and Ethiopia with a density of 24 ha/degree square, a mean of 99 giving an estimated population of 267 million. The Sudan is considered to have a larger quelea population than any other country in Africa, and this estimate suggests that the population is more than twice that of the rest of eastern Africa.

Magor and Ward (1972) showed that southern Africa has more confirmed breeding records than any other region of Africa. This may reflect on the number of active ornithologists who have recorded many 'small scattered colonies of perhaps a dozen nests' (McLachlan and Liversidge 1971). Such small colonies would have little impact on the total population, but their existence does extend the breeding range. Because it is unlikely that any major breeding colonies regularly occur south of 25° latitude, the 127 squares in southern Africa have, therefore, been reduced by the 39 occurring south of 25°, leaving a balance of 88 squares. This is still the largest distribution in any region in Africa. The colony density in the nearest measured area is Tanzania with 36 ha/degree square. Applying this to the southern African distribution gives a population estimate of 190 million birds. For the reasons described above, this is likely to be an overestimate because of migration within the region; for example, some quelea migrate from the northern Transvaal to Botswana (Jones 1972).

If the population estimates given in Table 3.1 (249 million) are combined with those of Sudan, Senegal, and southern Africa, the continental adult quelea breeding population would total 753 million. Because post-breeding quelea populations are considered to be the main cause of crop damage, this figure must be doubled to 1500 million birds to account for two chicks being fledged by each pair of adults in each colony per year. My estimate falls within the lower limits of the estimate made by Crook and Ward (1968), but probably is an overestimate given all the assumptions. Although no one has hazarded an estimate of the worldwide population of the House Sparrow *Passer domesticus* or the European Starling *Sturnus vulgaris*, the quelea does indeed seem to be the most numerous bird in the world. Its numbers are larger than the 500 million combined overwintering population of the four main pest species in North America, the Red-winged Blackbird *Agelaius phoeniceus*, the Common Grackle *Quiscalus quiscula*, the Brown-headed Cowbird *Molothrus ater*, and the European Starling (Meanley and Royall 1976). The quelea population also probably exceeds the 'more than 100 million' given for Wilson's Petrel *Oceanites oceanicus* (Campbell and Lack 1985).

It has been suggested that a correlation exists between serious damage outbreaks and regional population levels. Manikowski *et al.* (Chapter 12) mentions that quelea problems increased in West Africa following major flooding of the Senegal and Niger rivers. Unfortunately, no quantitative data are available to support this contention. In Zimbabwe, LaGrange (in press-*a*) found that in years with higher rainfall more birds were killed during control operations. No breeding data were available to show if this was due to more breeding in Zimbabwe or in the region as a whole during good rain years. The semi-arid nature of the quelea breeding habitat is such that it is unlikely that above-average precipitation will occur over the whole of its regional range. Therefore, regional population levels are unlikely to undergo major fluctuations, unless there is a major climatic change as has occurred in the Sahel.

Positive impact on the environment and man

The quelea as a study animal in ornithological research has contributed some important advances to the subject as a whole (Lack 1966; Jones and Ward 1976; Ward and Jones 1977) as well as over 400 published and unpublished papers on aspects of its biology and management. It also has aesthetic appeal to bird watchers who may marvel at its large numbers and flock manoeuvres and to aviculturists as a cage bird. Bruggers (1982) reported that about 25 000 quelea were exported as cage birds annually from Senegal between 1974 and 1978. Although illegal in the United States, pairs have been found in pet shops selling for $50 (R. Bruggers, pers. comm.), and for £20 in Britain (J. Parker, pers. comm.).

As a temporary food source, quelea bring positive benefit to humans (Chapter 23) and to many predators and scavengers (Chapter 16). The itineracy of the species usually prevents it from being exploited as food for more than a few weeks in any one place. For certain human tribes, it is obviously very welcome even to the extent of villagers collecting poisoned birds after a control operation (Chapter 23).

Vernon (in press) has speculated that quelea are involved in nutrient cycles, particularly that of the destruction and creation of acacia bush. He suggests that the guano deposited by colonies increases grass growth under the bushes, which promotes fire that destroys the bush. Overgrazing of the opened-up grassland causes bush encroachment and creates new suitable breeding habitat. Quelea guano is regularly collected at roosts in Nigeria (Ward 1965*a*; Plate 2); in Somalia, citrus farmers welcome birds roosting in their orchards even to the extent of hindering control operations. It is not known how useful quelea guano is for Sahelian agriculture, but it seems to be, at best, a local opportunistic benefit.

Quelea eat a variety of insects as a small component of their annual diet (Ward 1965*a*). Before breeding, they take mainly caterpillars, grasshoppers, beetles, and plant bugs (Jones and Ward 1976), some of which can be potential agricultural pests. Morel (1968) estimated that birds in a colony will consume 214 kg of insects compared to an average insect biomass around the colony of 0.7 kg/ha. However, because feeding usually occurs within 6 km of a colony (GTZ 1982), and colonies usually are several kilometres from the nearest cultivation, an immediate benefit to agriculture seems unlikely. Quelea have been observed feeding on American Bollworm *Heliothis armigera* larvae in ripening wheat (P. Ward, pers. comm.), Armyworm *Spodoptera exempta* larvae (Dyer and Ward 1977), and hopper instars of *Locusta migratoria* (H.F.I. Elliott, pers. comm.). These examples indicate that quelea could benefit agriculture, but they seemed to occur only under exceptional circumstances. Quelea sometimes feed in rice or wheat stubble. Farmers welcome these flocks because they consume the spilt grain and, therefore, clean the fields of seed that would germinate the following season.

In summary, it is clear that quelea can have a positive impact both on ecosystems in general and on agriculture and man, in particular. However, few of these benefits and aspects have been investigated in detail. Even so, it seems doubtful that any of them would economically outweigh the negative impacts of this species.

Negative impact on the environment and man

Apart from quelea's impact on agriculture, their negative effects seem limited to destroying trees and fouling waterholes. The former has been quantified only once; 20 ha of *Acacia arabica* plantations in the Sudan were 'flattened' by roosting quelea (Disney 1964) resulting in damage of $800 for reseeding and 15 years increment of timber. The absence of other records suggests that such occurrences must be rare. The quelea's habit of forming day-roosts near water leads to the fouling of waterholes in southern Ethiopia to the extent that the water becomes undrinkable for people and their stock (M. Jaeger, pers. comm.). This causes serious inconvenience to pastoralists who may depend on only a few sources of water as the dry season advances. The quelea's impact on cultivated cereals, however, is the main reason for its notoriety. I have analysed this impact at four levels; the amount of cereal consumed and destroyed on a daily basis by individual quelea, and the amount of damage caused locally, nationally, and over the species' entire range.

Table 3.2. Estimated quelea food consumption (g/day).

Situation	Consumption (g/day)	Grain	Source
None given	2.6	Sorghum	Da Camara-Smeets (1977)
Caged birds	4.6	Rice	Elliott (1979)
Counting crop/gizzard contents in wild birds	2.6	Rice	Elliott (1979)
Unknown	1.5	Rice	Mallamaire (1961)
Caged birds	5.0	Millet	Schildmacher (1929)
Unknown	2.0	Not specified	Serrurier (1965)
Extrapolation from energy studies in *Zonotrichia albicollis*	2.0–2.5	Not specified	Ward (1965*a*)

Individual daily damage

Table 3.2 gives the various estimates of daily food consumption that have been made for quelea. Feeding rates of caged quelea are greater than those of wild birds, and caged birds are often abnormally fat. Feeding estimates of wild birds suggest a consumption rate of about 2.5 g/day, but quelea are physically capable of eating up to 5 g/day. The daily diet can be composed entirely of cultivated grain. In estimating the damage potential of individual quelea, one must also estimate the amount destroyed but not eaten (Fig. 3.2).

When feeding on cereals, quelea mandibulate the grain to remove the husk, sometimes dropping whole or bits of grain. One sign of serious damage is the presence of husks and fragments of grain on the ground beneath damaged panicles. The amount of grain damaged is dependent on the stage of crop maturation. At the milky stage, quelea often leave large numbers of grain nipped or punctured. The punctured grain subsequently becomes dried or shrivelled (Jaeger and Erickson 1980) or can become infested with fungal or other diseases. It is impossible to count the number of grains eaten at this stage in dissected birds, because all that can be seen is an indistinguishable mush. Later, when grain begins to dry, it may shatter. Quelea flocks rising from and descending onto mature heads of rice and wheat, in particular, are likely to dislodge more grain that each individual would consume (Plate 3). Shattering also can be caused by wind making it difficult to distinguish from damage done by birds.

Opinions vary greatly as to the amount of grain destroyed but not eaten by individual birds. Most authors do not say how they reached their estimates,

(*a*)

(*b*)

Fig. 3.2. (*a*) Quelea can cling to the sides of grain heads and remove all the seeds without difficulty (photo: M-T. Elliott), leaving subsistence farmers like this Malian woman (*b*) with little or no grain to harvest (photo: R. Bruggers).

which may imply that they have merely guessed. Mallamaire (1961) and Elliott (1983*a*) doubled their estimates of daily food intake to reach figures of 3 g and 10 g, respectively, for the amount destroyed per bird per day; Mosemann (1966) reported 17–18 g/day. Bruggers *et al.* (1984*a*) reported a cage trial which showed that quelea have the potential to destroy 38 g/bird per day.

R. Luder (pers. comm.) attempted directly to measure consumption in wheat fields. He assessed damage (18.4 t) on one wheat farm, estimated the number of bird/days at 1.9 million and determined the proportion of wheat (55 per cent) in the birds' crop. He concluded that each bird ate an average of 1.65 g of wheat per day and would have had to destroy an additional 8.0 g/day to account for the estimated total loss. Jaeger and Erickson (1980) suggested that severe damage in Ethiopia (51 per cent over 35 000 ha—the highest ever measured in Africa) was caused by quelea destroying 100–600 g sorghum/bird per day. The loss was estimated by sampling only a total of 1000 sorghum panicles at five randomized sampling points. Although damage was undoubtedly severe, the amount of sampling in relation to total cultivation seems inadequate. Furthermore, M. Jaeger (pers. comm.) said that the quelea population may have been underestimated by at least five times, bringing the daily food consumption down to a more plausible 20–120 g/bird. My conclusion is that quelea damage to cereals may result in losses of about 10 g/bird per day. If damage occurs at the milky or early dough stage losses can be higher.

Local damage levels

Table 3.3 presents local damage data. By 'local' I refer to crops tied to a specific locality, which in most cases means tens or hundreds of hectares and sometimes several thousand hectares. I draw a distinction between these estimates and national estimates involving country-wide assessments of damage to particular cereals. Data presented in Table 3.3 have been chosen because they seemed to be reasonably reliable and were to some extent quantified or led to specific consequences.

It is clear that quelea can cause devastating local damage. This damage is not a new phenomenon, but occurred at the onset of large-scale cereal production in Africa—during the 1940s—and even as long ago as 100 years, when only traditional crops could have been involved. Some of the records illustrate that quelea can completely destroy small areas of crop in only a few days. Most losses amounted to tens or sometimes hundreds of tonnes; a few involved thousands of tonnes. The latter figures are restricted to Ethiopia, not because I think the quelea problem is so much more severe there, but because of the way in which Jaeger and Erickson sampled damage over wide areas.

Another aspect of locally severe damage is that it seldom occurs in the same area in consecutive years. Thus, in northern Nigeria sorghum was heavily damaged in 1961, but not in 1962. In the Ethiopian Awash damage was 14 per cent in 1976, but only 3.5 per cent in 1977. In Tanzania in 1979, 80–90 per cent of the wheat at West Kilimanjaro was damaged, but from then until 1985 losses have never risen above 5 per cent.

National damage levels

Many difficulties are involved in assessing damage caused nationally by quelea. These include the statistical decisions regarding the number of sampling sites and cereal heads necessary for valid samples. Such sampling logistics may be beyond the available resources of trained manpower. Reliable sampling is further complicated by the unevenness of damage among fields and, especially in subsistence farming, by asynchronous maturation (Elliott, in press).

Reports of bird damage by farmers often are exaggerated. Farmers know that bird damage can be devastating and unpredictable, and they regard bird pests in general, and quelea in particular, as a great menace to food production, even though serious losses may be infrequent (Manikowski 1981). Severe damage often begins with small flocks which gradually increase to sufficient numbers capable of destroying an entire crop. Farmers never know whether the small numbers present at the beginning of ripening will constitute a major or minor threat to their crops, but the mere presence of any birds often results in their requesting control action from local authorities. If the bird population remains low, its potential impact will be small compared to losses due to insects, weeds, lack of fertilizer, or poor management. Yet, birds often are the most vehemently accused. Such psychological factors make objective damage assessment very difficult.

The most comprehensive nation-wide survey of bird damage was carried out by a questionnaire submitted to the governments of 33 African countries in 1976 (FAO 1980*a*). The results showed that Botswana, Chad, Ethiopia, Kenya, Malawi, Mali, Sudan, and Tanzania rated quelea as major pests. Ten countries—Burkina Faso, Cameroon, Mauritania, Niger, Nigeria, Senegal, Somalia, Swaziland, Uganda, and Zambia—gave quelea an intermediate pest status. Outside the survey area, Mozambique, South Africa, and Zimbabwe can be included in the major pest category and Angola in the intermediate pest category (Elliott, in press). Recent developments suggest that the quelea should be regarded as a major pest in Cameroon and Somalia. However, given the psychological aspect to bird control, which can introduce a subjectivity into questionnaire-derived data (Dyer and Ward 1977), this survey probably reflects more the degree of concern of different countries about quelea than an accurate assessment of the quelea's pest status.

Some national or multinational damage estimates are presented in

Table 3.3. Quelea damage estimates at the local level.

Year	Area/Region/ Country	Estimated grain loss and/or value	Remarks	Source
1881	Ugogo-land; Dodoma, Tanzania	No data	Famine named 'kubwa-kidogo' (big-small); so named because birds responsible. 'Birds (probably Sudan Dioch) consumed most of the staple crop'.	Brooke (1967)
1942	Dodoma, Tanzania	$60 000	5081 t relief food imported. Scanty rainfall primary cause but 'an invasion from the north of Sudan Diochs contributed to low production in NE Dodoma'.	Brooke (1967)
1944	Ardai Plains, Tanzania	1215 ha wheat	Crop 'destroyed'.	Plowes (1955)
1953	Kenya	About 540 t (86%) wheat	Only 1000 bags harvested from expected 7000 after 'visitation of quelea'.	Plowes (1955)
1953	Near Richard Toll, Senegal	No data	Consideration was given to total abandonment of rice cultivation.	Morel (1965, 1968); Grist and Lever (1969)
1954	Near Richard Toll, Senegal	60 ha rice	Completely destroyed in a few days in September.	Morel (1968)
Before 1955	Filabusi, Zimbabwe	40 ha dwarf sorghum	Completely stripped of ripening grain in 2 days.	Plowes (1955)
1961/62	Northern Nigeria	No data	Dry season sorghum (massakwa) heavily damaged 1961 but no damage 1962.	Curtis (1965); Ward (1965*a*)
Before 1973	Lake Guier, Senegal	–	Rice cultivation abandoned because of birds.	Park (1973, 1975)
Before 1973	Senegal Valley	–	Millet abandoned because of birds.	Park (1973, 1975)
1973	N'Djamena rural zone, Chad;	15% of total sorghum crop	–	Manikowski and Da Camara-Smeets (1975*a*, 1975*b*)
	Maroua, Lere, Pala in Cameroon/Chad border	12% of total sorghum crop	–	

1975	Khartoum, Sudan	6.5% of irrigated wheat	–	Allan (1975)
1976	Jebel Sim Sim, Sudan	150 ha sorghum	Completely destroyed.	Bruggers *et al.* (1984*a*)
1976	Bongor, Chad;	354 t (12.7%) rice	–	Elliott (1979)
	Yagoua, Cameroon	38 t (0.9%) rice	–	
1976	Jijiga Plain, Ethiopia	51.4% of 35 000 ha sorghum	–	Jaeger and Erickson (1980)
1976/77	Awash Valley, Ethiopia	25 450 t (14%) sorghum in 1976; 3950 t (3.5%) sorghum in 1977; > $1.1 million	–	Jaeger and Erickson (1980)
1976	Zambia	3.2% of 583 ha of irrigated wheat	Three of eighteen farms had damage above 5%.	Jones and Pope (1977)
1977	Bongor, Chad;	358 t (14.2%) rice	–	Elliott (1979)
	Yagoua, Cameroon	722 t (25.8%) rice	–	
1978	Hawata, Sudan	44 ha (98%) sorghum; $18 000	Fields within 5 km of a breeding colony.	DWRC (1978)
1978	Nakuru, Kenya	42 000 ha (15.2%) wheat; $180 000	–	FAO (1981*a*)
1978	West Kilimanjaro, Tanzania	80–90% (2400 ha) wheat	–	FAO (1981*a*)
1978/79	Wanle Weyn, Somalia	31% of 122 ha sorghum	–	Bruggers (1980)
1979	Arusha, Tanzania	35% of 480 ha wheat	–	FAO (1981*a*)
1979/80	Niono, Mali	9.6% rice	–	Manikowski (1981)
1980	Jebel Sim Sim, Sudan	600 ha sorghum	Completely destroyed.	Bruggers *et al.* (1984*a*)
Before 1981	Office du Niger, Mali	3.5–15% rice	–	Manikowski (1981)

Table 3.4. Most of them can be harshly criticized. In several cases no supporting data are given; for example, Serrurier (1965) gives his methodology, but does not explain how he estimated the West African quelea population at 2000 million. My estimate for the same area, obtained as previously explained, would be 280 million birds after reproduction. National damage estimates not only need to be randomized within the sampling area, but also distributed throughout a country's production areas. None of the estimates achieved both these prerequisites. Bruggers (1980) randomly sampled damage in cereals in the Shebelle and Baidoa regions of Somalia in 1978/1979, but did not sample in the south or the north-west of the country. He found that in the main sorghum-growing inter-riverine area of Baidoa (71 000 ha) damage averaged less than 1 per cent. Near the Shebelle, where much less sorghum is grown, damage averaged 31–75 per cent. Bruggers very cautiously concluded that annual losses to birds in Somalia amounted to a minimum of $1 million.

In calculating damage in Sahelian West Africa, Manikowski (1984) sampled 443 fields, most of which were not randomly selected, and no account was taken of the total cereal distribution in the areas sampled. If sampling occurred only in the major bird pest areas, then the 8.3 per cent damage level should refer to those areas, rather than to all cereal-growing areas in the Sahel. Loss estimates in Sudan ($6.3 million; FAO 1980*a*) are the highest of any country, but are based mainly on questionnaire data (FAO 1980*a*) and are, therefore, unreliable. This unreliability was highlighted by Lenton (1981), who recalculated the estimates by updating cereal production figures and market values to reach a figure of $59 million for total losses per annum due to birds in Sudan. In retrospect, I believe that neither the original estimates I made, nor Lenton's updated estimates are credible. It would appear that insufficient information exists for Sudan to make a reasonable national estimate of bird damage.

In establishing quelea pest status by country, it would be useful to know how much damage occurs in the absence of lethal control measures. Unfortunately, in most countries where quelea are classified as a major pest, some form of lethal control has been done for many years. One exception would be Ethiopia, where damage was measured in 1976 before effective control began to be implemented. Losses in the Awash Valley were placed at 12–16 per cent (Jaeger and Erickson 1980). The Awash produces only about 26 per cent of the national sorghum production, and the damage levels that occurred in the remaining 74 per cent of the national production were not known.

Quelea damage varies annually. During 1976–1977, Ethiopia lost about $3 million to birds, but in recent drought years losses have been negligible (Hailu Kassa, pers. comm.). Even if all the correct randomized sampling procedures are followed, bird damage estimates made in 1 year in one area

Table 3.4. Quelea damage estimates at the national level.

Year	Area/Region/ Country	Estimated grain loss and/or value	Remarks	Source
1953	Kenya	\$1.5 million	Wheat.	Plowes (1955)
Before 1959	Senegal River Valley	*c.* 150 000 t; \$30 million	No supporting data.	Mallamaire (1959*a*)
Before 1962	Sahelian Region— Senegal to Lake Chad	*c.* 65 000 t; \$13 million	No supporting data.	Curtis (1965), in Drees (1980)
Before 1965	760 000 km^2, equivalent to West Africa	*c.* 94 000 t; \$18.9 million	Assumed quelea population of 2000 million birds eating 2 g/bird per day, 5–8% of which is cultivated grain.	Serrurier (1965)
Before 1968	Sudan	\$500 000	No supporting data; estimate for all granivorous birds.	Schmutterer (1969)
1974/75	Sudan	35 000 t sorghum; 9000 t millet; 16 000 t wheat; \$6.3 million	Based on FAO questionnaire and a few direct assessments.	FAO (1981*a*)
1975/79	Somalia	Minimum \$1 million	Extrapolated from direct estimates and from farm managers' reports.	Bruggers (1980)
1975/83	West Africa	8.3% of all cereals including maize	Average of 443 fields sampled, but including damage by all pest bird species.	Manikowski (1984)
1976/77	Awash Valley, Ethiopia	14 675 t sorghum; \$3 million	Random samples within Valley.	FAO (1981*a*)
1976–77	Senegal	\$3.9–4.9 million	Systematic sampling.	Bruggers and Ruelle (1981)
1978	Kenya	\$3 million	Extrapolation from direct measurements and farm managers' reports.	FAO (1981*a*)
1979	Tanzania	\$2.4 million	Extrapolation from many direct measurements on wheat, a few on rice, sorghum; for millet, extrapolation only from experience in other countries.	FAO (1981*a*)

cannot be considered representative of the situation. This suggests that even the better estimates probably are an insufficient statement of the bird problem. If more accurate and representative estimates of quelea damage are required, then more resources will have to be made available to achieve them.

Continental damage levels

Using a population estimate of 1500 million birds, 25 per cent of the total quelea population attacking cereals for 30 days/year, an average of 10 g/bird per day of cereal being destroyed, and an average cereal crop value of $200/t, then continent-wide damage would be about $22 million per annum. Production levels for the four main cereals vulnerable to quelea in the eleven countries which report quelea as a major pest (FAO 1980*a*) suggest that total production is 12 531 915 t, which at $200/t would be valued at about $2506 million (Table 3.5). If all the estimated $22 million quelea damage was restricted to these countries, which it is not, the loss would amount to only

Table 3.5. Production estimates in tonnes for four cereal crops vulnerable to quelea based on reports or personal observations in the eleven African countries where quelea are a major pest.

	Crop			
Country	Sorghum	Millet	Rice	Wheat
Botswana	15 000[a]	6000[b]	–	–
Chad	523 000[b]	580 000[a]	51 000[b]	–
Ethiopia	685 000[a]	190 000[a]	–	500 000[a]
Kenya	134 400[a]	130 000[a]	42 555[b]	215 000[a]
Malawi	146 000[a]	–	36 000[b]	–
Mali	699 600[c]	930 000[a]	30 000[c]	–
Sudan	2 849 000[a]	501 000[a]	–	150 000[a]
Tanzania	220 000[a]	150 000[a]	200 000[a]	68 080[a]
Mozambique	155 000[a]	5000[a]	100 000[c]	–
South Africa	268 000[a]	15 000[a]	–	2 440 000[a]
Zimbabwe	131 000[a]	190 000[a]	–	176 280[c]
Total [d]	5 826 000	2 697 000	459 555	3 549 360

[a]Source: FAO (1982*a*).
[b]Source: FAO (1981*a*).
[c]Source: Other information.
[d]Total cereal production = 12 531 915 t. Total value at $200/t = $2506 million.

about 0.75 per cent of these countries' total quelea-vulnerable cereal production. This suggests that viewed over the quelea's entire range the impact on food production is negligible.

Discussion

Ample evidence has been provided to show that quelea can cause serious local losses to cereal crops and considerable hardship to subsistence farmers. On the rarer occasions when larger farms suffer heavy damage, a significant economic loss must be felt by those particular farms, even if the impact of that loss on national production may be slight. Local damage can also have a number of indirect demoralizing effects on the process of crop production; for example, government policies to encourage development of high-yielding, drought-resistant millet and sorghum varieties to replace maize may be undermined if quelea attack the crop. Thus, efforts to distribute a high-yielding sorghum in Ethiopia were totally disrupted when, by chance, a 100-ha demonstration plot of the new variety was severely damaged by quelea (G. Brhane, pers. comm.). In some cases, damage may not be severe; yet, farmers may refuse to grow sorghum because of the birds. Such indirect effects must be taken into account when deciding national policies toward the problems caused by quelea.

Despite the sometimes serious local damage caused by quelea, the available evidence suggests that, if losses are viewed relative to national cereal production, they probably represent only a few per cent of total production even in the most seriously affected countries. In Senegal, losses due exclusively to quelea were put at 2.5 per cent of total vulnerable cereal production (Bruggers and Ruelle 1981). In Tanzania, the loss of cereals valued at $2.4 million in 1979 is only 1.9 per cent of production. In Ethiopia, the only damage estimates are from the Awash Valley, where most of the known quelea breeding occurs. Sorghum grown elsewhere probably suffers much less, so it would be surprising if overall losses for the whole country exceeded 5 per cent. These low proportional losses should also be viewed in the context of the present vigorous expansion of cereal production in Africa; for example, wheat cultivation in Tanzania has increased from 15 000 ha in 1978 to 35 000 ha in 1981/1982 (Mosha and Munisi 1983). This expansion and the encroachment of the growing human population on quelea breeding habitats will also eventually reduce the overall impact of quelea on national cereal production.

The conclusion that quelea have a negligible direct influence on crop production (< 1 per cent) in continental Africa and usually cause < 5 per cent losses to national crop production, places quelea firmly into the same bracket as bird pests elsewhere in the world. Stone and Mott (1973*b*) estimated that the 500 million U.S. blackbird population accounted for less than 1 per cent of the total maize crop. Damage within individual states was also not high, i.e. in Kentucky it was 0.48 per cent, a loss of $1.2 million (Stickley *et al.* 1979*a*). European Starling damage to freshly sown wheat in Britain did not affect overall yield (Feare 1984); Kentucky and Tennessee averaged only a 1.5 per cent loss (Dolbeer *et al.* 1978). These low overall damage levels include, as with quelea, serious local losses of grain. One per cent of the fields sampled by Dolbeer *et al.* (1978) suffered over 25 per cent damage. The low overall damage also does not stop the complaints of individual farmers in these developed countries who incur serious losses, nor does it interrupt research efforts to reduce or prevent it.

Although the pest status of quelea appears similar to that of other pest birds elsewhere in the world, the impact of quelea damage is totally different. Europe and North America produce vast surpluses of grain, while Africa faces famines and major food deficits. Farmers in the developed world can relatively easily absorb losses caused by birds, but the majority of African farmers are growing cereals for basic survival and have few other options. If birds cause serious local losses disaster can result.

In the past, many African governments have probably been led to believe that quelea are a direct constraint on cereal production and that, if quelea are controlled, a direct boost to production will occur. The conclusions reached here may come as a surprise. However, I believe that the combination of direct, serious local losses that can affect large cultivated areas and the indirect effects, especially the psychological constraint on farmers caused by quelea damage, justify the priority that governments have given to quelea control. It should also be pointed out that even low losses constitute a loss of millions of dollars' worth of production. Overcoming the other perhaps more important constraints on cereal production, such as poor agricultural practices, lack of fertilizers, and inefficient management and planning, may involve much more costly extension programmes that would have to go on for years and involve hundreds of extension workers. Provided that the cost of carrying out control remains well below the potential damage that quelea can cause, there should be little cause for concern about control measures. At the same time, the over-sensitivity of farmers to birds and their tendency to exaggerate the seriousness of the problem, and our lack of ability to predict the potential damage that quelea can cause, suggest that the pest status of quelea requires greater in-depth study. Our understanding of the timing of exactly when quelea pose a serious threat to cereals needs to be improved, if we hope to effectively and efficiently direct control operations at those quelea that genuinely threaten crops.

4

Monitoring the quelea

CLIVE C.H. ELLIOTT and
GRAHAM M. LENTON

Introduction

To monitor a living thing is to gather information regularly on its dynamic status; part of the monitoring process is then to record and interpret the data collected. Animals may be monitored both internally and externally in many different ways. We are primarily concerned with the quelea as a whole animal, the changes in its distribution in relation to cereal crops, the description of local populations, and the effects of different sorts of control on damage levels.

One aim of monitoring a pest species such as the quelea is to develop a system to forecast and prevent damage. We will discuss whether this aim can be realized by the National Bird Control Units (NBCUs) responsible for quelea control in the present state of knowledge. We hope also to familiarize readers with the standard measurements used in sampling quelea populations, concepts that will be encountered later in the book.

Surveying techniques

For a species that is reputed to darken the sky with its immense locust-like swarms (Plate 4), it might be thought that to locate the quelea would be relatively straightforward. Yet, it is possible to drive several hundred kilometres through typical semi-arid quelea habitat without seeing a single bird. This would certainly be the case if one's survey was in the Sahelian zone and one chose to look for quelea between 1000 and 1600 h local time in the hot dry season. Quelea collect in day-roosts in the shade near water as morning temperatures increase and do not stir until it gets cooler in the late afternoon. The singing-chatter, which is normally the main clue to the presence of a day-roost, cannot easily be detected from a moving vehicle.

In eastern and southern Africa, where quelea often occur at higher, cooler altitudes, daily activity is still concentrated in the first and last 2 or 3 h of daylight; small flocks may occasionally be seen at other times of day. Only during the breeding season, which coincides with the rains and with cloudy, cooler conditions, is quelea activity spread throughout the day. All surveys, therefore, have to take daily activity patterns into account.

Surveys must also of course take into account the distribution of the species; breeding distribution has been described in Chapter 3. The geographical range as a whole coincides with three distinct vegetation zones: grass steppe, dry savannah, and montane grassland (Ward 1966). These zones receive a minimum of 300 mm of rainfall and usually not more than 600 mm per annum. They normally have a short period of vegetative activity favouring the growth of annual grasses. The range of potential habitat is thus enormous. Quelea are 'liable to appear anywhere in Africa south of the Sahara, away from rain-forest regions and the wetter savannahs bordering the forest and from the higher mountains' (Magor and Ward 1972). In altitudinal distribution, quelea have, for example, been seen in Kenya from sea level (Haylock 1955) up to 2900 m on Mt. Kenya (M. Jaeger, personal communication).

The season should also be taken into account when planning a quelea survey. The migratory nature of the quelea may cause entire local populations to vacate large areas after the onset of the rains. Ward (1971) found no significant numbers after a wide search in Kenya and Tanzania in late November. Quelea were also completely absent from his weekly transect counts in Nigeria at 12°N from August to October, due to their absence first on early-rains migration and later by breeding in a geographically separate area from the transect site (COPR 1977). Of course a negative survey, caused by absence on migration, may be important for interpreting migration patterns, but the possibility of absence due to this cause rather than an unsuitable habitat should be considered. Having allowed for these constraints, the surveying options are limited to terrestrial or aerial methods, but the technique used will depend on whether the object is merely to locate quelea, census populations, or find roosts or breeding colonies.

Terrestrial surveys

The simplest and most reliable way to locate and identify quelea is on foot so that each flock can be checked with binoculars. If flocks of more than about 200 sparrow-like passerines are seen, they usually can only be quelea. However, the inexperienced observer can easily confuse smaller flocks with such species as the Pin-tailed Whydah *Vidua macroura* in non-breeding plumage and, especially in western Africa, the Golden Sparrow *Passer luteus*.

In eastern Africa, identification problems occur with the Chestnut Weaver *Ploceus rubiginosus*, the Grey-headed Social Weaver *Pseudonigrita arnaudi*, and with Red-headed Quelea *Quelea erythrops* and Cardinal Quelea *Q. cardinalis*.

The method used by Jones and Ward (COPR 1977) to survey quelea populations across their north-south distribution in Nigeria was to walk at dawn for 90 min along a prescribed pathway and count the quelea seen. Each transect was followed once a month at six different points between 13°30′N and 9°0′N. While this technique may be useful to assess changes in the 'centre of gravity' of the quelea population, the amount of country that can be covered on foot is too small either to locate the bulk of the population, to give reliable population estimates, or to locate breeding colonies.

Vehicle-based surveys have been most frequently used by Manikowski (1981). He drove along fixed census routes at regular weekly or monthly intervals. Vehicle speed was 30–35 km/h and the effective census distance was taken as 50 m on each side of the road. He stopped to verify species identification when necessary. It should be possible to drive about 100 km in the morning and again in the evening, enabling a sizeable chunk of quelea habitat to be covered. Surveys are limited to motorable roads, and this may be a serious limitation in uninhabited bush or during the rains when roads may be impassable. Manikowski (1980, 1981) used the data collected by vehicle survey to make population estimates of quelea in the Chari–Logone area of the Lake Chad Basin and to estimate population densities in the central delta of the Niger River in Mali.

Another important part of terrestrial surveying is to locate breeding colonies. Surveys of course have to be scheduled for the breeding season, if it is known, starting about 5 weeks after the onset of the rains. The presence of a breeding colony means many birds will be in the area since most breeding colonies involve at least several tens of thousands of birds, and sometimes hundreds of thousands. The activity of a colony is normally limited to an area with a radius of about 6 km (GTZ 1982), but radio-telemetry has shown that it can be as little as 2–3 km (Bruggers *et al.* 1983); other studies have shown that it can be as much as about 11 km (Jarvis and Vernon, in press-*a*). It follows that unless a survey road happens to pass within 2–11 km of a colony, the site may be missed altogether.

Compared to the tight, elongated clouds of birds moving to a roost, breeding quelea fly in looser formations on a broader front. Such flocks can be sampled and dissected for active breeding status before the effort is made to find the colony. Elliott and Manikowski (1976) described how the compass bearing of flocks flying purposefully in a consistent direction was plotted on a map to reveal the likely location of colonies in northern Cameroon. It may be impossible to reach the likely colony site by car if conditions are wet, so this usually has to be done on foot or by air. If on foot, the task can sometimes be formidable because up to 10 km of difficult, flooded terrain may have to be covered each way.

Quelea roosts are mainly found by terrestrial survey. Quelea only begin to move to a roost about 45 min before dark, and it is often only in the last 20 min that they begin to settle in to their final roosting site. The birds have, therefore, to be followed by car or on foot during this narrow time period. Roosts can be up to 30 km from a feeding site, although most often they are within 5 km of the last feeding area. Depending on the terrain, it may take several evenings to follow the birds to their roost. It is also essential that the position of the birds is checked until it is completely dark, because sometimes quelea shift position at the last moment. At dawn the departure of the birds is normally more rapid, taking only about 15 min, so that back-tracking the birds at dawn is hardly worth the effort.

Aerial surveys

Manikowski (1981) has been the chief practitioner of the aerial survey for locating quelea populations and for estimating their density. In Mali, he used a light aircraft flying at 30 m above the ground. The numbers of birds seen in a band of 50 m in width beneath the aircraft were counted, and the position on a map of each sighting was plotted. The area surveyed was divided into 50 × 50-km squares and the density observed during the flight within each square was presumed to be uniform throughout that square. The conclusions on the census were extrapolated from a sample of 0.1 per cent of the total area surveyed, assuming one flight only over each square. A similar technique was used for the quelea population in the Chari–Logone area of the Lake Chad Basin (Manikowski 1980).

In eastern Africa, aerial survey has been used to locate breeding colonies more for control purposes than for general censuses. Although this can be done with a fixed-wing aircraft, the most effective quelea survey is carried out by helicopter. In most parts of savannah Africa, given a reasonable rainy season, the potential breeding habitat is enormous with the result that colonies are seldom located by random survey. Information is, therefore, needed to home in on a potential breeding area, including the rainfall pattern which would produce optimal grass growth (Chapter 9), knowledge of past breeding sites, and quelea reports from field scouts, villagers, herdsmen, and cereal farmers. If the information received refers only to the presence of birds and no specific breeding sites are given, surveys can search for a likely looking breeding habitat such as dense patches of acacia adjacent to plains with luxuriant grass growth and nearby water. Flying height is usually about 50 m and survey speed 200 km/h.

Aerial surveys of colonies can be conducted at any time of day. Early morning and late evening are preferable because dry quelea nests seem to glow in the low rays of the sun, and colonies can sometimes be spotted from 3 or 4 km away. Having located the colony, the helicopter can land next to the site to inspect its stage of development. This is important for timing any subsequent control operations.

Aerial surveys can rarely be used to locate roosts because the noise of the aircraft or helicopter tends to disrupt the flights of roosting birds. It also becomes increasingly difficult to see the flocks against the ground from above as dusk falls, whereas from the ground the birds can still be clearly seen against the sky.

Surveying—the importance of co-operation with local people

In most countries that suffer from quelea damage, finding colonies near cultivation is an essential part of the control strategy. Therefore it is in the interests of local farmers to help locate the colonies. In Tanzania, a programme has been followed of encouraging villagers to report colonies to the nearest agricultural official and to support their reports with samples of nests and their contents. This enables the control team to arrive in a district, contact the local official, and inspect the nests. The unique form of the quelea's nest enables immediate identification, and the nest contents allow the colony to be aged. False reports of species of weaverbirds that are less harmful to crops than quelea or are harmless can be ignored.

Good co-operation with villagers can also help in the location of roosts. Their local knowledge may allow the roost to be found immediately instead of necessitating several evenings of back-tracking flocks from fields where damage is occurring. Farmers can also be encouraged to find the roosts themselves before sending in a report. If some degree of co-operation from the farmer in locating concentrations is not forthcoming, he cannot consider the bird damage losses to be a high priority.

Surveying—other techniques

Chapter 6 describes how radio-telemetry has been used to detect breeding colonies in Ethiopia and how it might be generally useful for this purpose and for finding roosts: searching time can be much reduced. However, the method may be too complicated to be incorporated into most quelea control operations at present.

The possibility of using infra-red detectors or photography to locate roosts or colonies at night has also been considered. Studies in the USA showed that Red-winged Blackbirds *Agelaius phoeniceus* are surprisingly well insulated

(Bray *et al.* 1978). More heat was radiated by the decaying vegetation in which the birds were roosting than by the birds themselves. Similar problems seem likely if the same techniques were to be tried on quelea. Suggestions have also been made that satellite imagery might help to detect quelea breeding sites, but the level of reception technology and expense required seems well above the resources that are presently available in most parts of Africa.

Sampling the quelea population

The authority for all concerned in sampling quelea is the 'Manual of Techniques Used in Research on Quelea Birds' (Ward 1973*b*). The manual has been followed in subsequent publications such as the Bird Scout's Handbook (FAO 1978) and most research papers have used the standardized indices. In almost every respect, the description of techniques and indices in the manual hold as good today as when they were first produced.

Quelea samples can be collected by shooting, mist-netting, or after control operations (Ward 1973*b*). For large samples, post-spray collections are ideal, and the only limit on their size is the time available to process them. The quelea is probably the most extensively sampled species in ornithological history, with samples of thousands of birds being common. The sampling technique used will depend on the data being sought. Ward (1973*b*) discussed the various factors involved, the kind of sample needed, when and where it should be collected, and what size it should be. For example, post-spray samples would be useless for diet studies (because the crops are empty), but ideal for mask index studies since several hundred males in breeding plumage can be quickly collected.

Standard measurements of the quelea

Ward's (1973*b*) manual also covers the standard bird-in-the-hand measurements applicable to all ornithological research, including wing, tail, bill and tarsus length, body weight, plumage and moult, sex and age (Fig. 4.1). For quelea body weight, the removal of the crop contents is essential, since the crop can contain as much as 2.9 g (wet weight) of food (Ward 1978) or 15 per cent of fresh body weight.

The manual gives indices for the black face mask polymorphism (Fig. 4.2) and for cranial pneumatization (Fig. 4.3), both of which are illustrated here as they are frequently referred to elsewhere in this book. An index is also given for mask-moult, and beak colour changes in females and juveniles are described. To see quelea flocks containing many females with bright yellow

Fig. 4.1. Monitoring quelea populations—samples of birds are taken and data collected in the field on morphological features such as moult, cranial pneumatization, mask indices, and gonad development (photo: R. G. Allan).

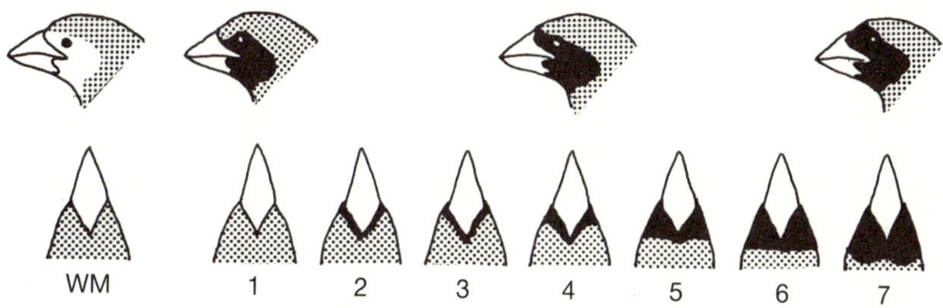

Fig. 4.2. Standards for mask index. Only the black-masked morph males are classified. The individual on the left is a white-masked morph; the extent of the mask is often difficult to see in such birds (source: Ward 1973*b*).

beaks is one of the clearest in-the-field indications that breeding is imminent. By contrast, the presence of males in full breeding plumage, but accompanied by females mainly with red beaks, usually means that breeding will not take place for many weeks, if at all. The manual describes how to age a quelea colony from its state of development and the size of the chicks. This guide is routinely used by control teams to establish the date of colony installation and to work out how many days remain in which to carry out control operations.

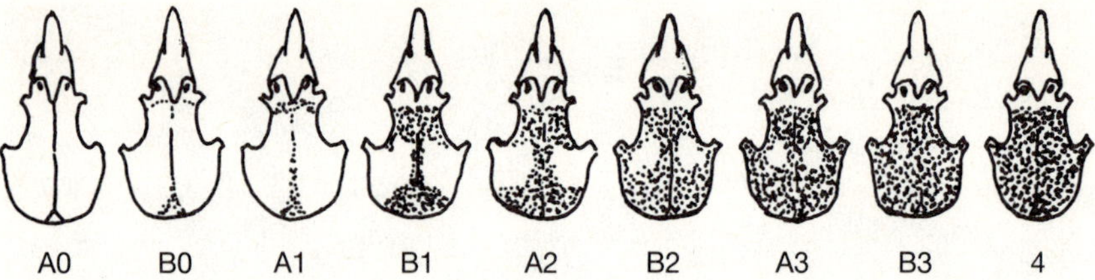

Fig. 4.3. Index of cranial pneumatization. The stippling shows pneumatized areas (source: Ward 1973*b*).

The only significant error in the manual is the use of a 0–4 scale for scoring primary moult instead of the internationally accepted 0–5 scale (Ginn and Melville 1983). No reference is made to the quelea's frequent suspension of moult, especially in the eastern African population. Suspended moult is a regular feature of some palaearctic migrant passerines in which the moult is completed in part before migration and part occurs in the winter quarters (Mead and Watmough 1976). In quelea, a similar system seems to operate in that moult is suspended some time after breeding when a local movement or migration to a new breeding area occurs. When the second breeding cycle nears completion, moult begins in the primaries either at the point of suspension or there *and* at the beginning again. Sometimes two waves of moult, even three, can pass descendantly through the wing, similar to the 'staffelmauser' or 'step-wise' moult described previously only in non-passerines (Stresemann 1965). For monitoring purposes, the importance of this pattern, apart from indicating the condition of the bird, is that it suggests on how many previous occasions individual quelea have bred in one particular season. The onset of post-juvenile moult is also a useful indicator: post-juvenile primary moult begins 9-13 weeks after hatching and is completed 23–30 weeks later (Elliott 1981*a*). Therefore juveniles found not to have started primary moult would have hatched less than 9-13 weeks earlier.

Estimating quelea numbers

This is an important part of the procedure for assessing potential impact on cereal crops. Counting the numbers of birds in a roost is done at dusk or dawn as the birds enter or leave. Because an overall impression of numbers can often be misleading, we have found it essential to count individual flocks before summing them for the total.

To estimate the number of quelea in a colony, a rough indication can be obtained by assuming that each hectare contains about 60 000 adults (S. Manikowski and C. Elliott, unpubl. data), but allowances must be made for colonies with exceptionally high or low nest density. If time allows, actual density can be measured in randomly selected quadrats. The hectarage itself may be estimated by mapping out the colony, but this is a laborious process (Elliott 1981*b*). An alternative is to estimate the area from the air by timing the distance between flags placed at each corner: this can be accurate if the colony has a regular shape, but if it is irregular, overestimation of the size is likely to occur (Elliott and Manikowski 1976).

Standardized forms for collecting quelea data

As part of the attempt to establish a monitoring network for quelea in eastern Africa, a number of standardized forms have been created (FAO 1981*b*). These forms can be used to record quelea data according to the standardized indices and methods, and easy comparisons with past studies can then be made. We will describe the use of all seven forms but actually illustrate only two of them (Fig. 4.4*a* and *b*).

The Bird Pest Survey Report is designed as the backbone of all incoming data of a monitoring system. It covers quelea and other common bird pest species and includes the main factors likely to influence or be involved in an outbreak of damage including the activity, composition and number of the bird pests, the cereal crop involved and its stage of development, the rainfall situation, the ground condition, and what sort of control has been used.

The Roost Report can be used to record details of any quelea roosting site including a rough sketch map of its shape and orientation to nearby landmarks. If an accurate map of the roost is drawn, using a compass and a pedometer or its equivalent, the degrees and distance of each bearing can be entered. The form provides information to pilots who might spray the area. The Breeding Colony Report is used to record details of quelea colonies.

Crop damage estimation is considered an important part of the evaluation of any quelea control operation (Elliott 1981*b*; Jaeger and Erickson 1980), and another standardized form provides the means to record damage levels. Two of the forms, which we feel are worth illustrating here, permit recording of all the standard measurements from sampling and dissecting quelea (Fig. 4.4*a*) and describing the critical standard parameters by which populations can be distinguished (Fig. 4.4*b*). We have designated the latter form as 'Quelea Population Finger-Print'. Finally, another form is used to record the necessary details of control operations.

These standardized forms were officially introduced in 1981 in eastern Africa. Since that time, their frequency of use has been extremely variable. The reasons for this reflect on the practicalities of using a monitoring system in present-day circumstances in Africa.

Reference ______________ Country ______________

Date ________ Time ________ Locality ______________

Method ______________ Recorder ______________

	QQ MALE			QQ FEMALE																		
		M		Follicle				Food			Body											
	M I	M I	Test (mm)	UN	Sze (mm)	Bk. Col.	Cran.	Wld Sd.	Clt Sd.	Ins.	Wt. (g)	Moult 1	2	3	4	5	6	7	8	9	Mlt Scr.	Remarks
1																						
2																						
3																						
4																						
5																						
6																						
7																						
8																						
9																						
10																						

(*a*)

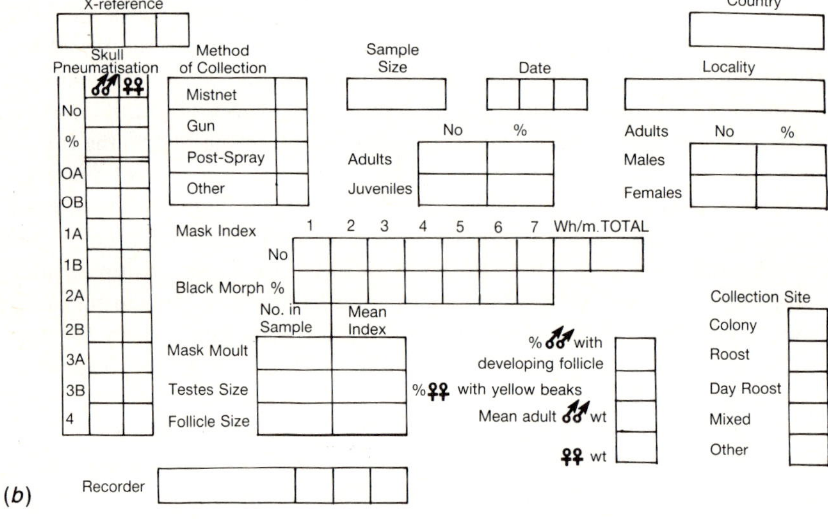
X-reference
Country
Skull Pneumatisation ♂♂ ♀♀
No
%
OA
OB
1A
1B
2A
2B
3A
3B
4
Method of Collection
Mistnet
Gun
Post-Spray
Other
Sample Size
Date
Locality
No %
Adults
Juveniles
Adults No %
Males
Females
Mask Index 1 2 3 4 5 6 7 Wh/m. TOTAL
No
Black Morph %
No. in Sample
Mean Index
Mask Moult
Testes Size
Follicle Size
% ♂♂ with developing follicle
% ♀♀ with yellow beaks
Mean adult ♂♂ wt
♀♀ wt
Collection Site
Colony
Roost
Day Roost
Mixed
Other
Recorder

(*b*)

Fig. 4.4. (*a*) Quelea dissection data sheet; (*b*) Quelea population 'finger-print'.

Potential purposes of monitoring

In the context of the quelea as a pest, it would seem that there are three types of monitoring that can be useful:

(1) Monitoring quelea in response to reports of quelea concentrations near farmland or causing damage, and then, in the event of a control operation, of the results of the control.

(2) Routine monitoring of known roosting and nesting sites at times when they may be occupied and in areas where damage has occurred in the past.
(3) Monitoring the quelea population on a country-wide regular basis through a network of observers with the object of forecasting outbreaks of damage.

Monitoring reports of quelea concentrations or damage starts from the moment that a report is received. Data on the birds and on the damage, if any, are then gathered until either the concentrations disperse, are successfully controlled, or the damage ceases. The data gathering initially provides the basis for a rational decision on whether or not to carry out control and, subsequently, on the effectiveness of the control in protecting crops.

A systems flow diagram of the different stages in this first type of monitoring is shown in Table 4.1. Some of the information is not necessarily easy to collect; for example, assessing the area of cultivation near the quelea concentration. Methods have recently been proposed for assessing crop areas by aerial survey (Otis 1984), but they have still to be thoroughly tested in the field. Alternatively, such data can be solicited from agricultural or other officials and may not be reliable. The field worker may be forced, therefore, to rely on a subjective impression of the importance of vulnerable cereal cultivation in the area under investigation.

Following a decision to carry out control, monitoring the results involves recording the technical details of the spray operation itself (flying time, volume of chemical used, etc.), the mortality inflicted on the birds and, most importantly, the effect on the numbers of birds attacking the crops for which protection is being sought (Elliott 1981*b*). Ideally, the final stage in this lengthy process would be to assess at harvest the damage levels in the area being protected, to show that damage had not significantly increased from the time of control up to the time of harvest. All these stages require a great deal of work and suitably trained manpower, and they may not all be possible to complete. However, the desirability of proper monitoring is clear because it provides a complete set of data by which the effectiveness of an NBCU can be objectively gauged.

The second type of monitoring is that of known roosting and nesting sites. This is an extension of the first in that instead of waiting for reports to come in, the NBCU attempts to speed up its actions by taking advantage of previous experience. For quelea colonies, the delay often involved in collating villagers' reports can make effective control more difficult, because colonies may have reached an advanced stage by which time high percentage kills are hard to achieve. If the NBCU can check sites known from previous years to have posed a threat to crops, at the earliest possible moment that the threat may develop, then efficiency of control will improve. Roosts and

Table 4.1. Flow diagram for monitoring quelea in response to reports of a bird concentration or of damage occurring.

Visit report site

↓

Observe birds; if they are quelea

↓

In non-breeding plumage or breeding plumage

↙ ↘

Non-breeding plumage	Breeding plumage
Time of year/condition of habitat suggests roosting. ↙	Time of year/condition of habitat suggests breeding.
	↓
Terrestrial survey to locate roosts.	Sample birds, dissect, and assess sexual condition to confirm breeding.
↓	↓
Count numbers of birds entering or leaving roost; sample birds in roost for per cent eating cultivated grain.	Aerial or terrestrial survey to locate breeding colony.
	↓
	Estimate size of colony, number of breeding birds.
↘	↙

Assess cereal cultivation near concentration: how many hectares, stage of maturity, how many days to harvest; are the quelea already attacking the crop: how many in the fields; if damage has already started, assess damage levels.

Decision on control operation (Chapter 18)

Yes ↙	↘ No
Monitor results: number and per cent quelea killed. Has the number of birds in the fields dropped to insignificant levels?	Continue monitoring quelea until some aspects change in favour of control or against (such as the birds disperse, the crops are harvested).

colonies occurring in areas and at times of year when the threat to crops is non-existent or negligible can be ignored by this type of monitoring.

The work involved is for each site to be visited at the appropriate time and, if occupied, for data on quelea to be collected that will assist in making the decision of whether or not to spray. For roosts, this would include monitoring the diet and feeding behaviour of the birds and the precise condition of nearby crops. For colonies, it would include monitoring colony development until the colony has stabilized in size and activity.

The third type of monitoring, a country-wide regular network monitoring of the quelea population, is much more general in its nature and objectives. It seeks to improve control strategy through a better understanding of why and when quelea damage cereal crops. The information gathered would, in theory, allow the identification of the specific subpopulation of quelea that would need to be controlled in order to prevent damage. Population monitoring is, therefore, geared toward discovering the source population from which the quelea causing the crop damage are derived. The entire quelea distribution within the country or region in question must be covered by a network of monitoring stations. These would provide the inflow of information by means of which influxes of birds from one area to another can be recognized, both by simultaneous changes in numbers and from the composition of the population as revealed by 'fingerprint' sampling data (see above). Thus, if a large decrease in quelea numbers at one monitoring station is accompanied by a large increase elsewhere and the fingerprints match up, evidence for the movement is quite strong. It may, of course, be strengthened further by using techniques such as ringing and fluorescent-marking (Chapter 5) and perhaps trace element analysis (Chapter 7). The most important aspect of this type of monitoring is that it should be systematic, regular, and widely distributed. If movements across national borders are suspected, then it must involve co-operative monitoring between neighbouring countries.

The positioning of sites within a population monitoring network must represent the types of habitat occupied by quelea at different times of the year and must reflect the known distribution of the species within an area, country, or region. The more survey areas and sample collection points there are, the more comprehensive will be the resultant interpretation. Accuracy will be further enhanced if surveys are made and samples taken simultaneously at several of the selected sites. Exploratory monitoring in areas where quelea are unknown, but could occur on evidence of habitat or gaps in distribution, would be included in this third type of monitoring.

Discussion

The status of the quelea problem varies considerably from country to country (Chapter 3). In some countries, there is concern about the damage to crops, but little information on its extent and the possibility of control operations being undertaken is only beginning to be considered. In others, quelea studies and control operations have been going on for many years. We suggest that the type of monitoring required in these two circumstances is different.

In countries where quelea control is well established, such as Tanzania, Ethiopia, and Sudan, it is important that a monitoring procedure similar to that described in Table 4.1 should be followed as closely as time and trained manpower allow. Registers of reports received and their fate will allow the performance of an NBCU to be evaluated. It is also clear that routine monitoring of roosts and nesting sites improves efficiency and, therefore, is desirable for a developed NBCU.

The situation is less clear in relation to the usefulness of population monitoring. From the point of view of the biological study of the quelea, much can be learned from routine and systematic monitoring of its populations. Because quelea has been studied more than any other tropical bird, enough is known about the quelea to pose all sorts of intriguing questions, some of which could be answered by a successful monitoring programme lasting a good number of years. But does this have any bearing on the quelea as a pest species?

Population monitoring would certainly seem to benefit an NBCU in its initial development stages. It would provide basic information on the quelea such as population size and distribution, breeding season, and the demography of the birds causing damage. Coupled with information on the damage, the distribution of vulnerable crops, their growing seasons, and the places where damage occurs regularly, an NBCU can formulate an effective control strategy. But having reached that stage, is it necessary for an NBCU to continue with wide-scale monitoring?

An intensive monitoring programme could reveal the origins of the quelea causing damage and the colonies they occupied before causing damage. If so, it would be far easier to control the birds while concentrated at source than when they disperse to smaller, unstable roosts. By source, we mean the breeding colony from which the adults or the progeny move on to cereal crops. Jaeger and Erickson (1980) succeeded in applying this control strategy in Ethiopia, although they developed the idea more by trial and error rather than by monitoring. In other regions where the geographical features less obviously channel the quelea towards the crops, monitoring might show a similar situation in which the number of source colonies was few. Alternatively, monitoring might show that birds causing damage derive from a wide catchment area of colonies: from each of many colonies perhaps only a small percentage of the population ends up attacking crops while the rest eat wild grain or migrate elsewhere. If this was proved, it would not make economic or ecological sense to control colonies as sources; it would only be necessary to control colonies that were themselves damaging crops or nearby crops.

Population monitoring in the broadest sense, including monitoring the condition of the habitat and the availability of wild food, might help to establish the ecological conditions prevailing either around source areas or around cereal-growing areas, which would lead to heavy bird pressure on crops. If such forecasting could take place early enough, control operations could be carried out against only those concentrations of quelea that posed a major threat to crops, probably saving a great deal of effort and expense. On the other hand, the ecological clues to serious quelea damage may prove so subtle and difficult to collect that a forecasting system could not in the foreseeable future be implemented by any NBCU.

The maintenance of a population monitoring network is an expensive undertaking for an NBCU. The data will be unreliable unless they are collected by highly motivated, well-trained staff, so heavy investment in training will be necessary. A properly run network requires that surveys are carried out at all times of the year and in areas where cereal crops are not widely grown, as well as during the growing season and in agricultural areas. This requires extra inputs of fuel, vehicles, and other costs. At present, it must be doubtful that any country in Africa has enough well-trained manpower to operate a monitoring network with sufficient scientific exactitude. It is equally doubtful that the resources exist to keep a network going year-round. Furthermore, the results obtained to date, at least in eastern Africa, suggest that the movements are complex, and making predictions would be difficult (Chapter 10).

Whether or not population monitoring should be carried out presents a developed NBCU with a dilemma. Opting for population monitoring means opting for extra costs, in the hope that such monitoring will eventually reveal ways that will significantly decrease the costs of effective control without any guarantee that such ways will be found. Opting out means giving up the chance of major improvements in control strategy and probably locking the NBCU forever into a fixed system of relatively expensive control.

5

Mass-marking quelea with fluorescent pigment particles

BRAD E. JOHNS,
RICHARD L. BRUGGERS, and
MICHAEL M. JAEGER

Introduction

An understanding of the annual or seasonal migration or daily feeding flights of quelea in relation to local and regional agricultural patterns is necessary for developing selective control strategies. In the 1950s in eastern Africa, ringing was initiated under the premise that 'when large numbers of birds have been ringed throughout East Africa, it will be easier to untangle some of the outstanding questions regarding the extremely complex migratory problems' (Haylock 1957). Ringing and recovery efforts in Tanzania (Disney 1960), South Africa (Chapter 11; Jones 1980; Ward 1971), and Ethiopia (Chapter 10) have provided examples of local and migratory movements and contributed to the overall migration theory of the species (Morel and Bourlière 1955; Ward 1971). However, this marking technique is extremely laborious during both ringing and recovery phases, and recoveries have been proportionally very few. The limitations of ringing and the desire for more precise knowledge of movements resulted in efforts to develop other identification or marking techniques. Populations have been intensively sampled for 'fingerprint' patterns (Chapter 4). Trace element profiles (Chapter 7) and radio-telemetry (Chapter 6) have been investigated or implemented as techniques to differentiate populations or follow local movements, respectively. As a result, movement patterns appear to be much more complex than Ward (1971) previously thought.

Another identification technique that has proven very useful has been the aerial or terrestrial application of fluorescent pigment particles to birds in nesting or roosting aggregations. This technique is compatible with the other marking methods, yet it permits marking literally hundreds of thousands of birds during one evening at relatively low cost using standard control equipment and procedures. It also offers the possibility of obtaining a reasonable number of marked recaptures, depending on the specific situation. Birds can be collected in mist-nets or after control operations and examined under ultraviolet (UV) light for the presence of fluorescent particles. This chapter describes the historical development, current methodology, and future direction and use of this unique capability.

Techniques development

Background

The development of the fluorescent mass-marking technique for quelea began when Bruggers and Bortoli (1979) conducted laboratory marking trials with various fluorescent dyes and paints in Senegal. At the same time, researchers at DWRC were studying ways to mass-mark Red-winged Blackbirds *Agelaius phoeniceus* with various particle markers. The combined efforts of personnel involved in these two initial approaches resulted in the development of mass-marking methodology for quelea. Mass-marking was first used in 1981 in Ethiopia to determine quelea movements in the Ethiopian Rift Valley (Jaeger *et al.* 1986). It was subsequently used in Kenya (Thompson and Jaeger 1984), in Tanzania (Luder and Elliott 1984), and in Niger in 1987 (R. Bruggers and J. Bourassa, unpubl. data) to delineate movement patterns relative to susceptible cereal crops. It was also used in the United States to study dispersal patterns from roosts of migrating Red-winged Blackbirds (Knittle *et al.* 1987).

Markers and formulation

In early attempts to mark wildlife, fluorescent dyes were dissolved in a liquid medium and sprayed on the animal (Evans and Griffith 1973; Taber and Cowan 1969). The dyes were translucent and fluoresced under UV light. For birds, these dyes presented problems when sprayed in small quantities because the liquid spread on the feathers, and even though it fluoresced, it was often quenched on darker plumage. In contrast, the new quelea marking technique described in this chapter uses particles that fluoresce when exposed to long-wave (360 nm) UV light. The highly fluorescent particles are not diluted or quenched on a feather; small flecks of markers, as minute as 10 μm in diameter (slightly larger than a red blood cell) can be detected on feathers in UV light with the unaided eye.

The markers are amorphous, transparent organic resin particles that contain dyes which are capable of fluorescing while in a solid state solution. DAY-GLO Color Corporation is the manufacturer of the markers we have used. They produce a variety of fluorescent particles for incorporation into various solvents and product bases to make paints, inks, plastics, or other products that require a fluorescent pigment. Most particles are 3.5–4.0 μm in size, but they range from <1.0 μm to 50.0 μm; 95 per cent of the particles fall between 2 and 13 μm. Ten colours are available, but depending on observer discrimination ability, only four to six colours or combinations of two colours work well. The A and AX series of particles have proven satisfactory for formulating avian mass-markers. The specific gravity of particles in these series is 1.36, which causes them to sink in the marker formulation but stay suspended when slightly agitated. The particles are essentially non-toxic; the oral LD_{50} in rats is >16 g/kg.

The marker spray formulation was developed to be compatible with current avicide application equipment and techniques used in Africa (Plate 5*a*). The components of the formulation are, by volume, 95 parts diesel fuel and 5 parts boiled linseed oil plus, by weight, 2 parts particle marker. Thus, a 100-l spray load would contain 95 l diesel fuel, 5 l boiled linseed oil, and 2 kg particle marker. The boiled linseed oil is the adhesive: it polymerizes to a hard, somewhat flexible film that holds the marker in place after the diesel fuel evaporates. Raw linseed oil is unacceptable because it does not harden but instead remains as a liquid on the feathers. The boiled linseed oil is first mixed with the diesel fuel, then the marker particles are mixed homogeneously into the liquid. The markers will slowly settle to the bottom if the mixture is not regularly agitated, and therefore the marker formulation should not be mixed and stored for later use. Storage of the formulated marker will also result in particle degradation. Initial examination of marker wear rates on quelea (Bruggers and Bortoli 1979) indicated that the wings held markers better than other body parts (Plate 5*b*). Additional examination of wear rates in Red-winged Blackbirds (Knittle *et al.* 1987) determined that wing collection was the most efficient means of detecting marked birds. Wing collection has been used for quelea mass-marking studies.

Fluorescent particles can be mixed into a toxicant spray formulation to determine the location of toxicant droplets. If the marker is used for this purpose, it is necessary to determine compatability between marker and toxicant formulation. Some toxicant chemicals will degrade or destroy the marker on contact. Marker particles in the GT series are more resistant to a wider range of toxicant chemicals.

Application equipment and methods

A variety of aircraft, spray systems, and ground-spraying equipment have been used to mark quelea. Both fixed-wing (De Havilland Beaver) and rotor-wing (Bell Jet Ranger 206) aircraft have been successfully used to mass-mark quelea roosts and nesting colonies (Plate 5*a*). They have been fitted with spray systems like Micronair[R] (AU 3000, AU 4000), or similar rotating-disc designs with two sprayers per aircraft, or a boom and nozzle system. Droplets with a volume median diameter (vmd) of 100 μm seem to mark birds well, but marking should still be successful with droplets in the range of 40–150 μm vmd. Droplets outside this range could be too fine to leave marks or too large to have a chance of impacting on a bird, either because they are too few or they do not remain airborne long enough. As a general rule, a boom- and nozzle-type spray system will require a greater volume of spray liquid than a rotating-disc spray system to mark the same number of birds. The boom and nozzle system produces a wide range of droplet sizes at a given setting, and thus is less effective in producing the desired droplet sizes. The suggested time and methods used in aerial spray control operations should be followed to mark the maximum numbers of quelea (Chapter 21).

The concept of mass-marking using ground application techniques is basically the same as for aerial application—bringing a cloud of droplets into contact with birds. During aerial spraying, spray both drops and drifts, while during ground spraying, spray movement relies mainly on wind drift with a slight amount of vertical movement. This difference in application requires that the ground-generated spray clouds have droplet diameters sufficiently small to hang in the air, as the wind drifts the spray cloud into the birds or the birds fly throught it. Droplets in the 40- to 60-μm range can be transported in light winds (2–5 m/s) and will mark birds well; droplets <40 μm are carried in the slightest breeze (<2 m/s), while droplets >60 μm will drop to the ground quickly unless the wind speed exceeds 5 m/s. Therefore, a spray cloud of droplets from ground sprayers with a vmd of 40 μm should provide acceptable marking under a variety of wind conditions.

In some situations, ground sprayers can offer a less expensive way to mark quelea than aircraft. Ground sprayers have been successfully used to mark quelea in field tests in Niger (R. Bruggers and J. Bourassa, unpubl. data) and in Kenya (Meinzingen and Latigo 1986). Both are transported to the marking site by vehicle. The ground sprayer used in Niger was a rotating disc unit that was mounted on a telescoping pole and driven by an air compressor powered by a generator. This unit is set in a fixed location upwind of the marking site (GTZ 1986). More than one unit may be required to obtain proper marking coverage of a site considering the size of the roost or colony and the wind conditions. The ground sprayer used in Kenya was a Micronair[R] AU 7000 rotating-disc type that can be attached to vehicles or mounted on a wand that is hand-held and powered by electricity from a portable generator. The process of matching droplet size, wind speed and direction, distance from marking site, sprayer height, spray volume and duration with the 'window' of best spraying time to mark birds requires the same skill and experience as that needed to aerially mark quelea. Any sprayer design that functions well for the purpose of controlling quelea should be adaptable to the delivery of the marker formulation.

It is very important to keep accurate records of all parameters during marker applications as this will help improve the success of future marking sprays. Table 5.1 illustrates the type of information that should be collected and typical spray parameters that should be determined and recorded for each marking operation.

Table 5.1. Example of aerial quelea mass-marking record sheet.

Spray site	
Concentration	Colony
Size (ha)	10
Mean no. active nests/ha	2380 ± 428 (± 1 SD)
Total no. active nests	23 800
% nests occupied	91
Installation date	21, 22 Mar.
Composition on spray night	24- to 28-day-old young and a few roosting adults
Total no. birds	
Nesting adults, nestlings, fledglings	50 000
Roosting adults and juveniles	5000
Spray parameters	
Date	25 Apr.
Time	1820–1830 h
Vol (l) marker formulation	60
No. passes above colony	4
Ambient temperature (°C)	24
Wind speed at 2-m height (km/h)	8
Flight height (m) above tallest canopy	15–20
Flight speed (km/h)	165
Flow rate (1/min per Micronair AU 4000 atomizer)	21.5
Droplet size (μm, vmd)	80
Estimated no. birds marked	
% adults	85
% nestlings, juveniles, fledglings	95

Mass-marking studies

Since the onset of mass-marking operations in Africa, several million birds have been marked using fixed-wing aircraft, helicopters, and ground sprayers. When marking has taken place in the context of a well-defined movement study and has been followed by conscientious and thorough post-spray bird collections, marked birds usually have been found and interesting and useful information has resulted.

Mass-marking efforts in Ethiopia (1981), Kenya (Dec. 1982/Jan. 1983, 1984 and 1985), and Niger (1987) have resulted in an estimated total of 6 million birds marked and 352 marked birds recovered of 12 760 sampled (Table 5.2). These studies documented the movements of birds from distant breeding grounds to agricultural areas and confirmed itinerant breeding in Ethiopia (Jaeger *et al.* 1986), provided a basis for recommending control strategies in Kenya (Jaeger 1984), and confirmed the impact of quelea on rice in Niger (R. Bruggers, unpubl. data). In general, mass-marking appears to be particularly valuable in circumstances exemplified by these situations (Table 5.2), when it is of interest to know when and where quelea aggregations are likely to threaten cereal crops. Besides permitting one to obtain information on breeding biology and bird movements, this information is necessary for making quelea control more selective and focusing it where it can be of greatest benefit (Thompson and Jaeger 1984).

Discussion

The success of the first field study in Ethiopia seemed to result in optimistically high expectations as to the utility of this technique in unravelling the movements of quelea. In some marking situations, birds were sprayed, collected in spray sites to determine marking effectiveness, and then little or no effort was made to collect them again. In other situations, birds were sprayed and collected in widespread localities without first having determined the effectiveness of the marking spray. In still other situations, large numbers of birds were sprayed, immediately collected to determine effectiveness, and conscientiously collected later without finding marked birds.

A number of factors seem to be responsible for variability in the results and the recovery rates of marked birds. These include (1) failure to estimate the number of birds marked, (2) deterioration of marks, (3) moult, (4) grouped dispersal of birds from particular spray sites, and (5) time and personnel available to collect and examine wings, a factor that greatly influences post-spray sampling strategies and, ultimately, sample size. The real dilemma, however, is interpreting negative data obtained from a properly designed and conducted study.

Table 5.2. Description of mass-marking operations of quelea in Africa between 1981–1987 from which marked birds were later recovered.

Country Location	Aggregation	Data	Marker[a]	Application	Birds marked No.	Birds marked %	Number of birds sampled	Number of birds marked	Location	Comments	Source
Ethiopia											
South-west	Colony A	June 81	FO	Aerial	655 000	64	200	6	Gewane I	93 days; 650 km; in colony	Jaeger *et al.* 1986
	Colony B	June 81	FO	Aerial	122 000	51	50	1	Gewane II	93 days; 700 km; in colony	Jaeger *et al.* 1986
	Colony C	June 81	FO	Aerial	463 000	44	89	2	Melka Sede	67 days; 650 km; in roost	Jaeger *et al.* 1986
							1000	1	Issa Plain	86 days; 650 km; in colony	Jaeger *et al.* 1986
							480	3	Lake Zwai	100 days; 500 km; in colony	Jaeger *et al.* 1986
Kenya											
Mwea	Roost	Apr. 84	FO	Aerial	1.2×10^6	82	940	29	Nanyuki	73 days; 95 km; in roost	Thompson and Jaeger 1984
							290	7	Timau	128 days; 95 km; in roost	Thompson and Jaeger 1984
Tsavo East	Colony A	Jan. 85	FO	Aerial	Several million (Colonies A–D)	39	2468, Tsavo (Colonies A–D)	5	Tsavo	58 days; immediate area; in breeding colony	Thompson and Jaeger 1984
	Colony B	Mar. 85	SY	Aerial		13–19	6075, Tanz.	1	Tsavo	27 days; immediate area; in breeding colony	Thompson and Jaeger 1984
	Colony C	Mar. 85	FO	Aerial		17	8543, Total	1	Dadoma	69 days; 500 km; in breeding colony	Thompson and Jaeger 1984
	Colony D	Mar. 85	FO and SY	Aerial		11		1,1	Tsavo; Dadoma	99 days; immediate area; in breeding colony	Thompson and Jaeger 1984
Niger											
Tillabery	Roost A[b]	Apr. 87	FO	Ground	400 000	64	1168 (Roosts A and B)	294	Local roosts and crop areas	20 days; 1–50 km; in roosts and crops	Bruggers *et al.*, unpubl. data
	Roost B	Apr. 87	SY	Ground	91 500	61				17 days; 1–50 km; in roosts and crops	Bruggers *et al.*, unpubl. data

[a]Day-Glo fluorescent particles were used individually [Fire Orange (FO) and Saturn Yellow (SY)] or mixed (FO, 75%; SY, 25%). Formulation ratios for 50 l consisted of 47.5 l diesel fuel, 2.5 l boiled linseed oil, and 1 kg marker.

[b]Roosts were comprised of Golden Sparrows *Passer luteus* and quelea.

Estimates of marked birds

The effectiveness of the marker spray must be determined at the spray site by collecting birds and estimating the number marked. Interpreting data from future collections of birds is based entirely on some idea as to the percentage of birds originally marked and the quality of the marks relative to detectability and longevity. These initial post-spray samples are particularly important if no marked birds are found in future collections.

Deterioration of marks

Marked birds may not be found if the marks have not initially adhered or if they have worn off by the time samples of birds are collected. Mark retention has not systematically been documented in the field, primarily because one can never assume a stable, local population of birds from which to sample over time. Population turnover further complicates any retention studies. Although R. Allan (personal observation) found it difficult to remove marks on primaries by simulated abrasion with a brush, and B. Johns (unpubl. data) demonstrated good marker wear on quelea wings subjected to weathering compounds in a paint weathering instrument, one must assume that markers are lost with the wear of primary feathers. Under laboratory conditions, caged quelea lose markers progressively over a period of several months (B. Johns, unpubl. data). In field studies, marked wings have been found up to 100 days post-spray when moult was not a complicating factor (Jaeger *et al.* 1986).

Moult

Post-breeding moult is an important concern when marking adults in breeding colonies. Unless post-breeding moult is arrested, adults usually complete it (which includes remiges and rectrices) about 4 months from its onset. If juveniles were present at the time a colony was sprayed, marks should be present on their feathers for considerably longer as they do not start their moult for 9–13 weeks after hatching. Likewise, marks could be on the feathers of adults which had begun moult and had fresh new feathers at the time of spraying (Luder and Elliott 1984). None the less, moult greatly reduces the chances of finding marked birds over time.

Group cohesion

Group cohesion appears to be another potentially important characteristic of quelea movements that affects marker recovery. Results from mass-marking suggest that quelea tend to remain together as a group during movements: group cohesion was evident in the Ethiopian study (Jaeger *et al.* 1986). Similarly, quelea marked at Mwea, Kenya, were recovered only at Timau, Kenya (Thompson and Jaeger 1984). Group cohesion may offer an explanation for not recovering marked quelea from suspected sites. Marked birds could be expected to be found if they disperse widely and are adequately sampled. However, a marked group of 100 000 quelea that migrate together can easily be missed in much of this species' inaccessible range without extremely widespread sampling.

Logistical constraints and sample sizes

In studies where marked birds were found after at least 1 month and away from the spray site, samples of at least 400–500 birds were needed to have a good opportunity to recover marked birds (Thompson and Jaeger 1984). Small collections of 10–100 birds can yield recoveries in restricted areas where birds have localized and apparently repeated movements such as the Tillabery region of Niger on the Niger River (R. Bruggers, pers. obs.), where 294 marked birds were recovered from a sample of 1168. However, obtaining large post-spray samples of quelea can be very time-consuming and tedious, unless they can be collected following lethal spray operations. Similarly, considerable time can be spent looking at wings for paint particles under UV light. C. Elliott (Luder and Elliott 1984) estimated that he spent 37 h over a period of several days checking 3716 wings following a spray in Tanzania.

The importance of thorough sampling cannot be overstressed as deductions about movements are also made on the basis of negative collections. Elliott (1983*b*) described the difficulty of interpreting negative data from colonies in Tanzania following a marking spray in Tsavo East, Kenya, in 1983. Negative results can fit so many hypotheses that using them to try to improve control strategies is confusing.

In contrast to the confusion that can exist in trying to interpret data from an inadequate post-spray sampling design is the clarity of information that can be generated from proper sampling. Knittle *et al.* (1987) wished to determine the spring dispersion of Red-winged Blackbirds in North America from spring roosts in the central United States. They estimated that they marked 10.6 million birds in roosts in 4 h. A post-spray collection network by investigators and co-operators was established in 19 states and 3 Canadian provinces in 1982 and 11 states and 3 provinces in 1983. They collected 8880 birds during 2- to 10-week sampling periods and found 687 were marked. Marked birds were found over an area of 1.8 million square kilometres. The general movement pattern of these blackbirds was according to predicted hypotheses. However, movements of a considerable number of other birds were unexpected. Prior to this work, the seasonal distribution of blackbirds had been determined on the basis of banding data, which, as we have mentioned, is a particularly laborious task. To recover sufficient numbers of birds over an appropriately large area to obtain a meaningful analysis had taken many years. This mass-marking study further demonstrates how ideal and efficient the technique can be for understanding complex distribution patterns.

The sampling thoroughness and overall success of these kinds of studies stimulate the desire for implementing more sophisticated studies and interpretations. Concerns relating to (a) the number of birds that must be sampled from a location remote from the spray site to ensure sufficient sampling to detect marked birds, and (b) proper sampling to determine the proportion of marked birds that migrate in different directions need to be addressed. However, accurate estimates of pre-spray colony or roost populations, the percentage of birds marked, and thorough post-spray sampling are needed. One must also make assumptions (1) of the percentage of marked birds that can be expected to migrate to a particular area, (2) that no loss of markers occurs during the study, (3) that markers are properly identified, (4) that birds collected on each occasion are a representative sample of the roost population each time, (5) that birds experience fates independent of each other, and (6) that mortality is negligible (Otis *et al.* 1986).

Technique improvements

Future use of the mass-marking technique could benefit from technical improvements in marking methodology. The marker particle size and the large quantity used during marker operations cause considerable concern about contamination and, ultimately, data interpretation. Care must particularly be taken to prevent colour contamination when two or more colours of particles are used in an operation. Wings collected for examination must be handled and examined in a way that prevents false positive marks from occurring. Any procedural innovations that reduce contamination will improve the technique: mixing procedures for the marker formulation and dedicated use of equipment and vehicles are possible areas of improvement.

The number of colour codes that can currently be differentiated on feathers is limited to between four and six, depending upon the use pattern. Improvements in the number of marker codes will be required if several groups or organizations wish to mark quelea several times at different sites. DWRC is developing a method to incorporate a new rare-earth elemental code in addition to the colour pigment in the mass-marking spray formulation. In this marking system, the fluorescent colour of the markers would indicate a marked bird and the new rare-earth code would identify the particular spray application. This marking system would provide an almost unlimited number of codes, but it would require a sophisticated electronic detection instrument to distinguish each of them. Also, improvements in the adhesive formulation and a better understanding of how the marks wear off the feathers are required to extend the marker retention for the lifetime of the feathers between moults. Finally, detecting markers on wings is very labour intensive. The screening process could be improved by using a detection device that would sense the presence of particular wavelengths of light emitted from the marker under UV light.

Conclusions

It has been clear to everyone involved in mass-marking activities that the technique has immense potential for understanding quelea ecology and for providing information upon which to develop efficient, selective control operations. The technique is particularly helpful when used in conjunction with 'fingerprint' profiles, to offer a much more valuable means of determining quelea movement patterns than either technique offers by itself. Information and conclusions arising from the marking studies in Ethiopia in 1981 and in Kenya in 1984 and 1985 were enhanced by the simultaneous development of 'fingerprint' profiles for birds collected at all locations. These profiles are particularly important when only a single marked bird is found in the sample. Obviously, interpretation of 'fingerprint' data is only as reliable as the ornithologists' expertise. However, finding fluorescent-marked birds in these samples provides indisputable proof for local and regional movements. As previously mentioned, the work must be conducted within the framework of a well-defined hypothesis and have adequate logistical support to determine the destination of marked birds.

6

Uses of radio-telemetry in quelea management

RICHARD L. BRUGGERS

Introduction

Radio-telemetry is a technique that is increasingly being used to investigate various aspects of the behaviour of pest birds in relation to agricultural damage, crop protection, and management operations (Besser 1978; Bray 1973; Bray *et al.* 1975; Heisterberg *et al.* 1984; P. Lefebvre *et al.*, unpubl. data). The technique has proved useful to locate roosts and nesting colonies of quelea in Ethiopia (Bruggers *et al.* 1983) and of quelea and Golden Sparrows *Passer luteus* in Niger (R. Bruggers and J. Bourassa, unpubl. data), follow daily movement patterns and dispersal of Black-headed Weavers *Ploceus cucullatus* in Ethiopia (Bruggers *et al.* 1985), and assess the impact of lethal avicide sprays of colonies on non-target birds (Bruggers *et al.*, in press). In addition, telemetry would seem to have other uses in quelea management.

For effective control of quelea breeding colonies, it is important to locate the colonies during their early stages. This gives maximum time to determine if control is justified and, if so, to make appropriate preparations. The difficulties associated with locating quelea concentrations are described in Chapter 4. The logistics involved are formidable whether overland vehicle or low-flying aircraft survey methods are used, given the vastness of the quelea breeding range. Most colonies are located more than 2 weeks after their inception, when the grass nests have dried and their tan colour contrasts with the surrounding vegetation. In Tanzania during 1985, the average time elapsed between the time 25 colonies were initiated and located was 16 days (C. Elliott, pers. comm.). The Tanzanian Bird Control Unit even had the advantage of using a helicopter, yet only seven of the colonies were found within 1 week of installation. Thus, Bird Control Units usually are faced with the dilemma of trying to control many nesting colonies in the last 3 weeks before the birds disperse. This insufficient time can lead to poorly executed operations, low kills, and inadequate evaluation of results.

Surveys and nesting colony detection in Ethiopia

In Ethiopia in 1981, 1.8-g radio transmitters with a battery life of between 2.5 and 3.5 weeks and a reception distance of 4–8 km at survey altitudes of 300–700 m and 31 km at 1525 m (Bruggers *et al.* 1981*a*), were attached to the base of the tail of two quelea males prior to nesting, and to 17 additional birds during advanced stages of the nesting cycle. The location and movements of these 19 radio-equipped birds were monitored between 17 May and 28 June in the Weyto and Sagon River valleys in the Rift Valley of south-western Ethiopia, using a Bell 47 helicopter.

Six quelea breeding colonies had been located there during 1979 and 1980 using about 50 h of helicopter time during several weeks of ground and aerial survey. Because of the remote, extensive area associated with the Weyto and Sagon rivers (*c.* 1200 km^2) and the Omo River (*c.* 2775 km^2), approximately 160 h of helicopter time costing US $350–550/h had been needed for the 1981 survey and related field activities. Less expensive fixed-wing aircraft were inappropriate for this area because of the absence of landing strips and the need to work in the nesting colonies. Roads were nearly non-existent in the area.

The two males that had been mist-netted on 17 May from a pre-nesting feeding flock and radio-equipped were relocated on 21 May during 2 h of aerial surveying in a 2-day-old 12.4-ha nesting colony of more than 1 million birds. This colony was only 0.5 km from a smaller, 0.40-ha colony in which the birds were fledging and which had been overlooked during the initial 10 h of pre-radio-tracking surveys of the valley. During the next 2 weeks, these two males (which did not nest, probably because of the radio package) travelled at least 25 km and were tracked to two other colonies, that were again only 2–3 days old. Four colonies were, therefore, located via the radio-equipped males, three of them within 3 days of installation. In a situation in which control might have been necessary, a maximum of about 39 days would have been available for a full evaluation of the potential impact of the colonies on agriculture and to prepare and execute control.

The information on daily movement obtained from the 17 other quelea shows the usefulness of the technique in survey operations. Bruggers *et al.* (1983) found adults that were feeding nestlings usually did not travel more than 2–3 km from the colony and foraged in a few preferred areas. Therefore, aircraft survey transects would have needed to be flown at about 5-km intervals, and ground surveys would have had to pass within 3 km of the colony to have a chance of detecting it. Flying such close survey transects is both expensive and impractical. Ground surveys must also overlook many colonies if roads do not pass within 3 km of the colony sites.

Post-spray behaviour

The behaviour of quelea after spraying influences efforts both to estimate control success and to evaluate the potential impact that dead or dying birds might pose to non-target wildlife. Elliott (1981*b*) reported that although the majority of quelea die within a few hours of a fenthion application, a few can take up to 24 h to die and some can be found dead up to 5 km from the spray site. Mortality estimates may, therefore, need to take deaths outside the spray site into account. The presence of dead and dying birds some distance from spray sites may also affect a much larger population of non-target wildlife, particularly raptors, than had been previously thought.

Radio-telemetry was one of the principal techniques used to investigate these problems in a study carried out in Kenya in 1985 (Bruggers *et al.*, in press). Twenty quelea were fitted with 1.1-g radio transmitters at two breeding colonies which were then sprayed (Plate 6*a*,*b*). Radio transmitters (some equipped with mortality sensors) were also attached to eleven Tawny Eagles *Aquila rapax*, three Bateleurs *Terathopius ecaudatus*, four Pale Chanting Goshawks *Melierax poliopterus*, two Gabar Goshawks *M. gabar*, two Pygmy Falcons *Poliohierax semitorquatus*, two Pearl-spotted Owlets *Glaucidium perlatum*, two Laughing Doves *Streptopelia senegalensis*, two Ring-necked Doves *S. capicola*, and one Taita Fiscal *Lanius dorsalis*. In addition, seven Black-backed Jackals *Canis mesomelas* and one Common Genet *Genetta genetta* were radio-equipped.

The radio-telemetry data showed that many raptors are highly mobile, are attracted to sprayed quelea colonies, and will sequentially predate colonies as the availability of young increases. In this manner, they are likely to suffer increasingly greater cholinesterase (ChE) depression and ultimately death. The fate of most of these birds other than the raptors is unknown because of the absence of mortality sensors in their transmitters. Most data produced by the radio-equipped quelea were inconclusive. However, several radio-equipped quelea apparently died outside of the colonies, one as far as 11 km away. Radio-telemetry made an important contribution to this study and would be a necessary component of any follow-up studies.

Additional management implications

Locating roosts

The ability of quelea to carry radio transmitters, and the success of these radio-telemetry efforts present other interesting uses related to increasing the efficiency and evaluating the effectiveness of control operations and crop protection efforts in Africa. One use may be the rapid location of roosts containing birds causing damage to crops. Roosts, which are transient, unpredictable, often composed of several species, and often not detectable from the air, can be even more difficult to locate than nesting colonies (Ward 1979). Unless they are in traditional, known locations, considerable time (often more than 1 week) must be spent trying to locate them. By catching quelea in the fields being damaged, radio-equipping them, then tracking them at dusk to the roost by aircraft or vehicles, it may be possible to reduce the location time to only a couple of days. The roost found in this way would also be certain to include the birds directly involved in the damage. This kind of information was obtained in April 1987 for mixed roosts of radio-equipped quelea and Golden Sparrows in Niger (R. Bruggers and J. Bourassa, unpubl. data).

It is possible that this technique would be particularly useful where damage is being caused by birds roosting in areas of rough or swampy terrain. Roosts near irrigated rice fields are often in small sections of large reed beds (as in Mozambique [M. Jaeger, pers. comm.]), in stands of sugar-cane (as in Kenya [C. Elliott, pers. comm.]), or on islands (as in Niger [R. Bruggers, pers. obs.]), but the control officer has great difficulty in locating the precise patch where the roost is situated. As a result, it is sometimes necessary to spray the whole stand of reeds or sugar-cane in order to achieve control. In such circumstances, radio-telemetry would not only locate the roost more quickly, but also help to define its precise dimensions, thereby reducing the amount of chemical needed for control. Likewise, because considerable roost interchange can occur (Heisterberg *et al.* 1984; R. Bruggers and J. Bourassa, pers. obs.), additional roosts could possibly be found by tracking radio-equipped birds.

Evaluating crop protection efforts

The technique of radio-telemetry might also be useful in evaluating the effectiveness of chemical repellents. Dyer and Ward (1977) suggested that birds repelled from one crop area might move to another and damage crops. By incorporating the techniques of radio-telemetry with repellent applications, the behaviour of the pest birds can be better understood. For example, Besser (1978) and Besser *et al.* (1979), using radio-equipped Red-winged Blackbirds *Agelaius phoeniceus*, found that on 43 per cent of 56 occasions, flocks frightened from vulnerable sunflower fields next fed in stubble fields, weed patches, non-vulnerable sunflowers, corn fields, and swathed wheat; 27 per cent of the flocks visited vulnerable cornfields, but inflicted only negligible damage. Radio-telemetry could be similarly used in African bird pest damage situations to determine the movements of repelled birds and understand the relationship of alternative food sources and adjacent cropping areas to the outcome of crop protection efforts.

Conclusions

The use of radio-telemetry in studies of quelea (and other African bird pests) has important implications for future investigations. The information that could be collected on local daily movements could improve our understanding of the quelea's crop-damaging behaviour. The use of the technique to locate roosts and colonies at early developmental stages, could greatly increase the efficiency of survey and control operations. Although radio transmitter weight may not decrease much below 1.0 g for some time, it is possible that further technological advances will be made that will permit the reception distance to be increased, the battery life to be extended, and the overall reliability to be improved (L. Kolz, pers. comm.), characteristics that will increase field effectiveness. However, radio-telemetry techniques must be used in the context of a well-defined avian agricultural problem to achieve their maximum potential.

7

Identifying quelea populations by trace element analysis of feathers

JAMES O. KEITH,
IWAO OKUNO, and
RICHARD L. BRUGGERS

Introduction

Damage by the Red-billed Quelea *Quelea quelea* to ripening cereal crops is often more intense in some areas than in others. Localized damage is predictable because it is caused by concentrations of quelea that annually arrive in an area as crops ripen. Simultaneously, other nomadic groups of quelea in non-crop areas feed on maturing native grasses. This is their natural behaviour, and quelea probably prefer native grass seeds to cereal crops (Ward 1973*a*). Because some groups cause crop damage while others do not, protecting crops can best be accomplished by reducing those groups of quelea that predictably feed on crops. Therefore, the annual movements and nesting locations of groups that cause damage must be identified.

Identifying quelea populations

Various attempts have been made to identify quelea populations by plumage and morphological features (Ward 1966, 1973*b*; Chapter 4) and to study their long-distance movements by ringing birds (Disney 1960) or marking them with fluorescent particles (Jaeger *et al.* 1986; Chapter 5). Here we examine another technique, the amounts of trace elements in feathers, that may be helpful in identifying various quelea populations.

For more than three decades, wildlife biologists have used trace elements in birds' feathers to identify discrete populations. The earliest work was with Ruffed Grouse *Bonasa umbellus*, where regional differences were found in the mineral composition of feathers (Grant 1953; McCullough 1953). Considerable research followed on waterfowl to define natal areas (Devine and Peterle 1968; Hanson and Jones 1968; Kelsall and Burton 1977; Kelsall and

Calaprice 1972; Neth 1971). Hanson and Jones (1976) examined soils and plants as well as goose feathers to relate minerals in habitats with those in feathers. Later work showed that feathers from Kirtland's Warblers *Dendroica kirtlandii* had large differences in mineral composition among individuals (Means 1981), which made identification of groups in the population difficult. However, Parrish *et al.* (1983) readily distinguished subpopulations of Peregrine Falcons *Falco peregrinus* by their feather minerals.

The levels of different elements in feathers are believed to be influenced by the relative availability of those elements in habitats where feathers are produced. The mineral level in the diet largely determines the kind and amount of elements in feathers, although this can be modified by differences in the absorption and excretion of elements by the animal. The levels ultimately deposited in growing feathers create a mineral profile that can be similar for all birds of a species that grow feathers in a particular area.

Mineral profiles are most uniform among individuals of species, such as waterfowl, that moult all of their flight feathers at one time. The new flight feathers produced in the area where birds are flightless tend to contain the same kinds and amounts of trace elements. Likewise, young birds raised in an area should have the same profiles in their juvenile feathers because they all eat similar foods and ingest comparable minerals while growing their feathers. In species that move while sequentially moulting feathers, mineral profiles become more variable among individuals, but birds moving together and moulting in unison should have similar levels of minerals in their feathers.

Most workers have found that some differences in mineral profiles are related to sex and age of birds, age of feathers, feathers examined, and methods chosen for washing and analysing feathers. Variations among sex, age, and individuals can result from differences in food habits, physiology, and metabolism (Kelsall *et al.* 1975). Kelsall and Burton (1979) found that trace element profiles of individuals from different groups of Snow Geese *Chen caerulescens* could show similarities, but large sample sizes ($n \simeq 40$) were necessary to separate groups. They felt the usefulness of the technique depended on how finely it was necessary to subdivide a population, but concluded that with adequate sample sizes some separation of groups within a population was possible.

A considerable and common problem in using feather mineral profiles to identify groups within a bird population is that feather composition can change over time. Kelsall and Burton (1979) noticed significant changes in feather profiles of captive geese between October and May. Other work with Canada Geese *Branta canadensis* suggested that some ions in the environment were adsorbed on to feathers with time (Edwards and Smith 1984). When geese were moved to different environments, ion exchange created new mineral profiles in feathers. Nevertheless, group affinities and profiles were maintained and, with adequate sample sizes, birds that moulted together could later be identified.

Collecting and analysing quelea feathers

In 1981, primary feathers were collected from *Q. q. quelea* in Senegal, *Q. q. intermedia* in Tanzania, and *Q. q. aethiopica* in Ethiopia to compare mineral profiles among these three subspecies of quelea. In addition, primary feathers were obtained from different groups of quelea within Ethiopia and adjacent western Somalia to examine differences among more local bird concentrations. Details of feather collections are given in Table 7.1. In Somalia, birds had suspended their moult. This offered the opportunity to compare 1-year-old feathers with newly developed ones from the same birds. Samples of old and new feathers were selected on the basis of their availability. In 27 cases, samples of old and new feathers were obtained from the same bird.

Feathers from each bird were analysed as individual samples. The first six primaries taken from the right wings of birds weighed at least 30 mg and provided enough feather material for analyses. To remove oils and adhered particles, feather samples were sonicated twice in water, once in acetone, and then rinsed twice in acetone. Samples were ashed in a muffle furnace and submitted for analysis by emission spectroscopy. Analyses were conducted by Spec Resources, Denver, Colorado. This technique provided only semi-quantitative data, but it permitted multi-element analysis at a reasonable cost. In this study, we wished to determine the levels of trace elements in quelea feathers and the ones that varied sufficiently to be of help in discriminating between populations. More quantitative methods can be used in future work once the diagnostic elements are identified. The emission spectrographic analyses were automated so that 24 samples were analysed as a group. As the method is semi-quantitative, variations were expected between each run of 24 samples. To reduce the bias of differences between runs, some samples from each collection were analysed in each run.

Analyses were conducted in three lots. The Ethiopia Colony B collection was analysed first to determine the results that could be expected. Samples were analysed in two runs of 21 and 22 samples. Techniques used in preparing and analysing Colony B samples were exploratory and somewhat different than those for other samples; therefore, the results are presented separately. The second lot consisting of collections from Senegal, Tanzania, and Ethiopia (Colony D) was analysed to see if feather mineral profiles differed among quelea subspecies. Each of these collections of 40 samples was divided into five groups of eight samples each. One group of samples from each country was combined into a run of 24 samples. It took five runs to analyse all 120 samples from the three countries. An identical approach

Table 7.1. Description of feathers collected from quelea in Africa during 1981. Collections consisted of primary feathers from 40 individuals in all areas except Colony B ($n = 43$).

Location	Concentration	Date (1981)	Age	Sex	Wing	Primary nos. analysed	Collector
Tanzania							
Arusha	Colony	15 June	Juveniles	M and F	Right and left	1–3	C. C. H. Elliott
Ethiopia							
Lake Chew Bahir	Colony B	22 May	Adults	M	Right	1–6	M. M. Jaeger and R. L. Bruggers
Lake Chew Bahir	Colony D	8 June	Adults	M	Right	1–6	M. M. Jaeger and R. L. Bruggers
Melkassa	Colony	30 Sept.	Adults	M	Right	1–6	M. M. Jaeger and R. L. Bruggers
Somalia							
Hargeisa[a]	Roost	15 Sept.	Adults	M and F	Right and left	As available[b]	H. Musa
Senegal							
Richard Toll	Colony	23 Sept.	Adults	M	Right	1–6	Ph. Ruelle

[a]Birds killed by chemical sprays.
[b]Old feathers—21 males, right wings; 19 males, both wings. New feathers—35 males, both wings; 5 females, both wings.

was used to analyse the final lot, which consisted of feathers from Melkassa in Ethiopia and the old and new feathers from Hargeisa in Somalia.

Levels of trace elements are reported in parts per million (p.p.m.) based on the dry-weight of feathers. Graphic presentations of results are similar to those used by Hanson and Jones (1976) and depict a mineral profile for feathers from each population of quelea. Data obtained on trace element levels in feathers were not normally distributed; many results were reported as zeros or traces. A zero meant either that elements were not present or were present at undetectable levels. In other samples elements were present but levels were too low to be quantified (trace levels). In the latter case, values equal to the lowest levels that could be quantified were assigned to elements. These values differed for each lot of samples (Table 7.2).

Kruskal-Wallis tests were used to analyse data on trace elements in feathers from Senegal, Tanzania, Ethiopia, and Somalia (old feathers). Non-parametric multiple comparisons were made to separate means. Assessment of differences in mineral levels between old and new feathers from Somalia was made with Wilcoxon's signed rank test. Finally, non-parametric, nearest neighbour, discriminant analyses were conducted to evaluate how well mineral levels permitted identification of quelea groups and which minerals were the most useful. All minerals except magnesium increased discrimination. Values were obtained on the percentage of samples from each popula-

Table 7.2. Lowest levels (p.p.m.) of quantitation for different lots of samples and amounts of trace elements found in feathers from Colony B, Ethiopia.

	Lowest levels (p.p.m.)			
Trace elements	Ethiopia (Colony B)	Senegal, Tanzania, and Ethiopia (Colony D)	Melkassa and Hargeisa, old and new	Levels from Colony B, Ethiopia ($\bar{x} \pm$ SE)
Iron	90	100	70	180 ± 22
Magnesium	30	40	40	76 ± 5.8
Calcium	100	90	80	179 ± 9.4
Titanium	4	4	2	66 ± 7.5
Manganese	2	2	1	26 ± 1.9
Boron	2	2	2	38 ± 1.5
Zinc	20	40	30	71 ± 4.6
Silicon	2000	1800	1400	2000 ± 0.0
Sodium	90	130	100	206 ± 9.6

tion that could be correctly classified by using the eight elements other than magnesium.

Quelea trace element profiles

Feathers from Colony B in Ethiopia were analysed first to determine which trace elements were present, and which elements varied sufficiently to help possibly in differentiating between groups of quelea. Analyses were made for thirty-five elements in Colony B feathers. Thirteen of these elements, if present, occurred at levels below the detection limit for emission spectroscopy. Analyses for two elements, aluminium and potassium, gave highly variable results; emission spectroscopy apparently was a poor method for measuring these elements in feathers. Eleven elements either were present in trace amounts or varied only slightly among samples, both of which limited their diagnostic value. For the remaining nine elements (iron, magnesium, calcium, titanium, manganese, boron, zinc, silicon, and sodium), levels varied over a large enough range to permit some recognizable differences among samples. Average levels for these latter elements in Colony B feathers are listed in Table 7.2, and a mineral profile is shown in Fig. 7.1.

Levels of these nine elements were then determined for the feather samples from Senegal, Tanzania, and Ethiopia (Colony D). This was done to compare feather elements in birds from populations known to be discrete because of geographic separation. Mineral profiles for these three populations were different (Figs. 7.1 and 7.2). Amounts of titanium and boron were different ($P \leqslant 0.05$) in each of the populations, while at least one population could be separated from the other two by using levels of iron, manganese, zinc, silicon, and sodium (Table 7.3). All three populations had similar amounts of magnesium and calcium in feathers.

Discriminant analyses indicated that 75.0, 97.5, and 80.0 per cent of individual quelea from Senegal, Tanzania, and Ethiopia (Colony D), respectively, could be correctly assigned to their place of origin by using levels of trace elements found in their feathers. All minerals except magnesium increased discrimination. Thus, separation of the subspecies of quelea in Africa was possible by using trace element analyses. Because birds lived in distinctive habitats and were isolated from each other, this was expected.

Analyses subsequently were conducted to determine if trace elements were distinctive for groups of quelea within Ethiopia. Collections of feathers from Melkassa in eastern Ethiopia and from Hargeisa (old feathers) in north-western Somalia were compared with each other and with those from Colony D in south-western Ethiopia (Figs. 7.1, 7.2, and 7.3). Levels of iron in feathers were different in all three areas ($P \leqslant 0.05$). For each of the other eight trace elements, levels from one area always differed ($P \leqslant 0.05$) from those in

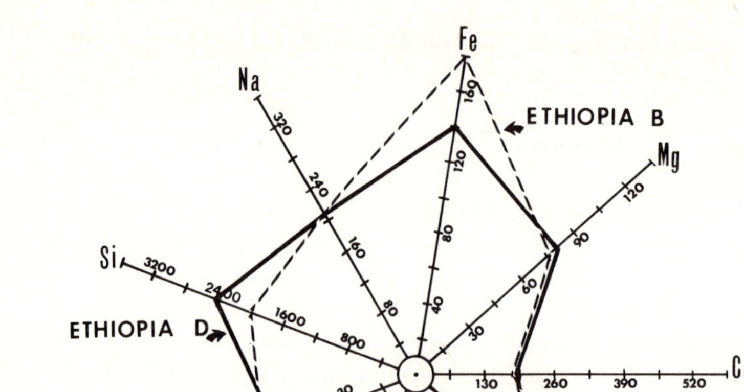

Fig. 7.1. Polygons depicting mean levels (p.p.m.) of nine trace elements in quelea feathers from Ethiopia (Colony B and D). Profiles of the colonies appear similar except for differences in boron.

one or both of the other areas (Table 7.3). Discriminant analyses of these data allowed correct assignment of individual birds to their groups in 85 per cent of cases for Hargeisa and 80 per cent for Melkassa and Colony D. Thus, groups of birds from the Awash Valley and adjacent areas seem to be separate and distinct from one another and can be identified by trace elements in their feathers.

The trace element composition of old and new feathers from the wings of the 40 birds collected in Hargeisa, Somalia, were quite different. Mineral profiles differed more due to age of the feathers (Fig. 7.3) than did other collections among populations. For all elements, except zinc and sodium, mineral levels appeared higher in old feathers than in new ones (Table 7.3). In 27 cases, samples of old and new feathers were obtained from the same bird and statistical comparisons of those samples suggested real differences existed among all minerals except zinc ($P \leqslant 0.05$).

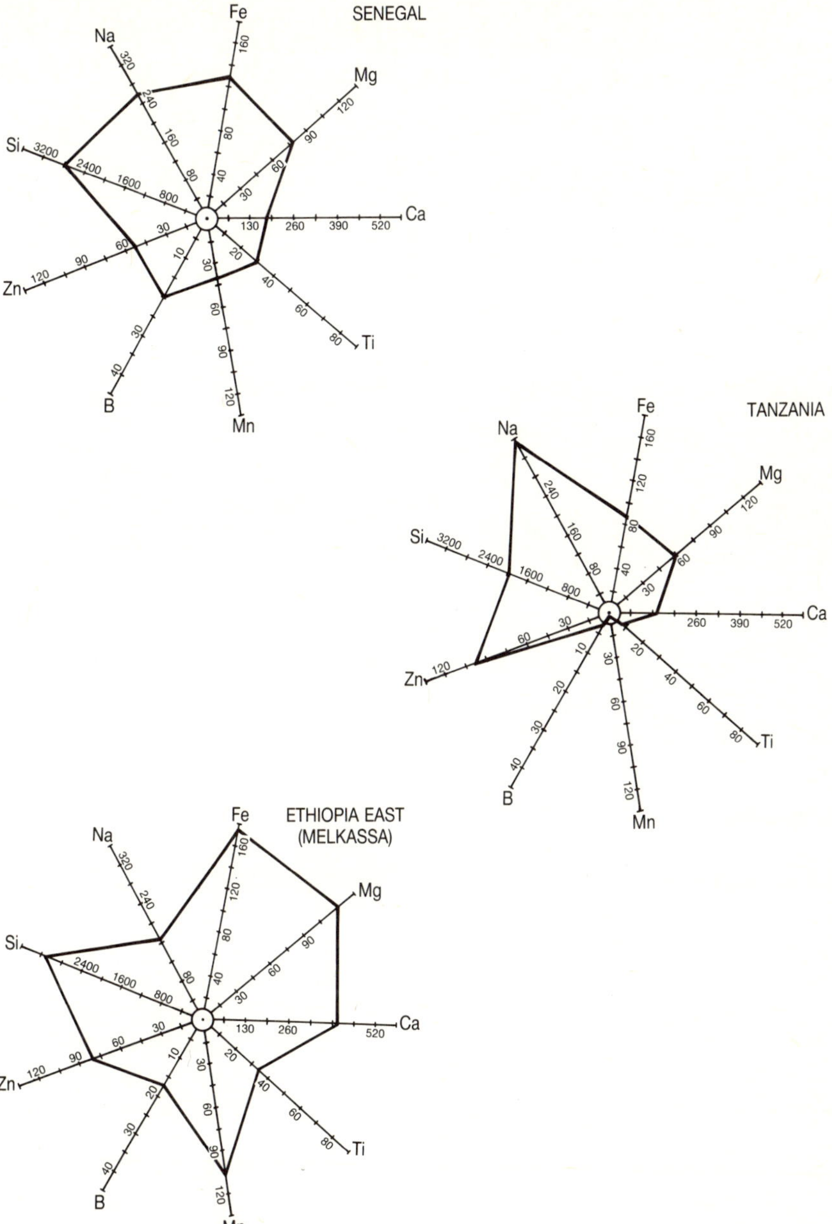

Fig. 7.2. Polygons depicting mean levels (p.p.m.) of nine trace elements in quelea feathers from Senegal, Tanzania, and Ethiopia (Melkassa). The profiles appear different from each area.

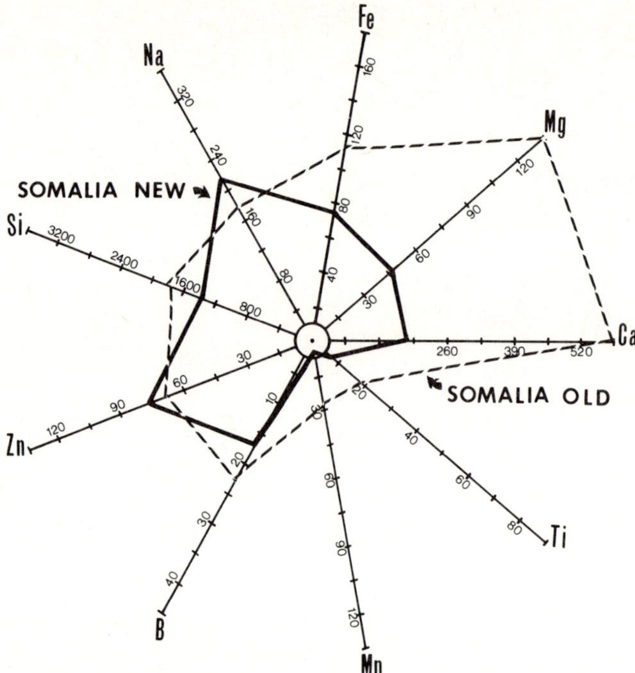

Fig. 7.3. Polygons depicting mean levels (p.p.m.) of nine trace elements in old and new quelea feathers from Somalia (Hargeisa). The profiles are distinctly different.

Discussion

In 1981, Jaeger *et al.* (1986) undertook intensive searches for quelea breeding colonies in the vast flood plain above Lake Chew Bahir, Ethiopia. This area, inhabited by nomadic herdsmen, contains little agriculture, but the grasslands offer ideal breeding habitat for quelea. At three colonies (B, D, and E) in south-western Ethiopia quelea were sprayed with fluorescent particles of different colours to mark them for later identification. Three quelea marked in Colony E were found 100 days later and 600 km away in the breeding colony near Melkassa, where birds were collected for trace element analysis.

Colonies B and D in Ethiopia were adjacent to each other and birds from the two colonies had similar mineral profiles in feathers (Fig. 7.1). Only levels of titanium and boron differed ($P \leqslant 0.05$), and differences were large only for boron. In contrast, feathers of birds from the Melkassa colony, where the marked birds from Colony E were found, had trace element levels that differed ($P \leqslant 0.05$) from those found in Colony D for all elements except titanium and silicon (Table 7.3).

These findings suggest several possibilities. If the Melkassa colony was

Table 7.3. Levels of trace elements (p.p.m.; $\bar{x} \pm SE$) in feathers of quelea from five areas of Africa. For each element, means that do not share the same letter in parentheses are different ($P \leqslant 0.05$).

Trace elements	Location					
			Ethiopia		Somalia (Hargeisa)[a]	
	Senegal (Richard Toll)	Tanzania (Arusha)	Colony D	Melkassa	Old feathers	New feathers
Iron	129 ± 8.6 (b)	88 ± 5.3 (c)	139 ± 9.9 (b)	175 ± 13 (a)	111 ± 12 (c)	73 ± 1.6
Magnesium	78 ± 5.4 (b)	58 ± 2.3 (b)	80 ± 5.7 (b)	122 ± 9.7 (a)	135 ± 13 (a)	47 ± 2.9
Calcium	181 ± 8.5 (b)	143 ± 5.5 (b)	186 ± 10 (b)	411 ± 28 (a)	583 ± 42 (a)	182 ± 14
Titanium	30 ± 2.0 (b)	7 ± 1.1 (d)	45 ± 2.6 (a)	34 ± 2.0 (a,b)	19 ± 1.7 (c)	7 ± 1.6
Manganese	41 ± 5.5 (b)	2 ± 0.02 (c)	25 ± 1.5 (b)	107 ± 8.0 (a)	28 ± 2.6 (b)	5 ± 0.70
Boron	20 ± 1.5 (a)	3 ± 0.30 (c)	10 ± 1.0 (b)	17 ± 1.5 (a)	23 ± 2.9 (a)	17 ± 2.4
Zinc	53 ± 1.7 (d)	97 ± 4.7 (a)	65 ± 3.2 (c,d)	81 ± 3.0 (a,b)	69 ± 2.9 (b,c)	76 ± 3.7
Silicon	2780 ± 202 (a)	2000 ± 0.0 (b)	2439 ± 119 (a)	3100 ± 206 (a)	1747 ± 110 (c)	1400 ± 0.0
Sodium	261 ± 8.7 (a)	365 ± 14 (a)	210 ± 7.5 (b)	165 ± 6.7 (c)	178 ± 8.0 (b,c)	217 ± 10

[a] Means for old and new feathers were analysed separately; means where different ($P \leqslant 0.05$) for all elements except zinc.

composed largely of Colony E birds, then Colony E birds appear quite distinct from quelea in Colonies B and D, which were only 56 km away. Thus, the many colonies found in 1981 throughout south-western Ethiopia and northern Kenya may have been composed of quelea from several different subpopulations. Their common use of this area for breeding may be a response to the excellent nesting habitat there, rather than an indication of any particular affinity with each other. Jaeger *et al.* (1986) found that quelea colour-marked in Colonies B, D, and E were segregated from each other when recovered 1-3 months later between 500 and 700 km to the north-east. Differences in trace element profiles among quelea in the Awash Valley support their findings that discrete groups of birds move and breed throughout the area whereas, at one time, all birds were thought to be part of a single homogeneous population.

Levels of specific trace elements sometimes varied considerably among the feather samples within a collection. The range of values for some elements was large, but often values were clustered around several different levels. This suggested that even individual collections may have contained birds from different sources. Several approaches were used to see if birds in collections were composed of different groups that could be recognized by trace element levels in feathers. First, trace element profiles were compared to mask classifications of males (see description of masks in Chapter 4). At the time they were collected, the 43 birds from Colony B were classified by mask type. Most birds (23) were classified as type 1, while others had mask types 2 (10), 3 (1), 4 (2), 5 (3), and 6 (4). No consistent similarities were found to suggest that those with the same mask type had similar mineral profiles. Next, groups of samples from Colony D were examined that had unusually high or low levels of specific elements. These groupings also did not show any consistent patterns for levels of other elements, and no distinctive mineral profiles for these groups were obvious.

Apparently, variations in feather elements within collections were not related to any unique history of the birds. They were either normal variations in mineral content or an artefact resulting from the semi-quantitative method used for chemical analyses. We have established which elements are of greatest diagnostic value, and now a more quantitative analytical method can be used to reduce variation within collections and provide more precision in distinguishing birds from different populations.

Kelsall and Burton (1979) also found that identification by mineral profiles was practical for separating subpopulations of birds, but not for separating individuals within collections. Because values vary within and can overlap between subpopulations, there is a need for large samples in order to establish reliable population means. Kelsall and Burton (1979) concluded that with adequate sample sizes and sufficiently different populations, variations due to sex, age, and feather year became unimportant.

Edwards and Smith (1984) showed that trace element profiles could change within a year when metal ions from the environment are adsorbed onto primaries. Thus, feathers exposed to different environments can attain different mineral profiles. They found, however, that profiles still maintained group affinities. In our work, for instance, the old and new quelea feathers obtained from Somalia had different mineral profiles. Still, the profile of the group was distinctive, even though it differed with the age of the feathers. A change in the profile over time does not negate its utility as long as the profile for the group remains unique.

We concluded that the populations we examined could be differentiated by trace element content of feathers. If profiles for populations of adult birds remain constant or change predictably during the year, this technique could be used to distinguish subpopulations at any place or time. Our results from Somalia show that profiles for a population can change over time. The nature of this change may differ depending on the specific habitats used each year; due to ion adsorption, adults can end up with quite different profiles than those with which they began the year. Further studies will be needed to find out how these factors influence the utility of feather minerals to identify specific populations.

Levels of trace elements were consistently lowest in the feathers of juveniles from Tanzania and in the new feathers of adults from Somalia. In addition, the mineral profiles for these two groups were quite similar. This suggests that new feathers may be somewhat alike in mineral content even though they are grown in different areas. Also, the total levels of minerals in or adhered to feathers seem to increase throughout the year. These aspects also deserve further study.

It may prove possible to identify the breeding areas of birds that cause crop damage by monitoring trace elements in feathers of juvenile birds. This would enable population control efforts to be directed at breeding colonies of birds known to cause damage. Young birds grow their primaries at colonies and all birds, regardless of sex, should acquire similar profiles in feathers. This is the only time that birds of different sexes and ages might be expected to grow all of their primaries while feeding on the same foods in the same area. In young, a post-juvenile moult, which is a complete moult and includes the primaries, begins about 9 weeks after fledging (C. Elliott, pers. comm.). This moult continues over the next 3–4 months (Morel and Bourlière 1955; Ward 1973*a*). It is during this period that juveniles often move into grain crops and cause damage. Trace element analyses might be useful in determining the origin of young that predictably cause crop damage during the months following dispersal from colonies. Such an approach might be the quickest way to utilize trace element analyses in quelea management. Additional research will be needed to test the feasibility of using mineral profiles to identify specific populations of quelea at different times and places throughout the year.

8

Damage assessments—estimation methods and sampling design

DAVID L. OTIS

Introduction

Quelea damage estimation is important for two main reasons. Firstly, it is the necessary initial step of problem definition before management strategies can properly be applied (Dyer and Ward 1977). Secondly, it provides the means to evaluate the success of control operations or other damage avoidance techniques in terms of crops saved (Elliott 1981*c*). In common with bird damage elsewhere in the world (Chapter 3), quelea damage is extremely variable in space and time, presenting many difficulties to the field worker attempting to obtain realistic and statistically valid damage estimates. This chapter reviews these difficulties and makes suggestions as to how they may be overcome.

Damage assessment methods

There are two components to any estimation method of crop loss: sampling design and damage assessment. Both must be statistically sound to produce useful final estimates. Sampling design determines sample size and sampling locality. It may involve a single stage, as in locating random points in an experimental unit, or several stages, as in choosing geographical regions of a country, then choosing fields within these regions, and, finally, sampling points within the fields (Stickley *et al.* 1979*b*). Damage assessment is what is done once the sampling points are reached and may have several components, e.g. measurements taken on the plant and on the size and shape of the sample plot, if plot sampling is being used, and a search strategy, if distance sampling is being used.

All assessment methods are to some degree subjective, depending on the quality of the decisions made by the assessor. These decisions will include precise measurements with a tool or the eye, and accurate discrimination between bird damage and other causes of losses. Assessor bias, defined as a consistent deviation between true loss and recorded loss, can best be eliminated by training. Jaeger and Erickson (1980) used correction factors for individuals after they had been through a training programme for quelea-damage-to-sorghum surveys. Alternatively, assessors can be randomly assigned to sampling sites so that bias is not associated with a particular area (LeClerg 1971).

The three parameters commonly used to describe crop damage are percentage loss, absolute loss, and incidence (the percentage of plants damaged more than a stated threshold value). The optimum sampling design may differ according to which parameter is of primary interest (Otis *et al.* 1983). Although the recorded data can be either continuous (e.g. area of seed removed, weight) or discrete (e.g. a percentage scale with intervals of 10 per cent), the same principles and suggestions set forth in this chapter apply. Efforts should be made to record data on a continuous scale, since these data will contain maximum information. In those instances where a discrete scale is used (most often in visual estimation of percentage loss), it is wise to strive for as many intervals as possible, while at the same time maintain a level of practicality and reality.

Plot size and shape

Quelea damage assessment methods usually have used plot sampling to produce loss estimates; all plants within one or more sample plots within the field are assessed. The plots may be of fixed size (e.g. 1 m^2) or may vary in size, as when a fixed number of plants is sampled at each location (Bruggers and Ruelle 1981; Jaeger and Erickson 1980; Kitonyo and Allan 1979). Jackson (1979) made recommendations for sample size based on numbers of plants per hectare as opposed to a number of sampling points per hectare. He did not mention the need to record plot sizes at each sampling point. Without knowing the actual area sampled, it is impossible to estimate directly the yield of the sampled field. Therefore, these considerations lead to the recommendation that damage assessment methods should use sampling plots of fixed size.

Plots can be of different shapes, the most obvious being rectangular, square, or circular. Although sampling efficiency influences this choice (Ghosh 1945), the most logical, practical plot shape would be rectangular. When sampling row crops, a rectangular plot is easily defined in terms of row length × row spacing. For example, the sampling plot is defined as those plants contained in a portion of a row K metres long; the actual size of the plot is then $K \times$ row spacing. (This plot will be referred to as a K-row metre-plot.)

Extensive statistical literature exists on determining optimum plot size

(Cochran 1977; Federer 1955), but these theories have rarely been applied to sampling situations in vertebrate pest research. Perhaps the most straightforward approach is to conduct damage assessments for several different values of *K*. Variances of estimates can then be used, together with time:cost factors (amount of time required to sample different plot sizes), to calculate the relative net precision of each method. Cochran (1977: 234–6, 243–4) described two methods for handling such data and presented an example. Otis *et al.* (1983) used a similar approach to determine plot size for sampling bird damage to sprouting rice and found smaller plots to be preferable. Generally, I believe the smaller the plot size the better, particularly in those situations in which damage is likely to be clumped or heterogeneous in the field, and the size of the clumps is large relative to practical plot size. It usually is best to sample many small plots in a field rather than a few large ones.

Crop types

Rice Most field surveys of quelea damage to rice have estimated percentage loss from the difference in average weight between damaged and undamaged panicles collected in plots (Bruggers and Ruelle 1981; Jackson 1979). This approach requires that the percentage of damaged panicles be used to adjust the difference in average weight. This method is subject to a potential bias because the assumption must be made that birds select and damage panicles at random. If, in fact, birds do show some preference for larger and heavier panicles, then there will be a sampling bias resulting in underestimation of loss. Comparison of the histograms of variables such as length or area of head on selected (damaged) and unselected (undamaged) heads (such variables being correlated with weight, yet unchanged by the occurrence of damage), will reveal if a bias exists (L. McDonald, pers. comm.). The bias could be avoided by comparing weights of random plots exposed to birds and random plots that have been protected, usually by use of exclosures (Bruggers *et al.* 1981*b*). However, such a procedure is totally impractical in large-scale surveys, and in most situations is not feasible even for small experiments.

Although bird damage research on ripening rice in the US is over 50 years old (Kalmbach 1937), reliable and standardized methods have not been developed (Meanley 1971). Recently, F. Crase and R. DeHaven (pers. comm.) evaluated seven assessment methods and concluded that either visual estimates of percentage loss, or estimates based on actual counts of missing and remaining spikelets were most feasible. P. Lefebvre (pers. comm.) recommended that visual estimates by trained assessors be given further consideration for large-scale surveys, and that the only practical alternative for use in small-scale experimental work may simply be incidence of damaged panicles.

For quelea, it appears that there are two available choices for rice damage estimates:

(1) If data are available that indicate random selection of panicles by birds, comparison of weights of damaged and undamaged panicles can be used. The appropriate formula is:

Estimated per cent loss =

$$1 - \left(\frac{\text{Av. wt(g) damaged panicles}}{\text{Av. wt(g) undamaged panicles}}\right) \times \left(\frac{\text{No. damaged panicles}}{\text{Total no. sampled panicles}}\right)$$

A minimum of 50 row centimetres should be used as the plot size.

(2) If assessors can be trained visually to estimate percentage loss, then use of this simplest assessment method is justifiable. If training is impractical and assessor accuracy difficult to verify, this method cannot be recommended. Minimum plot size can be reduced to perhaps 20 row centimetres. In addition, a random sample of plot weights should be taken from the field so that a direct estimate of yield is available, if the actual yield cannot be obtained from the farmer.

Fig. 8.1. Damage in bullrush millet can be assessed by measuring the length of the panicle eaten relative to the total length of the panicle, and extrapolating to the amount of grain lost (photo: R. Bruggers).

Millet The most accurate technique developed thus far to evaluate millet damage by quelea measures the length of damage on four axes of the ear and compares the average length of damage to the length of the total ear to estimate percentage loss (Manikowski and Da Camara-Smeets 1979*a*). These authors also developed a prediction equation to estimate percentage loss from incidence of damaged heads only, but state that such a technique would probably only be useful in planning or preliminary studies, and not in detailed, presumably experimental work. Considering the relatively easy length measurement technique, this procedure, with minor modifications, would be appropriate in all millet damage research (Fig. 8.1).

The accuracy of this method depends on the assumption that the width of the length of damage measured is fully one-quarter of the head's circumference. If the average width of the damage is less, then percentage loss is overestimated. To avoid this potential bias, it is necessary to measure the average width of damage along each axis and the circumference of the head. Then an estimate of percentage loss becomes:

$$\text{Estimated per cent loss} = \sum_{i=1}^{4} w_i l_i / WL$$

where

w_i = average width of damage on the ith axis
l_i = length of damage on the ith axis
W = circumference of the head
L = length of the head.

In the absence of real data on the efficiency of various plot sizes, any recommendation is *ad hoc*, but I suggest that for millet, plot sizes averaging approximately five plants per plot be used.

Wheat Most bird damage surveys in wheat in Africa have used weights of damaged versus undamaged panicles (Elliott and Beesley 1980; Kitonyo and Allan 1979). The same potential bias described for rice is applicable for wheat. Allan (1975) visually classified heads into five categories of per cent loss. Dawson (1970) estimated wheat damage in New Zealand by actually counting damaged and undamaged seeds on each sampled head, but stated that accurate counts of missing seeds were difficult because 'the bracts tend to hide the gaps and the number of grains per spikelet varies'. This observation implies that visual estimates could be subject to significant negative bias. Dawson (1970) accommodated the difficulty of counting missing seeds in the presence of a gap by estimating numbers of seeds present on adjacent spikelets, a method requiring 1 h to assess approximately 75 heads, even with low bird damage. Therefore, as with rice, counting missing and remaining seeds appears impractical.

Comparing weights and visual assessments seem to be the best techniques for wheat. If a visual method is used, the plot size used could be less than 20 row centimetres, because Dawson (1970) concluded that sampling four plants per sampling plot was the most time-efficient method. Also, if damage categories are used, I suggest that there be at least 10 categories. The necessary training programme for assessors should, therefore, strive for accuracy with this degree of refinement.

Sorghum Most surveys of quelea damage to sorghum have used a visual estimation method (Bruggers and Ruelle 1981; Jackson 1979). Assessors can be trained by use of known-damage heads (Jackson 1979). Alternative methods have been investigated. D. Otis (unpubl. data) protected sorghum heads from bird damage in the US by enclosing them in paper bags and comparing the yields of protected plots with paired, unprotected plots. Yield differences were quite variable, necessitating large sample sizes to achieve reasonable precision. This approach would probably not be feasible in large surveys, because a trip to each field to be sampled is necessary before the onset of bird damage. A linear regression equation for predicting percentage loss from incidence of individual damaged heads was developed by Manikowski and Da Camara-Smeets (1979*a*), but their subsequent discussion states that the amount of error in the model precludes its use in experimental studies. I remain sceptical of indirect approaches such as theirs because the method is usually developed from studies conducted in one general location in 1 year, and the consistency of the relationship over space and time is not verified.

For experimental research, visual estimations made by trained assessors probably is the most viable approach (Fig. 8.2). Visual estimates should also be used in large surveys, if it is believed that the behavioural feeding patterns of quelea could be different from those investigated by Manikowski and Da Camara-Smeets (1979*a*). Different feeding patterns and, hence, different quantitative relationships between incidence and per cent loss, could be caused by variation in crop culture, weather, habitat, and other factors.

Data on optimum plot size for sorghum apparently are not available. I suggest that a plot size large enough to contain five plants on average be used. This recommendation, as for all recommendations for row crops, assumes that plants are sampled down rows, not across rows or any other configuration. This choice is justified basically because of its simplicity and ease of plot definition.

One of the drawbacks of visual damage estimates in sorghum, as in other

Fig. 8.2. Sorghum damage in Somalia is usually measured by making a visual estimate of the percentage lost from each head (photo: R. Bruggers).

crops attacked by quelea, is that they provide no information on actual yield loss. If crop yields vary significantly among fields, and if research objectives dictate, it becomes desirable to collect data that can be used to adjust for potential yield differences. A quick, simple approach would be to devise a simple scale for rating the yield of sampled plants. This scale could be used to assign scores to each sampled plant, which could then be used to weight the estimates of percentage loss. For example, loss on higher yielding plants would receive more weight in the final estimate of loss for the field than would loss to low-yielding plants. An example of this technique is illustrated in the section 'Large-scale survey design.'

Field sampling design

In field sampling design, a field is defined as a sharply delimited, continuous planting of grain, e.g. a 1-ha experimental unit in a protection trial, a 0.24-ha plot within a larger planting, or a 100-ha commercial planting. The assessment technique is influenced by the size of the area to be sampled and the precision required of the results. Our studies on blackbird damage to sprouting rice (Otis *et al.* 1983) showed that the variance contributed by an individual field to the overall variance of the percentage damaged for an entire area was negligible compared to the variance in loss among fields. We chose, therefore, not to sample individual fields heavily but to sample as many fields as possible.

Precision of loss estimates for individual fields is an important consideration when designing experimental field trials. Firstly, precise loss estimates are usually necessary if compensation is to be paid to farmers for use of the experimental fields. Secondly, sampling of experimental units, as opposed to a complete census of damage within the unit, results in a loss of information, i.e. the power of the experiment in detecting differences is reduced (Yates and Zacopanay 1935). The amount of loss depends upon the relative sizes of the within unit variation in damage (sampling error) and the between unit variation (experimental error), and the optimum choice for numbers of experimental units versus level of sampling in each unit can be determined if preliminary information is available (Federer 1955). In the typical experimental situation in Africa, experiments must be performed with very limited resources (Bruggers and Jackson 1981), including numbers of experimental units available. Often, the researcher will not have the flexibility to adjust the number of experimental units based on expectations of the relative sizes of sampling and experimental error. The necessary requirements of adequate bird numbers and farmer co-operation severely restrict the range of choices for number of units. Therefore, we should attempt to minimize, within practical constraints, that component of the experimental error due to variance in individual unit estimates.

The above considerations lead to the following three general principles of field sampling designs. All designs subsequently considered in the chapter adhere to these principles.

Principle: The level of sampling effort in individual fields will be greater in an experimental context than in a large survey context.

Principle: It should be possible, at least conceptually, to construct a list or 'frame' of every sampling unit (plot) in the population (field).

For example, if a sorghum field is to be sampled, using one row metre as the sampling unit and the field has 100 rows, each 50 m long, then the description of each sampling unit in the field could conceivably be written down. There would be 5000 such sampling units, beginning with the first metre in Row 1 and ending with the last metre in Row 100. Of course, it is not necessary to do this exercise in practice, but it is helpful to be able to think of the field as a collection of definable sampling units when constructing a design that specifies how units are to be chosen.

Principle: Each sampling unit in the population, and therefore, every plant in the field must have a positive (not necessarily equal) probability of being chosen for assessment. This property is guaranteed and the probabilities determined by a specified method of random selection of the units.

This principle is the foundation of probability sampling. It allows measures of variability of estimates, such as standard errors, to be calculated from the sample data, and hence, leads to valid statistical inferences concerning the population sampled. Alternatives such as 'haphazard' or 'expert choice' sampling depend on the validity of broad assumptions about damage distribution which are difficult to evaluate.

Field sampling in large-scale surveys

Both the suggested field sampling designs and the actual methods used in surveys of quelea damage have been diverse. They vary from the random selection of points from a grid superimposed on the field (Anon. 1979; Kitonyo and Allan 1979) to sampling at regular intervals along diagonals (Church 1971) or along transects placed to cross the field adequately (Jackson 1979; Jaeger and Erickson 1980) or systematically (Bruggers and Ruelle 1981) (Fig. 8.3). If neither the initial transect location nor the starting location along transects is randomized, the design will not result in every

Fig. 8.3. Wheat damage is assessed in West Kilimanjaro, Tanzania, by cutting samples and comparing the yield of damaged spikes relative to the yield of undamaged spikes (photo: M-T. Elliott).

portion of the crop having a positive probability of being sampled. These designs, therefore, violate the principle of probability sampling.

For large-scale surveys, the correct procedure would be as follows (see also Fig. 8.4):

(1) Construct a rough diagram of the field and establish a baseline, preferably perpendicular to the direction of rows.
(2) Divide the estimated length of the baseline, e.g. L, by 4, and let the result be denoted W and select a random number between 1 and W, e.g. R. Four transects running perpendicular to the baseline, i.e. along rows, will be sampled at distances of R, $R + W$, $R + 2(W)$ and $R + 3(W)$ along the baseline. Simply stated, four rows, located at regular intervals of W starting from a random point R, are chosen to be sampled.
(3) Estimate the total length of the four rows to be sampled, say M, and divide by the number of samples to be taken, say N. Choose a random number between 1 and M/N, say T, and take the first sample a distance T up the first row. Continue to take samples at regular intervals of M/N. When the end of a row is reached, simply continue the count to the next row.

Of course, the question still remains of how large the number of sample plots (N) should be. A specific answer is difficult in the absence of any

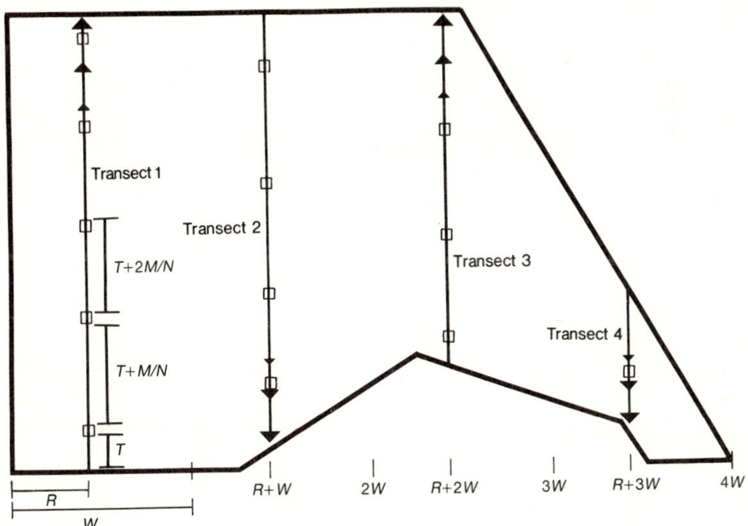

Fig. 8.4. Illustration of the placement of four transects and sample plots (□) across the region of interest. Note that the transects are parallel and equally spaced, but have different lengths due to the irregular shape of the region.

preliminary estimates of distribution or variation of losses, but I suggest a range of 20–60 sampling locations of any size per field. The reasoning is to devote more effort for the given amount of manpower available to increasing the number of fields that are sampled and therefore, to minimize (within reasonable limits) the amount of sampling done within a field. Modification in this sampling design will undoubtedly be necessary as specific applications are encountered. The researcher should strive to ensure that such changes in design do not result in violations of the basic principles of sampling design that have been discussed.

Field sampling in experimental trials

In experimental work, it is more desirable to have unbiased estimators of precision for each experimental field, and therefore, a different sampling design is necessary. The size of the experimental unit is of course a major consideration. In African field trials, this size has ranged from <1 ha (Bruggers 1979*a*,*b*; Funmilayo and Akande 1977), to 2 or 3 ha (Martin 1976) to occasional use of large (30–40 ha) fields (Bruggers 1979*a*). As a generalization, Bruggers and Jackson (1981) stated that experimental trials in developing countries are often conducted at agricultural research stations with experimental units of <0.25 ha. In experiments with units of $\leqslant 1$ ha, a simple sampling design should be used for locating sample plots. The steps to construct this simple random design are:

(1) Construct a diagram of the field. If the field is irregularly shaped, circumscribe the field with a rectangle or a simple polygon.
(2) Select N random locations (row, pace) in the field and sample damage at these locations. If the field is irregularly shaped, it is convenient to select a few extra locations. Once in the field the assessor may discover that some of the samples fall outside of the field boundaries due to inexactness in the field diagram. When this occurs, the extra random locations can be substituted.

The above design would obviously be too time-consuming in larger fields. In addition, when placing a relatively small number of locations completely at random in a large area, a satisfyingly uniform coverage of the field may not be achieved. To guard against this possibility, and possibly to increase efficiency of sampling effort, stratification can be used.

In stratified random sampling, the population (field) is divided into mutually exclusive and exhaustive parts called strata, and a random sample of plots is taken within each stratum. Efficiency is improved if sample plot damage within the individual strata is more homogeneous than the entire population of sample plots. Strata boundaries are established at the discretion of the researcher, but I suggest that there be from two to five strata. Populations may be stratified by simply splitting the field into equal areas to ensure uniform coverage or by preliminary guesses of where the most variation in damage in the field is expected, and then establishing strata in which low, moderate, and high variation (which is usually proportional to the amount of loss) is expected. I would recommend that stratification schemes be kept simple with the major objective of achieving more uniform coverage. Within each stratum, sampling locations could be located randomly if the individual strata are sufficiently small or along at least four transects (rows). However, transect locations along the baseline should be selected randomly and not systematically. (Random selection will permit valid estimation of variance of the loss estimate.) An example of this sampling design is illustrated in Fig. 8.5.

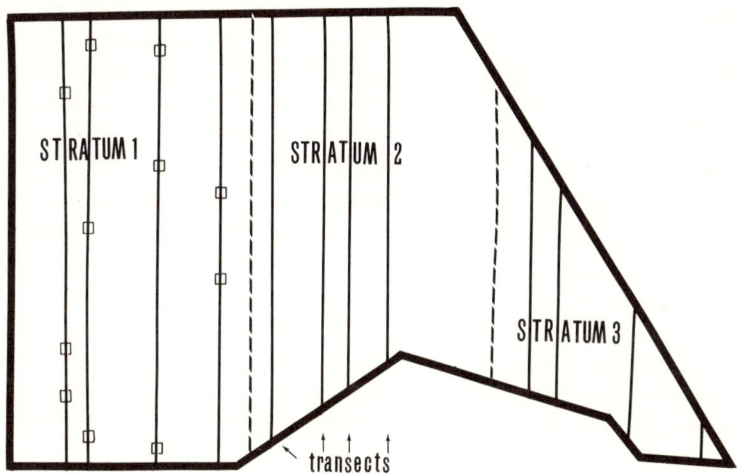

Fig. 8.5. Illustration of the use of stratification. The field has been divided into three strata of equal width but unequal area due to the irregular shape of the field. Within each stratum, four random locations for transects are selected along the baseline. Sample plots (□) are located at random on the transects.

A third sampling strategy involves the technique of post-stratification (Cochran 1977:134). This approach could be used when the original sampling locations are placed in the field at random and without stratification. After data have been collected, the researcher may be able to examine the damage pattern and delineate boundaries that produce strata which are homogeneous within. It is important for the researcher to know the exact size of the resulting strata. These exact strata sizes can then be used as weights, as if stratification had been used originally. For example, consider a 1-ha field in which 50 plots have been placed at random. Upon examining the resulting damage data, the researcher concludes that damage within the outer 10 m perimeter of the field is much greater than in the interior. Therefore, the data are post-stratified into interior and perimeter strata, which receive weights of 0.64 $=(100-2(10))^2/10\,000$ and 0.36 ($=1-0.64$), respectively, to calculate losses for the entire field. Note that the stratum weights do not depend upon how many plots actually fell within each of the strata. Post-stratification could be a useful way to increase precision in situations in which a few obvious, easily-defined strata can be created after examining the data.

The choice of the number of plots to be sampled (N) should guarantee the desired precision in our estimates. However, 'the decision [about N] cannot always be made satisfactorily; often we do not possess enough information to be sure that our choice of sample size is the best one' (Cochran 1977). This information essentially is a quantitative estimate of the expected variation within field damage. In the absence of previous research involving similar conditions and parameters, a somewhat arbitrary decision must be made. It has been my experience that variation in damage within any field with a significant amount of damage is large, and that sample sizes must, therefore, be relatively large, perhaps on the order of 80–100 plots. This suggestion is made with the goal of achieving a coefficient of variation of 10–15 per cent. With some standard available, as well as previously collected data, determining necessary sample size (and perhaps changes in sampling design) can be made accurately and objectively. Good sampling designs are the result of a sequential process of data collection.

Sample size recommendations are independent of field size. This is contrary to many researchers' intuition; in fact, sample size recommendations are often made in terms of plots per hectare or some other unit of measurement (Jackson 1979). However, in this context, the most important parameter in determining N is the variation in damage among the sampling units in the population (field) and not the size of the population. Usually the number of potential sampling units in any field is practically infinite. In a 2-ha field with 1-m row spacing, there are 100 000 sampling units of 0.2 m^2 (20 row centimetres). Thus, for any practical sample size N, the sampling fraction ($N/100\,000$) will be negligible, i.e. very close to zero. In a field one-half as large, this fraction will remain essentially zero; thus, variances of the estimates will be the same in both instances, given that the variation among sampling units is the same in both fields. For example, using eleven 0.37-ha experimental units of sunflower planted in Sand Lake, South Dakota, C.E. Knittle (unpubl. data) evaluated the efficacy of several treatment forms of Mesurol^R in repelling Red-winged Blackbirds *Agelaius phoeniceus*. In each unit, twenty 5-head plots were located using a simple random sampling design, and square centimetres of seed removed were measured using a plastic template. Below is the average loss ($\bar{x}$, cm^2) per plot and the coefficients of variation (CV) for each unit, assuming a fixed plot size:

Experimental unit	$\bar{x}$ (cm^2)	CV($\bar{x}$) (%)
1	1365.90	9.50
2	1125.35	8.86
3	1259.35	7.14
4	1157.35	6.55
5	1132.45	5.59
6	650.35	19.54
7	679.15	11.78
8	1448.15	8.03
9	1049.60	7.51
10	1152.50	7.77
11	1115.80	5.60
Average	1103.27	8.90

These data indicate that estimates of loss for the units are of adequate precision. In fact, if the number of plots per units were reduced to 11, the average CV ($\bar{x}$) would increase only slightly, to an acceptable 12 per cent (calculated by solving the equation $\sqrt{\frac{20}{N}} \times (8.9) = 12.0$).

We may also examine the adequacy of sampling units from the viewpoint of the sensitivity of the analysis of variance in detecting differences among the treatments. That is, how much information (sensitivity) was lost by having to sample the plots as opposed to measuring or harvesting them completely? The key to this approach is in the ANOVA (Analysis of Variance):

Source	DF	MS
Total	219	
Treatments	3	787 402
Experimental error	7	$1\,396\,875 = \hat{\sigma}_e^2$
Subsampling error	209	$179\,672 = \hat{\sigma}_s^2$

The estimate of the percentage loss of information is (Federer 1955:80):

$$L = 100 \left[1 - \frac{k}{n}\right] \frac{\hat{\sigma}_s^2}{\hat{\sigma}_e^2}$$

where k = number of sample plots/unit and n = total number of plots available in the unit. Stated differently, the number k/n merely represents the proportion of the unit that was sampled. This fraction is almost always very close to zero; thus, an approximation to L may be taken as $100 \times \hat{\sigma}_s^2/\hat{\sigma}_e^2 = (100)$ (179 672)/1 396 875 = 12.7% . Thus, about one-eighth of the information contained in the experiment was lost due to sampling of the units. If the same rate of sampling is maintained in future experiments, the size of the experiment (number of experimental units) would have to be increased by a factor of 1.145 = 1/ (1–0.127), relative to an experiment in which complete enumeration of the plots was planned. Finally, it can be shown that if the level of sampling were decreased to 11 per unit, as suggested above, the corresponding information loss would increase from 12.7 to 20.5 per cent. This is probably an acceptably small increase, considering the savings in manpower associated by nearly halving the amount of damage assessment required.

Large-scale survey design

Direct assessment

The most challenging of the sampling design problems encountered in quelea research is that of constructing a design for surveys over large geographical areas. The objective is to produce an unbiased picture of the level and distribution of damage in the area and to obtain estimates of the precision with which the losses are assessed. As with any sampling problem, a design is easily created given unlimited manpower, but of course the situation in Africa is one of very limited manpower and a plethora of logistical problems. These obstacles have, unfortunately, led some researchers to discard the above objective. Ash (1981) stated that 'the gathering of crop damage data is tremendously dull work, very time consuming, and no one in Somalia has much interest in undertaking it'. Elliott's (1981*c*) opinion was that 'a definitive, fixed assessment of the problem in East Africa is not considered possible' because of variations in climate and rapidly changing agricultural practice. As a substitute for objective surveys, extrapolations of farmers' opinions or of a limited tour of some known damaged areas have been used to assess the magnitude of the problem. I agree with Lenton (1981) who, in talking about his estimates of crop loss in Sudan derived from sketchy data, stated that 'until good objective damage assessment data are produced over wide areas many of these estimates of loss are meaningless.' Moreover, such substitutes, even if they produce reasonably accurate estimates of country-wide crop losses, do not provide much information on damage distribution, which is often of more interest.

In spite of all the difficulties, several large-scale objective surveys of quelea damage have been conducted, each with its own particular survey design (Plate 7). In Kenya, Kitonyo and Allan (1979) used 0.5° grid maps (55 km^2) and Land Registration numbers to construct a sampling frame of wheat farms. They then selected farms in proportion to the acreage planted in the 0.5° blocks making up the defined region of interest. Allan (1980) subsequently altered this scheme by using the political division of a ward instead of 0.5° blocks. Both of these designs seem to satisfy the need for a theoretical sampling frame from which to randomly select fields, a defined geographical area (population) of interest, and a known, positive probability of selection for each field.

Thus, using appropriate estimation formulas (their design is really a two-stage cluster design), valid estimates of loss and associated precision could be calculated. In 1978, Bruggers (1980) assessed crop losses in Somalia using a network of randomly located sampling locations on government and private farms. He later stated, however, that the resulting figures should be used cautiously since 'it is uncertain whether the area sampled is representative of the district'. Implied is the fact that all fields did not have a chance to be sampled and, therefore, valid inferences about district losses are not possible. Cereal crops in Senegal were surveyed by Bruggers and Ruelle (1981). Fields in all crop-growing regions were sampled at systematic distances along paved or dirt roads (although the distance between stops actually varied somewhat based on finding a mature crop). Usually, only fields within 500 m of the road were sampled, so that inferences to the entire crop are not possible. As a result, it is difficult to assign a probability of selection to a field (a sampling frame does not exist) and, therefore, to construct the appropriate estimation formulas.

I wish to emphasize that the method of field selection (i.e. the survey design) determines how loss estimates are actually calculated. There is no one formula (e.g. the average of all sampled fields) that is appropriate for every design. Therefore, it is important that the investigator is aware of the type of design actually being used (e.g. cluster, stratified, multi-staged) and knows the proper formulae associated with such a design.

With this brief review of past surveys, let us consider the essential properties of a properly designed survey:

(1) The population for which estimates of crop loss (or any other parameter) are desired must be clearly defined. In this context, a population could be a

country, a region, or a type of farm (e.g. governmental). Notice that this property is closely related to an earlier stated principle involving conceptualization of a sampling frame. That is, once the population of interest is defined, a theoretical listing of all the members (fields) of the population is possible.

(2) Clear objectives for the survey should be identified, usually in terms of desired loss estimates in an area or in several areas, e.g. estimates of dollar loss in the country, of the percentage of fields with any quelea damage in each of several regions, of kg/ha loss on private farms, etc. A second type of objective, statistical in nature, must also be specified in terms of the desired reliability of the estimates, i.e. it is desirable to have estimated per cent loss with standard errors of less than 2 per cent, or estimates of dollars lost in a region within limits of $500 000.

(3) After points (1) and (2) have been established, the most critical element of the planning process occurs. An estimate must be made of the cost, in terms of manpower and money, associated with achieving the desired objectives. If the resources available are not sufficient, then either the objectives must be revised or the project dropped. The point to be made is that if the survey cannot be done well, with a reasonable chance of satisfying the stated objectives, then it is not worth doing. Effort has been mostly wasted if, upon completion of the survey, estimates of reliability of the loss figures are not available, or extrapolation to the population of real interest is not valid. Of course, cost will vary with the proposed sampling design, both field and survey, but the range in cost will probably be small compared to the average cost.

(4) Choice of the survey design should be influenced by both the objectives of the survey and the availability of resources. A variety of questions must be answered. If regional estimates are desired, what are the boundaries of these regions, and how shall effort be allocated to the strata within regions? Does the within field sampling design require a team of assessors, or can individuals handle the sampling alone? What type of statistics concerning the amount and distribution of the crop are available to help define sampling units at each stage of the design? What are the appropriate estimation formulae? Of paramount importance is that a probability sample of the fields (or whatever the basic sampling unit is) is achieved. Ultimately, every plant of every field in the population of interest must have a chance to be sampled.

The following example of survey design is a summary of the methodology proposed by Otis (1984) for surveying relatively large sorghum-growing areas in Tanzania. Discussion of this design will serve to illustrate the application of principles stated above.

The primary objective was to estimate the number of hectares of sorghum lost to birds in a large ($\simeq$10 000 km^2) region of central Tanzania. Most cultivation in this area consists of small, scattered, subsistence farm holdings. Because no statistics were available on the amount and distribution of cultivation, it was decided that the stated objective would require two different, independent surveys. The first was to estimate the hectarage of sorghum cultivation and to create a crude map of the distribution of crops within the region. The second would produce an estimate of the percentage of sorghum lost to birds; the product of this estimate and the number of available hectares produces an estimate of the number of hectares lost. Let us consider the design of these two surveys in more detail. The aerial survey described was conducted, but no data were collected using the ground survey design. Therefore, this part of the example is hypothetical.

A fixed-wing aircraft, capable of carrying at least three passengers and maintaining a speed of 200 km/h, was used to produce a survey estimate of hectarage. The idea was to fly a series of parallel transects across the area at an altitude of 100 m. One passenger served as a navigator, helping the pilot to locate landmarks, time the length of transects, and time the counts of two observers seated on either side of the aircraft. A small sighting guide, such as a 0.5-cm circle, was placed on each observer's window in a position that allowed the observer, at any given instant, to fix a location on the ground. In effect, the eye was taking a snapshot of a point on the ground. Once the navigator determined that the aircraft was on the transect, he signalled the observers in 15-s intervals until the transect was completed. At each signal, each observer took a sighting and recorded a 1 if the sight was on maturing sorghum at that instant, and a 0 otherwise. The resulting sequence of 1's and 0's for each observer for each transect represented the basic data collected in the survey.

At the planning stage, we had to define precisely the area to be surveyed and we chose to sample 4.5° × 0.5° blocks within the Singida region. To determine transect locations, a baseline was established along one side of each block. Transects were flown perpendicular to this baseline that was divided into N intervals of equal width, e.g. W, and N was the number of transects flown. For each block, a random number between 0 and W, e.g. R, was chosen and the first transect was begun at this distance down the baseline from the starting corner. Subsequent transects were flown at equally spaced intervals at distances $R+W$, $R+2W$, . . ., $R+(N-1)W$.

The appropriate formulae for estimating the number of hectares in sorghum cultivation within each block, and its associated variance, require the following notational definitions:

N = number of transects;

M_i = number of observations taken by each observer on the ith transect, $i=1, 2, \ldots, N$;

L_i = length of the ith transect, $i = 1, 2, \ldots, N$;
A = area, in hectares, of the sampled region; and
b_i = total number of 1's recorded by both observers on the ith transect, $i = 1, 2, \ldots, N$.

The estimate of the proportion of the ith transect devoted to sorghum cultivation is then $\hat{P}_i = b_i/2M_i$, $i = 1, \ldots, N$, and the estimated proportion for the block is

$$\hat{P} = \sum_{i=1}^{N} L_i\hat{P}_i / \sum_{i=1}^{N} L_i.$$

A simple estimate of the total number of hectares in sorghum in the entire block is then $A\hat{P}$, with estimated variance

$$\hat{\mathrm{Var}}\,(A\hat{P}) = A^2\sum_{i=1} L^2_i\,(\hat{P}_i - \hat{P})^2/\bar{L}^2N\,(N-1),$$

where $\bar{L} = \sum_{i=1}^{N} L_i/N$.

The estimate for the total hectarage in all four blocks was obtained by adding the four individual block estimates. The variance of this estimate is the sum of the individual block variances. A rough map of the pattern of cultivation in the region was made using the ordered sequences of 1's and 0's recorded by the observers. As mentioned previously, this map can assist in the planning of the damage assessment phase of the survey.

There are two essential components to a design to estimate percentage bird loss: (1) an accurate map of the target region that contains all available information about the distribution of the crop and the locations of all possible roads, and (2) a clear idea of manpower availability and associated operating resources or the desired precision of the final survey loss estimates. Ultimately, the design will reflect a compromise based on both specifications, because rarely are both compatible. These considerations are important for determining sample size, but not to the fundamentals of the design itself. Let us first discuss the basic outline of this design.

Any random sampling design involves a random selection of sampling units, and often subsampling units, which must be conceptually defined. In large-scale crop surveys, the primary sampling unit is usually defined as a single cultivated field, and plots of a specified size, e.g. 4 m^2, within the field are defined as the subsampling units. These definitions dictate that some method be constructed for randomly selecting specific fields. However, in Tanzania we were not concerned with the distribution of damage to individual fields and, thus, to the individual farmer, but only with the regional loss. Therefore, the concept of a field as a primary sampling unit was inappropriate in this sampling design. Rather, the target region was viewed as a single mosaic consisting of two classifications—sorghum cultivation and other cultivation or land use. (The same concept was involved in the aerial surveys of cultivation.) The idea would be to sample this mosaic by selecting locations at random and assessing damage within a specified area around these locations. The key to specifying the sampling locations is a road system within a region. On the regional map, roads are divided into 2-km intervals, and each interval is assigned a unique number. A random selection of these intervals is then made and each of the selected intervals would be sampled for sorghum damage. Note that this method represents a simple random sample of locations within the region, and not a systematic approach, in which all roads are travelled and assessments made at equally spaced intervals. The statistical efficiency of both of these sampling methods should be approximately equal, but the recommended approach should save a substantial amount of travel time.

Survey teams of a driver and two assessors then assess damage at the identified sampling locations. When reaching the sampling location, the first assessor chooses a side of the road via a coin flip, and begins to walk a 0.5-km transect perpendicular to the direction of the road. Upon encountering sorghum cultivation along this transect, the assessor samples circular plots of 1-m radius placed at predetermined intervals. The length of this interval should range between 20 m and 50 m and vary inversely with the density of the cultivation in the region. In practice, the assessors should be allowed temporarily to leave the transect to sample cultivation within perhaps 50 m on either side. If this is done, the circular plots should be sampled in a direction parallel to the direction of the original transect and should begin at a random distance along the dimension of the cultivation that is perpendicular to the transect. If possible, the assessor should then return to the original transect after reaching the end of the cultivation. The important point is that the assessor continue until he is at a point 0.5 km in perpendicular distance from the road. He then executes a 90° turn in the direction of travel in which the vehicle approached the sampling location and walks another 0.5-km transect, proceeding to sample circular plots as described. After this transect is completed, he then executes a second 90° turn towards the road, and walks the 0.5-km transect back to the road, again sampling, as described. In the meantime, the driver will have dropped off the second assessor at approximately this same location, 0.5 km down the road from where the first assessor began. This second assessor will proceed exactly as described for the first, except on the opposite side of the road. Figure 8.6 illustrates the resulting pattern of sampled plots. Note that, in effect, the sorghum within a 1-km^2 block of land has been sampled at this location. Also, this description

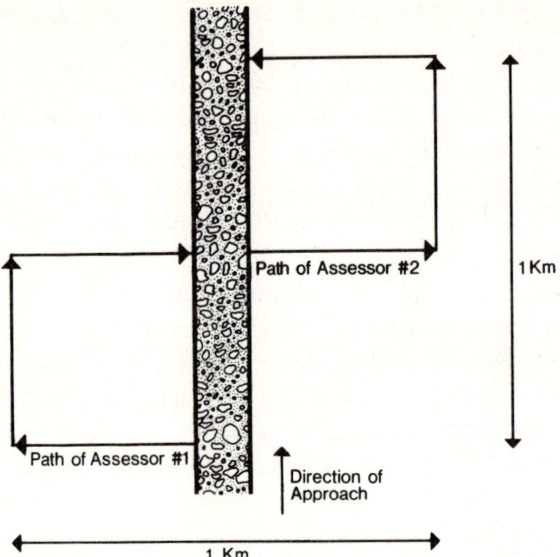

Fig. 8.6. Illustration of the ideal transect path followed by two assessors, beginning at a randomly selected location on the road. In practice, the assessors may vary from this path in order to encounter cultivation near the transect.

assumes that the 2-km stretch of road involved is relatively straight. Curves will cause the pattern of samples to be asymmetrical, which is permissible, although intervals containing sharp curves should probably be avoided.

This design for locating plots is not the same as that recommended and described earlier for use in large-scale surveys. The key difference is that the concept of a field as a primary sampling unit is not appropriate in this situation. The very small average size of a planting and the basically unknown distribution of these plantings precludes the use of a systematic system of parallel transects in each sampled 'field'. Note, however, that the present design involves locating sampling plots at systematic intervals and that straight-line transects are used, although they are not parallel. Also realize that the actual population being sampled does not include all cultivated land. Only the sorghum within 0.5 km of a road is eligible for sampling, and thus the inferences made from the data collected under this design pertain only to the collection of 1-km-wide strips centred on the area roads. If the investigator makes inferences to the entire area, he assumes that areas adjacent to roads are representative of the entire area. Obviously, this situation is less than ideal because all sorghum in the target area should have a chance to be sampled. The practicalities of the situation in Tanzania, i.e. lack of easy access to all agricultural lands, forced this compromise into our design.

Each plant within each sample circular plot is visually examined by the assessor and assigned two values. The first value is a head size category, e.g. 1, 2, or 3, that indicates if thc hcad is in a below average, average, or above average size class. The definition of these categories should be specified before the ground surveys are begun, and the assessors should be trained accurately to classify heads according to the defined size classes. The purpose of recording this variable is to provide some information about the quality of the assessed crop, and together with an estimate of plant density obtained from the average number of plants per plot, a crude index to the potential yield of the crop can be calculated. No direct information about the yield will be available, because it is simply too time-consuming and impractical to collect and weigh heads. Instead, the assessor records an estimate of the percentage of seed removed, in a 5 per cent incremented scale, using a visual examination of the head (Fig. 8.7). It must be remembered that the validity of the results of the entire survey ultimately depends upon the competence of the assessors.

Summarizing and analysing damage data are straightforward. The following notation is used to define the estimators. This notation does not distinguish between data collected by the two different observers in the same sampling location (1-km^2 block), i.e. the data from both observers are pooled for each location.

Region Singida, Tanzania   Date 17 June 1985   Beginning time 0800

Road location 43   Assessor Otis   Ending time 1100

Plot Interval 30 m   Direction of initial transect 240°

Plot number	Size class 1	Size class 2	Size class 3
1	0, 10, 20	0	0
2	15, 0, 0	5, 5, 5, 0, 40, 0, 0	
3	0, 10, 15	35, 50	15
4	0, 0, 0, 0, 0, 0		
5		10, 15, 25, 60	10, 5
6	0, 0	0, 10	0, 0
7	0, 15, 20	20, 20, 35	
8		60, 40, 20, 0	0, 0
9	0, 0, 0		
10		0, 0, 0, 0, 0, 0	

Fig. 8.7. Example of a completed data form for recording bird damage within circular plots of 1 m radius. Each entry represents per cent loss estimated visually on a single head, and categorized by estimated size class of the head.

Y_{ijk} = Estimated per cent loss of the kth plant in the jth plot at the ith sampling location, $i=1, \ldots, T, j=1, \ldots, R_i, k=1, \ldots, S_{ij}$,

X_{ijk} = Size class of the kth plant in the jth plot at the ith sampling location,

$$\bar{Y}_i = \sum_{j=1}^{R_i} \sum_{k=1}^{S_{ij}} Y_{ijk} X_{ijk} / \sum_{j=1}^{R_i} \sum_{k=1}^{S_{ij}} X_{ijk}$$

= Weighted (by size class) average per cent loss of all plants assessed at the ith sampling location,

$$\hat{Y} = \sum_{i=1}^{T} \bar{Y}_i / T$$

= Estimated per cent loss in the region,

$$\hat{\text{Var}}(\hat{Y}) = \sum_{i=1}^{T} (\bar{Y}_i - \hat{Y})^2 / T(T-1).$$

Note that, because of the weighting by size class, larger heads have greater influence in determining the overall loss. It is true that there is a slight statistical bias and potential loss of efficiency inherent in the estimator $\hat{Y}$, because the $\bar{Y}_i$ have not been weighted by the relative amount of sorghum cultivation contained in the 1-km^2 areas sampled. The reason for this is simply that these amounts are unknown, i.e. the data collected by the assessors cannot provide an accurate estimate of the amount of cultivation in the area sampled. If the assessors were required to adhere strictly to the transect line, then such an estimate would be available by using the proportion of the transect length that intersected sorghum. However, it seems impractical to expend resources to get to a location and then not have the flexibility to collect a reasonably large sample, and it is this flexibility (allowing departure from the transect line) that invalidates an attempt to estimate the relative cultivation of the area. Thus, the choice was made to make the trade-off between obtaining weights and collecting a satisfactory number of samples at each location, with the assumption that the unweighted estimator $\hat{Y}$ will be satisfactory in practice.

Bioenergetic models

Theoretical estimates of the amount of a specific crop likely to be consumed by a specific bird population can be produced by constructing a bioenergetic model for the particular ecological system involved. The bioenergetic requirements of each age and sex class in the population are estimated by creating a model that uses parameters such as digestive efficiency, temperature, population density and age structure, reproduction energy requirements, and produces a resulting population energy demand (Wiens and Dyer

1975). This output, combined with knowledge of the food habits of the species and the density and distribution of the crop, can then estimate the impact of the population on the crop. Wiens and Dyer (1975) used such an approach to estimate the amount of maize eaten by Red-winged Blackbirds in northern Ohio. Their energetics model was modified by Weatherhead *et al.* (1982) who were also interested in the impact of Red-winged Blackbirds on maize in Quebec. The authors also conducted a field study that measured actual consumption of maize by captive blackbirds during the damage season, and these figures were compared to their model predictions. The agreement between their model estimates and field data was quite good, although they differed markedly from estimates produced by the Wiens and Dyer energetics model.

Both studies emphasized the need for objectively evaluating the magnitude of the pest problem before further research or management strategies are put in place, and I am in full agreement with this attitude. They suggest that a modelling approach could replace large-scale damage surveys because it is less expensive and because of the highly variable estimates that often result due to the large variation in loss within the surveyed region. However, modelling also has its problems. First, the models require that values be specified for a large number of bioenergetic and pest population parameters, which may or may not be available. Second, the models are deterministic, so that a measure of the stochastic variation in the system, i.e. measures of variation in damage among fields or strata, are not available. These measures of reliability are valuable not only from a statistical point of view but can also be used as information concerning the distribution of loss within the region. Finally, of course, as with all modelling, the results will be subjective in the sense that no two modellers are likely to create the same model.

Despite these drawbacks, a model can be of value in that it requires the researcher to think in the context of an ecosystem with all its interactions and complexities, as opposed to a narrow focus that considers a pest and a crop isolated from the system. Also, if reliable estimates of parameters critical to the model are available from previous research, then bioenergetic estimates of loss may help to put the depredation problem in context. Although I am unfamiliar with the status of the data base involving quelea biology, I suspect that information of bioenergetic and population dynamics sufficient for the construction of reasonable model outputs does not presently exist. Considering the urgent need for information of problem definition on quelea, direct estimates of regional losses are presently more appropriate.

9

General aspects of quelea migrations

PETER J. JONES

Introduction

The quelea was long thought to be the textbook example of an entirely nomadic species. The early literature generated a popular image of the 'avian locust', roaming opportunistically in marauding swarms and homing in to fields of ripening grain to cause untold damage. In order to unravel the quelea's seemingly erratic movements, massive ringing campaigns were instituted in eastern Africa and southern Africa in the 1950s and 1960s. Despite many long-distance recoveries no immediately obvious seasonal pattern of movement emerged from these results, and it was concluded that there seemed to be no definite migration (McLachlan 1966). Nevertheless, it was appreciated that this apparent nomadism reflected the very patchy availability of the quelea's main food, the seeds of annual grasses. The production of such seeds varies markedly from place to place, depending on soil type, topography, and local differences in rainfall from year to year.

It was not until Ward (1971) attempted to analyse the seasonality of grass growth and seed production in relation to rainfall patterns over the African continent as a whole, that the ringing data and many casual observations of sudden appearances and disappearances of quelea fell into place. Two important results emerged from his study. First, the nomadic movements can be seen to be part of regular seasonal migrations of considerable complexity, whose differences of detail from year to year and from region to region make it extremely difficult to appreciate the underlying pattern. Second, the highly abbreviated breeding schedule of quelea may be seen as an adaptation to locally ephemeral conditions suitable for breeding. An important consequence of the short breeding cycle is that several consecutive breeding attempts are possible within the same breeding season. Occasionally, rainfall is prolonged and conditions allow repeat breeding in one locality, though not necessarily all birds breeding in the second round may have previously bred. More often quelea migrate, as soon as the previous brood is independent, to areas of later rainfall and breed again there.

Quelea migrate in response to changes in the availability of their principal food—the seeds of annual grasses, e.g. *Panicum*, *Setaria*, *Urochloa*, and *Echinochloa*. These grasses grow and set seed only during the rainy season, or in some situations during the flood cycle on the flood plains of rivers. Changes in seed availability are thus governed by movements of rainfronts across the continent.

Every part of Africa where quelea live experiences highly seasonal rainfall lasting several months and separated by long, hot, dry seasons. Rainfall is brought to tropical Africa north and south of the equator at different times of the year in accordance with the slow, seasonal movement of the Intertropical Convergence Zone (ITCZ). This is the 'meteorological equator', where the northern and southern tropical air masses meet and generate rainfall across a wide latitudinal band. The ITCZ moves north and south across the geographical equator following the zenith position of the sun, so that in the northern tropics rain falls between March and November and in the southern tropics between September and May. Equatorial eastern Africa receives two rainfall peaks each year around November/December and April/May. In general, the further away from the equator, the later the rains begin and the sooner they end, with a corresponding decrease in the amount of rain falling. Although rainfall patterns in sub-tropical southern Africa are more complex, the basic patterns north and south of the equator are mirror images of one another, with seasons being 6 months out of phase (Fig. 9.1). The slow movements of the rainfronts cause important regional differences in the timing of grass growth and are responsible for quelea migration patterns.

The 'early-rains' migration

During the entire dry season, quelea mainly subsist on fallen seeds of annual grasses produced during the previous rains (Ward 1965*a*). They remain in any one area only for as long as seeds are available. The enormous roosts, to which the birds return every evening, may be disbanded as quickly as they are formed. Some roosts persist throughout the dry season where there is abundant food, and as the dry season progresses, they become larger as they attract birds from exhausted feeding sites elsewhere. Thus, while some quelea may remain in one area for the entire dry season, there is considerable interchange of others among nearby and distant roosts. Because these movements do not have any set pattern and are governed only by the local depletion of food stocks, they may be termed nomadic. They differ greatly from the movements that take place at the start of the rains.

The beginning of the rainy season is heralded by increased cloud cover, humidity, and increasingly frequent local showers. These early showers are usually insufficient to cause seed germination except in depressions where

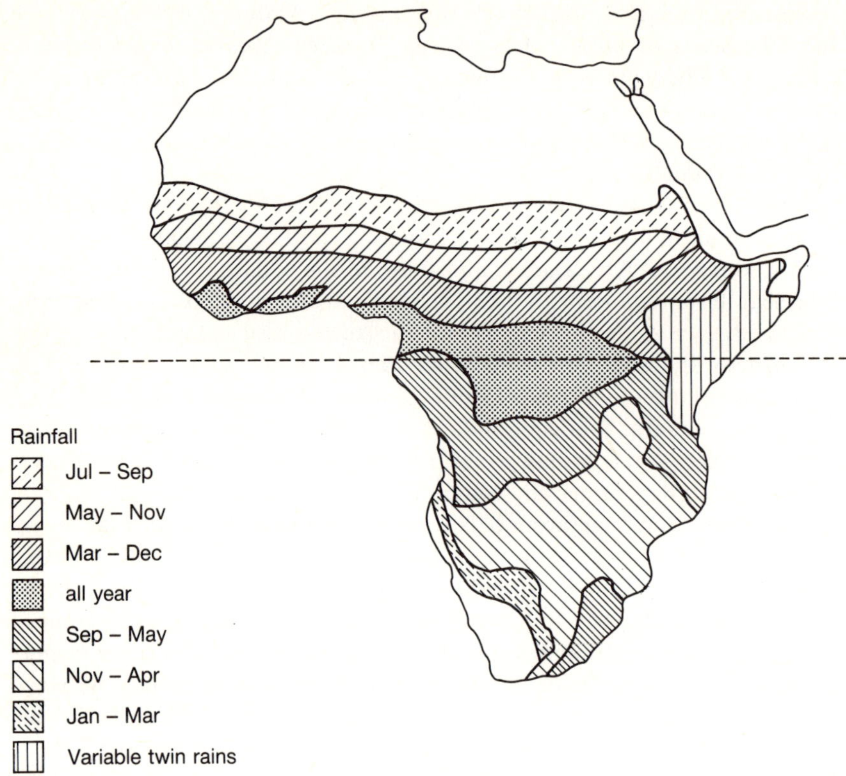

Fig. 9.1. Rainfall patterns in sub-Saharan Africa.

water may collect or in the shade of trees where soil remains damp. As the showers become more frequent, more seeds germinate, and the remaining seed stock becomes progressively more depleted. Finally, widespread and continuous rain occurs, resulting in a sudden and widespread germination of virtually all the remaining seed, which leaves the ground carpeted with growing grasses.

Almost overnight, quelea are faced with an acute shortage of food. Although insects, such as moths or grasshoppers, are plentiful at the start of the rains, they are mostly in active adult stages and difficult or impossible for quelea to catch. The larvae and nymphs of these insects that quelea eat later in the rains, are either too small or not yet available. However, one insect food source is available at this time, the clumsy, winged reproductive stage of termites that swarm in vast numbers to mate and establish new colonies. Termites soon shed their wings and are easily caught by quelea and other birds that feed opportunistically on them. Winged termites are rich in fat, which comprises some 30 per cent of wet weight and 50 per cent of dry weight

(Jones and Ward 1976), and they provide an important alternative food that is abundant just when grass seeds are fast disappearing through germination. Termites are not available for long, as the swarming period usually lasts only for a few days, and in some cases quelea can put on fat without eating termites at all (Ward and Jones 1977).

Faced with this acute food shortage, quelea must migrate elsewhere. They have basically two options: first, to seek places where it has not yet rained and where dry, ungerminated seed still exists; and second, to fly to regions where it has rained sufficiently long for the germinated seed to have grown, flowered, and set new seed. For many species of annual grasses this growth period is about 6–8 weeks. Thus, for quelea to be able to take advantage of this second option, there must somewhere be an area of suitable habitat where the rains began 6–8 weeks earlier.

Rainfronts normally advance quite slowly and fairly predictably across Africa, so that the rains begin at successively later dates along the track of the rainfront. Obviously, the first option is closed to quelea living in the last region of suitable habitat to receive rain; there will be no remaining dry areas to which they could move. The second option is closed to birds living in the first areas to receive rain; it is still too early for fresh seed to be available anywhere else, and their only recourse is to fly ahead of the rains to areas that are still dry. However, a retreat to dry areas ahead of the rains can only be a delaying tactic. Sooner or later it will rain there also, and these birds must then follow the second option, together with all the local birds that had spent the dry season in the late-rains areas (Fig. 9.2).

Birds that fly ahead of the rains find dry seed as they go and may encounter new supplies within a short distance. In contrast, those that fly over the rainfront to areas where the rains began earlier must cross a wide zone where the grasses are at all stages of growth, but may not yet be bearing seed. The width of this zone will vary depending on how quickly the rainfront has traversed it. If it has done so rapidly, the birds must fly a considerable distance to reach the zone where the rains began 2 months earlier. If the front moved slowly, then the distance will be relatively short. Probably nowhere in Africa is the journey likely to be less than about 300 km, and it will be a journey on which the birds are unlikely to find food. To accomplish this movement, which Ward (1971) termed the early-rains migration, quelea lay down deposits of pre-migratory fat, just as do palaearctic migrants that undertake the much longer journeys to and from Africa in autumn and spring. The progress of fattening in the southern African race, *Quelea quelea lathamii*, is shown in Fig. 9.3.

The environmental stimulus that triggers fattening is not known, but is unlikely to be a simple response to rainfall itself, since the early showers produce no fattening in most individuals (Ward and Jones 1977). It is also unlikely to be a response to the sight of green grass, since fattening has been

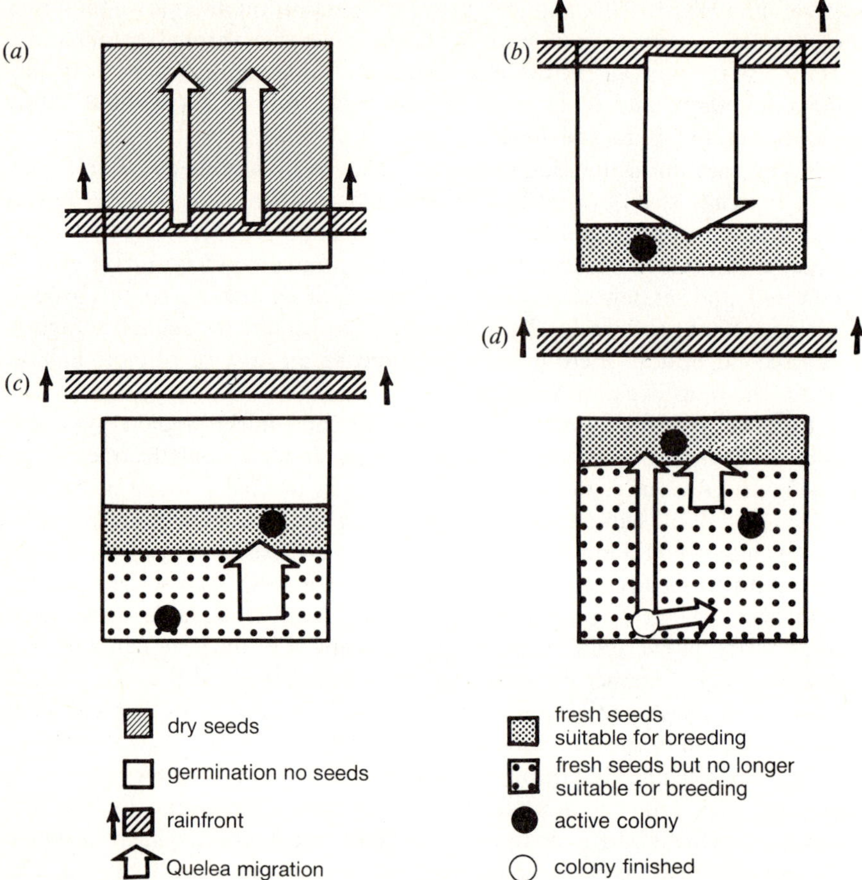

Fig. 9.2. Schematic representation of migration patterns in *Quelea quelea*: (*a*) quelea in early rains areas forced ahead of rains; (*b*) early-rains migration; (*c*) breeding migration; (*d*) breeding migration and itinerant breeding.

observed before widespread germination occurred. Possibly birds respond to changes in the overall meteorological conditions at the start of the rains, such as overcast skies and high humidity, or perhaps they respond to the changes in their food, since many seeds show obvious signs of germination and termites are readily available. Whatever may be the proximate stimulus to fattening, it is remarkable that quelea can do so on a rapidly diminishing food supply.

The amount of pre-migratory fat laid down by quelea is small compared to that deposited by palaearctic migrants. Nevertheless, it is interesting that the average amount laid down by the different quelea populations is propor-

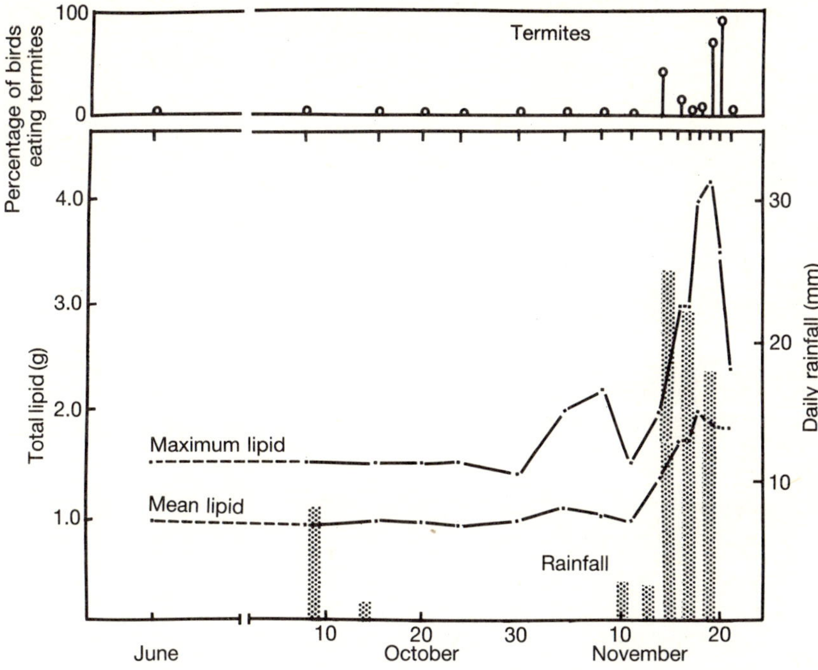

Fig. 9.3. Pre-migratory fattening in *Q. q. lathamii* before the early-rains migration (source: Ward and Jones 1977).

tional to the minimum distances that the birds must expect to fly before encountering freshly maturing seed (Ward and Jones 1977). In western Africa the rainfront moves rather slowly, advancing northward about 300 km between April and June. To fly this distance, the local subspecies, *Q. q. quelea*, lays down only 1.5–2.0 g of fat, barely double the amount that normally is taken to roost each evening for overnight metabolism. In contrast, in eastern Africa the rainfront moves more rapidly southward across the equator in the latter half of the year, so that the distance the local subspecies, *Q. q. intermedia*, must fly may be as much as 1200 km. These birds accumulate 3.0–4.4 g of fat before departure. In southern Africa the rainfall spreads northwest across the subcontinent at an intermediate rate. Consequently, the local subspecies, *Q. q. lathamii*, fly an intermediate distance of about 550 km, for which they deposit 2.0–2.5 g of fat (Fig. 9.4*a*,*b*).

It would appear to be a simple matter to calculate the energy requirements of migratory flight from Fig. 9.4b, but unfortunately we cannot. We do not know at what time of day quelea actually depart on the early-rains migration, nor whether they fly only by day or also at night. The migration is short; for example, *Q. q. quelea* migration in western Africa could probably be

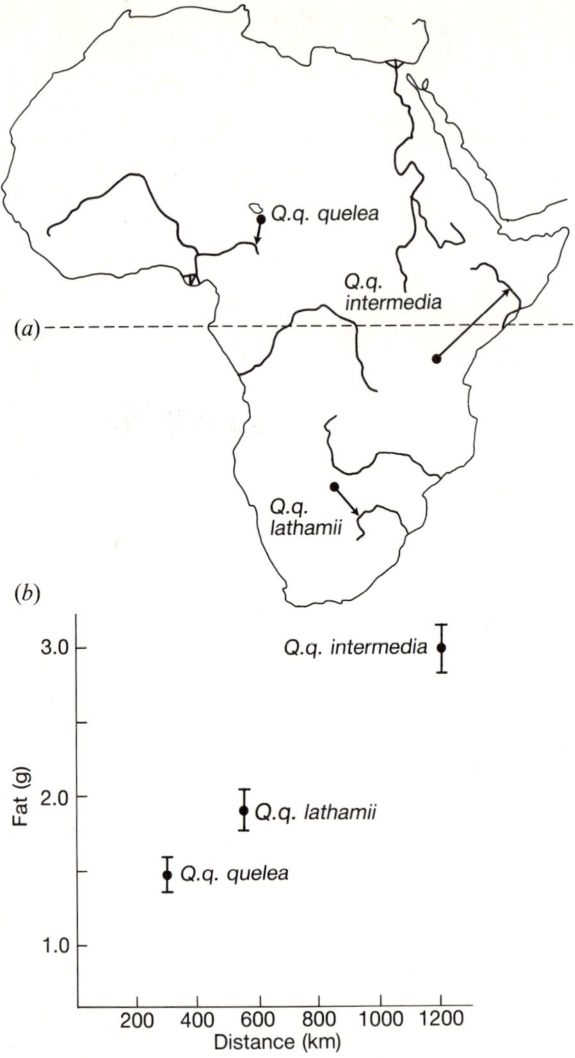

Fig. 9.4. (*a*) The early-rains migrations of three subspecies of *Quelea quelea* and (*b*) the relationship between the expected length of the migration and the degree of pre-migratory fattening (source: Ward and Jones 1977).

accomplished in 8 h non-stop flight, but because the amounts of pre-migratory fat deposited are so small, any night spent resting, either prior to departure, during the course of the journey (if it is too long to be completed during daylight), or after arrival (if this is too late in the day for birds to feed), will appreciably deplete the fat reserves available for flight and extend the period in which the birds cannot feed.

Although we do not yet fully understand the circumstances of the early-rains migration, pre-migratory fattening occurs at the start of the rains in all three populations that have so far been studied, and the extent of the fattening correlates with the distance the birds must expect to fly. This, therefore, provides compelling evidence that true migration is a regular feature of quelea ecology throughout Africa.

The breeding migration and itinerant breeding

Quelea sampled during the early-rains fattening period were at all stages of breeding readiness (Ward 1971; P. Jones, unpubl. data). A few males were already in full nuptial plumage with enlarged gonads, whereas others had not begun the pre-nuptial moult and their gonads were regressed (Fig. 9.5). The most advanced would undoubtedly have been able to breed very soon after arrival in the zone where fresh green seed and larval or nymphal insects were plentiful. This region has been called the early-rains quarters (Ward 1971), and the earliest breeding colonies of the season are established there. However, most birds would not come into breeding condition straight away. By the time they do, the optimal period for breeding in the early-rains quarters would have passed, for in any one locality optimal environmental conditions last for only a short time (Plate 8). By now the best environmental conditions would be found some distance away, along the track of the

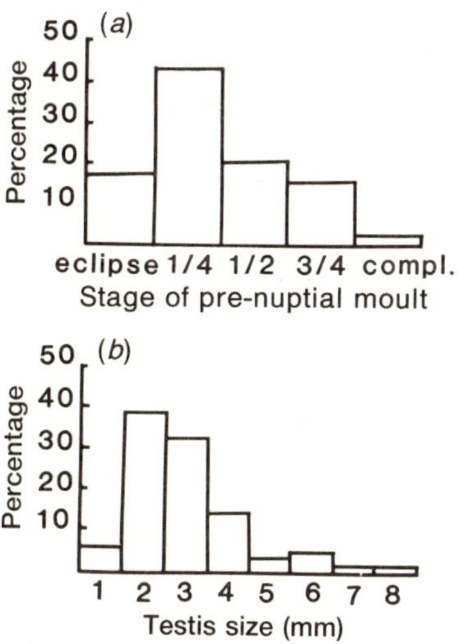

Fig. 9.5. Variation in (*a*) stage of pre-nuptial moult and (*b*) testes size in male *Q. q. lathamii* in the same sample of birds undergoing pre-migratory fattening.

advancing rainfront in areas of later rainfall, where grasses are beginning to produce seed (Fig. 9.2). Quelea that were not among the first to breed have two options. Either they may remain in the early-rains quarters until ready to breed and then make a long flight to catch up with the zone of newly seeding grasses, or they may move slowly along the same heading as the rainfront, but remain within the seeding zone as it advances. As soon as they are ready to breed they stop migration and do so. Whichever option they follow, they would not need to deposit pre-migratory fat for this breeding migration. They may feed on fresh seed at any point, and there is no urgency for the journey to be made non-stop. Eventually, the breeding migration will bring birds back to the areas where they had concentrated at the end of the dry season and from where they departed on their early-rains migration. By this time, some 6–8 weeks later, this region also offers conditions suitable for breeding.

Meanwhile, the earliest birds to breed in the early-rains quarters will have successfully completed their breeding attempt. The breeding cycle of quelea is very short. It is completed in 5–6 weeks, and the adults often disappear from the area immediately, leaving the fledglings alone (Ward 1965*b*, 1973*b*) (Plates 9,10). Quelea have never been recorded attempting to rear two broods in the same colony, nor do they even attempt to replace lost clutches. Occasionally, new colonies may be found in the immediate neighbourhood, just as the young from earlier colonies are fledging (Chapter 14; Bruggers *et al.* 1983), but they may not involve the same birds. It seems that the environmental conditions suitable for breeding do not often last long enough in any one locality to support two consecutive, successful broods. Nevertheless, in some colonies up to 20 per cent of the females begin to develop a new clutch when they are still feeding almost fully grown nestlings and could lay again within a few days of leaving the colony (Ward 1971). Ward (1971) suggested that these birds, and particularly the ones that breed first in the early-rains quarters, now continue the breeding migration by flying along the track of the rainfront to catch up with the zone of seeding grasses and breed again. Such a strategy of itinerant breeding may involve two or perhaps more breeding attempts in the same rainy season, each time with a new mate and each hundreds of kilometres apart. For example, quelea in breeding colonies in southern Ethiopia were colour-marked in June. Colour-marked birds from these colonies were later collected in August/September 500–700 km north in the Awash River Valley (Jaeger *et al.* 1986). Some 44 per cent of adult birds showed an interrupted primary wing moult consistent with a 2- to 3-month interval between breeding attempts, and they were accompanied by an older class of juveniles that had also already begun wing moult. Colonies in the Awash were distributed over more than 300 km and were established over a 2-month period that coincided with local differences in the timing of seeding in grasses (Chapter 10).

Post-breeding dispersal

Newly independent juveniles do not normally follow their parents on the breeding migration. They remain behind in the vicinity of their natal colony, often using the site as a roost for some weeks. Their food supply must become greatly depleted during this time, both in absolute amount (Morel 1968) and in its relative availability. As wild grass seed matures, it falls from the seed head and becomes hidden on the ground among the tangle of dying vegetation. Young quelea evidently experience difficulty in finding this seed, since it is at this time that they are seen feeding on burnt areas and ground trampled by animals where seed is exposed, or they turn to the more vulnerable, but less preferred cultivated cereal crops.

How long juveniles remain near the colony into the dry season probably varies. Quelea that stay in the early-rains quarters for the dry season are likely to be mostly juveniles, since their parents would be expected to have departed to follow the rains and possibly breed again. However, some juveniles certainly move when quite young and have been seen in colonies and roosts in the late-rains areas (Jaeger *et al.* 1986; P. Jones, pers. obs.), where they can be distinguished from any locally hatched young by having begun their post-juvenile moult (at about 9 weeks old). It is not known if any early-hatched young could achieve sexual maturity early enough to breed the same season, but several instances are known of young birds moulting directly from juvenile plumage into adult male breeding plumage without any intervening eclipse plumage, for example, in Tanzania (Ward 1971) and in Botswana (P. Jones, pers. obs.).

In regions where quelea have been breeding itinerantly, most will complete their final breeding attempt of the season in the area that was the last to receive rain. The end of the breeding migration leaves quelea in areas that become the major concentration zones for the dry season and where they will resume their nomadic life until the next rains. In some areas, as in the Awash River Basin of Ethiopia, adult females disperse much more widely than adult males after breeding, so that local concentrations of quelea may have significantly male-biased sex ratios (Jaeger *et al.* 1979). The same study also reported continued segregation of juveniles and adults through the dry season. The adaptive significance of such segregation is unclear, but it may serve to reduce competition for food.

The significance of migration in quelea taxonomy

The migration patterns of different quelea populations, some of which have been formally accorded subspecific recognition (Ward 1966, 1971) are shown in Fig. 9.6. These patterns must be a major factor in maintaining sufficient

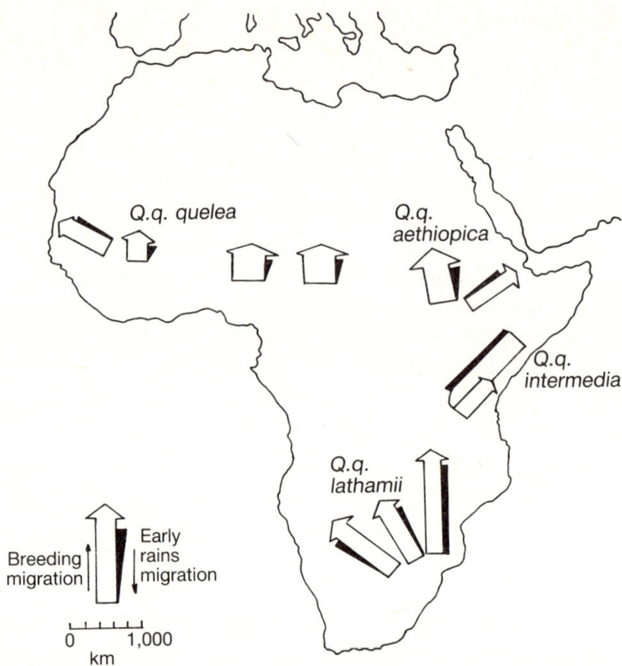

Fig. 9.6. Map of Africa showing diagrammatically the suggested migration patterns of different quelea populations (source: modified from Ward 1971).

genetic isolation for subspeciation to have occurred. For example, *Q. q. lathamii* in southern Africa is separated from *Q. q. intermedia* of equatorial eastern Africa by a zone of unfavourable habitat comprised of dense *Brachystegia* woodland. Presumably this woodland could easily be overflown by either population if it were advantageous to do so. Isolation occurs because the seasonality of rainfall and grass seed availability in each of their ranges is such that neither has cause to cross into the range of the other, and their early-rains migrations take them in opposite directions to breed. Similarly, quelea populations living in the northern tropics immediately south of the Sahara migrate north and south in parallel. There is no advantage in moving east or west along the rainfall contours, and so each population is effectively isolated from the one adjacent to it (see Chapters 10, 11, and 12 for further details).

10

Distribution, populations, and migration patterns of quelea in eastern Africa

MICHAEL M. JAEGER,
CLIVE C.H. ELLIOTT,
RICHARD L. BRUGGERS, and
RICHARD G. ALLAN

Introduction

Developing effective strategies for selective control of the Red-billed Quelea *Quelea quelea* depends on understanding their seasonal distribution in relation to that of susceptible cereals (Dyer and Ward 1977). Based on this information, control can be focused on only those quelea concentrations likely to do damage. Where possible, control can be directed against nesting colonies before quelea disperse and move into cereal-growing areas.

Biogeographical research has been an important part of quelea projects in eastern Africa since 1968. Four geographical races of quelea were proposed for the region with only *Q. q. aethiopica* (Sundevall) and *Q. q. intermedia* (Van Someren) posing important threats to cereals (Ward 1971; Fig. 10.1). The *Q. q. intermedia* race was believed to migrate back and forth between central Tanzania and southern Somalia (Figs. 10.2 and 10.3) following the rainfront as it moves north and south across the equator (Ward 1971). It was considered that breeding occurred wherever favourable conditions were encountered, that locations could change from year to year depending on the pattern of favourable rainfall, and that in years of widespread rain the same birds could breed as many as five times in five different areas.

Ward (1971) proposed the same general model for *Q. q. aethiopica* migrations to the north (i.e. migration following the rainfront), but acknowledged that little was actually known about these quelea, particularly in Ethiopia and northern Somalia.

In 1979, a UNDP/FAO regional quelea project was established to monitor the seasonal distribution and movement patterns of *Q. q. aethiopica* and *Q. q. intermedia*. Timely reporting was considered essential for coordinated re-

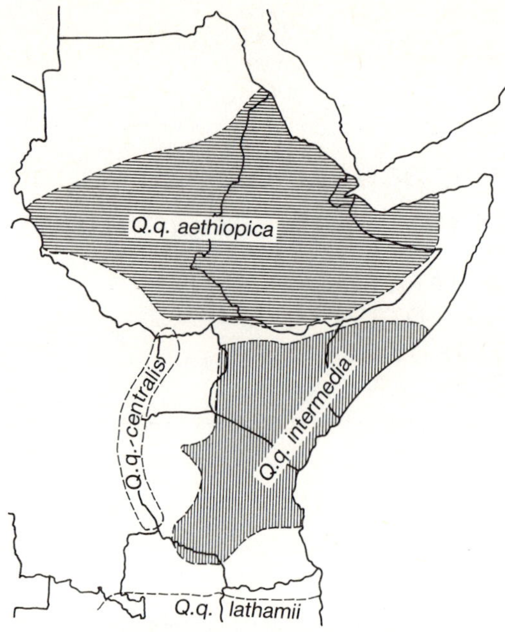

Fig. 10.1. The approximate distribution of races of *Quelea quelea* in eastern Africa, as proposed by Ward (1971).

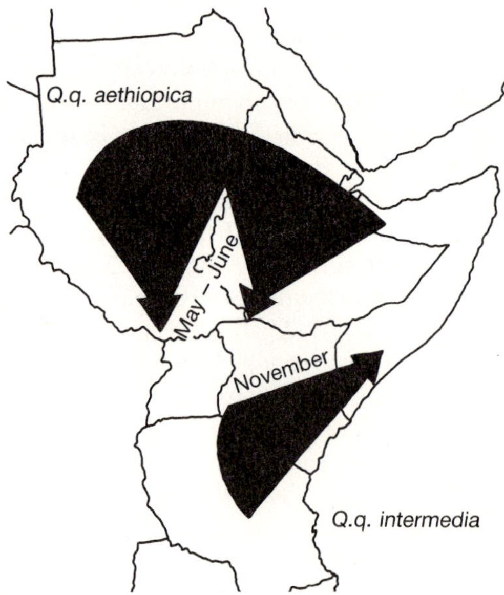

Fig. 10.2. The 'early-rains migration' routes of the *Q. q. aethiopica* and *Q. q. intermedia* races in eastern Africa, as proposed by Ward (1971).

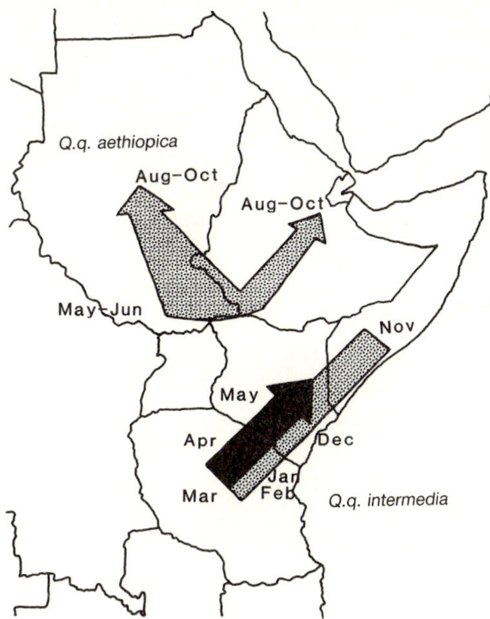

Fig. 10.3. The 'breeding migration' routes of the *Q. q. aethiopica* and *Q. q. intermedia* races in eastern Africa, as proposed by Ward (1971). The black arrow illustrates the return 'breeding migration'. The route shown for Ethiopia was predicted by Ward (M. Jaeger, pers. comm.) subsequent to the 1971 publication.

gional control of these two populations before quelea outbreaks in areas of ripening cereals. The quelea problem in eastern Africa was viewed as being analogous to that of the Desert Locust *Schistocerca gregaria* swarms where regional control could be restricted to clearly defined source areas.

An operational programme for regional monitoring and control at the population level, however, has not been realized, owing to the complex nature of quelea movements within the region. Present information suggests that there are three or more populations that intermix within the range of Ward's *Q. q. aethiopica* and *Q. q. intermedia*; that within populations nesting is widely scattered in both time and space; and that migratory movements are fragmented as opposed to massed. We will review the evidence upon which these observations are based, discuss how this information has changed our perception of the quelea problem, and suggest ways to approach its solution.

Surveys and methods

Quelea have been sampled over much of eastern Africa since 1950, but in general, this sampling has not been systematic. Consequently, quelea populations, their migratory movements, and the occurrence of breeding are unknown in many areas. This is due, in part, to the difficulties involved in getting to remote areas during the rains. Since 1978, when helicopter surveys began, considerable information has been obtained on the location and timing of quelea nesting. Quelea have been most thoroughly sampled in the Ethiopian Rift Valley (Bruggers *et al.* 1983; Erickson 1979; Jaeger *et al.* 1986; Jaeger and Erickson 1980; Jaeger *et al.* 1979) and in Tanzania (Disney and Haylock 1956; Disney and Marshall 1956; Elliott 1983*a*; Luder 1985*a*; Luder and Elliott 1984; Vesey-FitzGerald 1958; Ward 1971; Ward and Jones 1977). Data also are available from eastern Kenya (Allan 1983; J. Thompson and M. Jaeger, unpubl. data) and from southern Somalia (Ash and Miskell 1983; Bruggers 1980). The data collected through 1981 have been tabulated in Jaeger *et al.* 1981. Potentially important areas from where little or no information is available include southern Sudan, northern and south-eastern Ethiopia, western and north-eastern Kenya, south-western Somalia, western Tanzania, and Uganda.

For each sample taken since 1980, a population 'fingerprint' was prepared (Jaeger *et al.* 1986; Chapter 4) for male mask index, stage of cranial pneumatization, progress of post-juvenile or post-breeding primary moult, interrupted primary moult, and gonad condition (see Ward 1973*b* for a description of these methods). Quelea are polytypic for the extent of the black facial mask of males (Ward 1966, 1973*b*). 'Fingerprint' data can permit determination of (1) the occurrence and timing of nesting within the previous 6 months, and (2) the geographical population to which the sample is affiliated. When taken together with knowledge of the spatial and temporal pattern of nesting, it can be used to determine when and from where a particular group of quelea is likely to have come.

Distribution

Spatial

The known distribution of quelea habitat in eastern Africa is illustrated in Fig. 10.4. It includes central Sudan, the Rift system, and a broad swath extending from southern Tanzania north-east through Kenya to southern Somalia (Allan 1983; Ash and Miskell 1983; Bruggers 1980; Bruggers and Jaeger 1982; Bruggers *et al.* 1983; Disney and Haylock 1956; Disney and Marshall 1956; Elliott 1983*a*; Erickson 1979; Hall and Moreau 1970; Jaeger *et al.* 1986; Jaeger and Erickson 1980; Jaeger *et al.* 1979; Magor and Ward 1972; Urban and Brown 1971; Vesey-FitzGerald 1958; Ward 1971; Ward and Jones 1977; Williams 1954). In Sudan, the quelea's range is retreating to the south because of desertification and the destruction of thornbush for large-scale farming. Figure 10.5 gives the distribution of quelea breeding

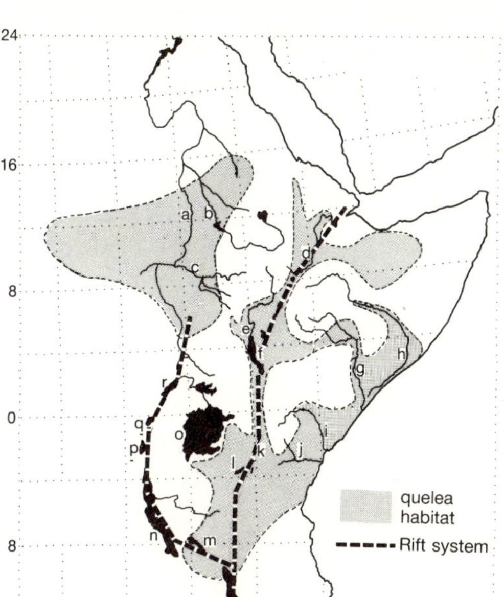

Fig. 10.4. The known distribution of quelea habitat (shaded areas) in eastern Africa. Major rivers and lakes include *a*, White Nile; *b*, Blue Nile; *c*, Sobat; *d*, Awash; *e*, Omo; *f*, Turkana; *g*, Juba; *h*, Webi Shebelli; *i*, Tana; *j*, Galana; *k*, Natron-Magadi; *l*, Iyasie; *m*, Ruhkwa; *n*, Tanganyika; *o*, Victoria; *p*, Kivu; *q*, Idi Amin; *r*, Mobutu.

records from within the region. The greatest number of nesting colonies reported each year are from control operations in Sudan and Tanzania. Sufficient data are not available to allow statistically reliable estimates of population numbers, as surveys for nesting colonies are neither comprehensive nor systematic.

Temporal

The seasonal movements and breeding of quelea are closely tied to movement of the rainfront (Intertropical Convergence Zone) and the subsequent production of grass seeds (Chapter 9; Elliott 1979; Jaeger *et al.* 1986; Jaeger *et al.* 1979; Ward 1965*a*,*b*; Ward and Jones 1977). The array of physiographic features within the region, such as the mountains and lakes associated with the Rift system, together with proximity to the equator and Indian Ocean results in a complex pattern of rainfall (Brown and Britton 1980; Brown *et al.* 1982). Consequently, quelea nesting can occur somewhere within the region throughout most of the year and simultaneously at two or more widely separated locations.

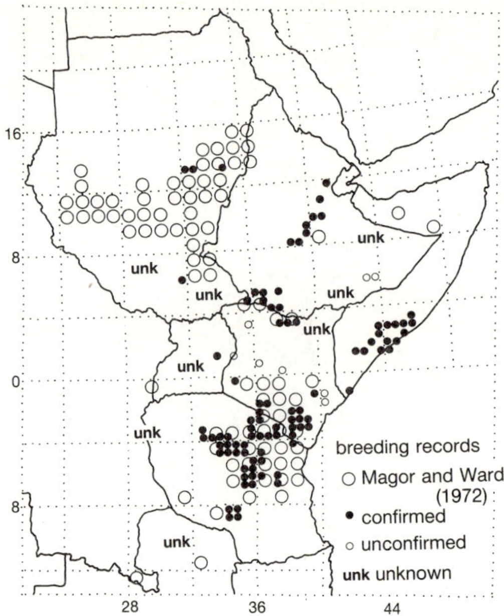

Fig. 10.5. The distribution of breeding records of *Quelea quelea* in eastern Africa. Small circles are records with 1/2° grid squares, while larger circles represent records by 1° grid squares.

Figure 10.6 illustrates the months and general areas where evidence of quelea nesting has been found since 1978. The general trend is for two separate nesting seasons nearer the equator (May–June and Dec.–Jan.), particularly in eastern Kenya and southern Somalia where a bimodal rainfall pattern is more evident (rainfall regions D and E; Brown and Britton 1980). Nesting has also been found during August and September near the coasts of southern Somalia and northern Kenya; but it is not known how regularly this occurs. In the remainder of the region there is normally a northward wave of nesting beginning earlier to the south (southern Tanzania, Feb.–Mar.) and ending later to the north (central Sudan, Sept.–Oct.). This scheme, however, does not include the Rift Valley lakes of western Uganda and western Tanzania where little is known about quelea nesting.

In eastern Africa, the same quelea can breed two or more times in a year. Opportunities for multiple breeding can occur both within and between breeding seasons. Within a nesting season successive breeding can occur in two ways. The more common way probably is itinerant breeding (Ward 1971) where adults depart from a successfully completed colony, follow the rainfront, and re-nest hundreds of kilometres away where later-maturing grasses are green and ripening. The best evidence for itinerant breeding is

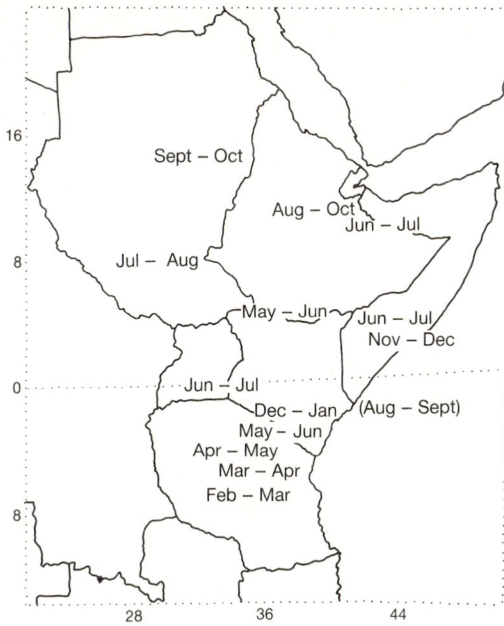

Fig. 10.6. The known temporal distribution of quelea breeding in eastern Africa. This is not to suggest that nesting occurs at these times and places each year, but that it regularly occurs when rainfall is sufficient. Parentheses indicate that nesting at this time and place may not be a regular occurrence.

from the Ethiopian Rift Valley (Jaeger *et al.* 1986) where adults, colour-marked at colonies in the southern Rift during June 1981, were recovered in September at colonies 500–700 km to the north in the Awash River Valley. Indirect evidence from 'fingerprinting' samples suggests a similar occurrence in Tanzania where quelea nesting during February and March in central Tanzania can move north to re-nest during May and June in north-eastern Tanzania.

A second form of successive breeding occurs with re-nesting in the same general area in response to prolonged rains. Strong evidence for this has been found in south-eastern Kenya where nesting colonies were found in the vicinity of Tsavo East National Park from December 1984 to June 1985. Adults, colour-marked at one nesting colony in January, were recovered at another colony 125 km to the east in late March (J. Thompson and M. Jaeger, unpubl. data). In central and southern Tanzania, breeding sometimes occurs over a 3-month period (C. Elliott, pers. obs.), allowing the possibility for successive breeding by the same individuals. However, this has yet to be substantiated. At present, there is no compelling evidence of three or more nestings in succession, although the opportunities seem to exist in the south-east (Feb.–Mar., May–June, Aug.–Sept., and Dec.–Jan.; Fig. 10.6). Here quelea can also nest in the same area in two separate seasons (May–June and Dec.–Jan.). Figure 10.7 illustrates possible opportunities for multiple breeding within the region.

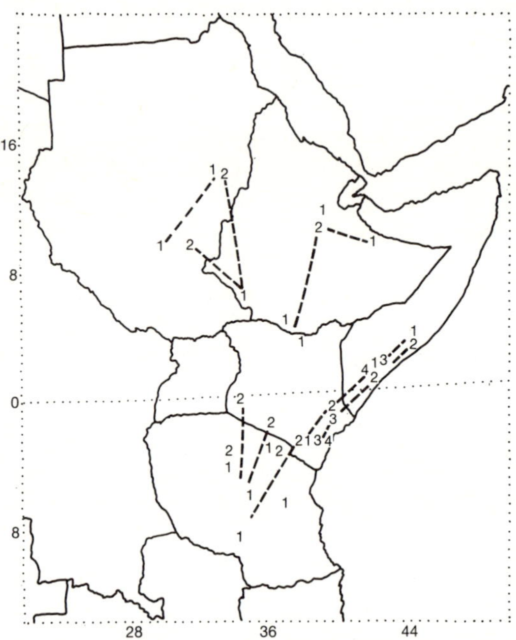

Fig. 10.7. The known opportunities for multiple breeding by quelea in eastern Africa within the same year. Numbers 1–4 represent first to fourth breeding by the same birds. In some situations, birds nesting for the first time may do so in an area with birds that have nested earlier.

Populations

As a basis for understanding quelea movements in eastern Africa it was first necessary to determine whether discrete populations or races of quelea exist, and if so, to determine the limits of their distribution. Ward (1966) compared geographical variation in several characters of the colour pattern of males in breeding plumage. He concluded that in eastern Africa only one valid race *Q. q. aethiopica* existed, distributed through the savannahs of Sudan, Ethiopia, and northern Somalia and that the rest of the region was populated by a hybrid swarm of *Q. q. aethiopica* and the *Q. q. lathamii* (Smith) race of southern Africa. Ward decided at that time that the *Q. q. intermedia* race, proposed by Van Someren (1922) to describe the quelea of Tanzania and adjacent areas, should be rejected because of its high variability. For similar reasons he also rejected the *Q. q. centralis* (Van Someren) race of the Rift Valley of western Uganda, Rwanda, Burundi, and western Tanzania. Later (Ward 1971) larger samples from Kenya and Tanzania convinced him that *Q. q. intermedia* was after all valid because it was so distinct from both *Q. q. lathamii* and *Q. q. aethiopica* despite its variability. Data supporting this conclusion were never published.

The *Q. q. aethiopica* and *Q. q. intermedia* races postulated by Ward (1971) now seem to be an oversimplification of the pattern of quelea distribution and intermixing in eastern Africa. These distinctions were based largely on the composition of male mask types from relatively few samples of restricted distribution. For instance, large samples had not been collected from Ethiopia or northern Somalia. Since then, more than 26 000 masks from a much wider variety of sampling sites have been examined. Figure 10.8*a* shows the percentage of males with little or no black frontal band (mask types 1 and 2) from pooled samples collected from 1978 to 1985. A geographical gradient in the composition of mask scores is apparent, with the highest percentage of frontal bands to the south-east and the lowest to the north-west (Sudan). This follows the general pattern of the cline suggested by Ward (1966), with increasing percentages of frontal bands to the south and to the west, so that samples from South Africa and Senegal, respectively, are almost exclusively of males with broad frontal bands (mask type 7).

The distribution of mask types illustrated in Fig. 10.8*a* suggests that the cline is broken into at least three local populations, or demes (Endler 1977; Mayr 1971) within eastern Africa: (1) Sudan, (2) the Ethiopian Rift Valley and adjoining areas, and (3) Tanzania, eastern Kenya, southern Somalia. The population in Sudan (95–97 per cent mask types 1 and 2, Fig. 10.8*a*) has been the least sampled, particularly in the south; nevertheless, it seems to be physically separated from the population in the Ethiopian Rift Valley by the highland plateau forming the Rift. Regular exchange probably takes place across south-western Ethiopia.

The population centred in the Ethiopian Rift Valley (85–87 per cent) includes quelea to the east in the northern Ogaden (84 per cent) and to the south into Kenya (79 per cent). An unconfirmed report of quelea nesting colonies in the Kenya Rift Valley at Lodwar on the south end of Lake Turkana (Fig. 10.4) and at Lake Baringo in the central Rift in 1986 (F. Kitonyo, pers. comm.) may explain the different mask index found in central Kenya (72 per cent from Fig. 10.8*a* and 77 per cent from Fig. 10.8*b*). This may be a fourth population distinct from that in the Ethiopian Rift; further sampling is needed to clarify the matter. The third population from Tanzania (53–55 per cent), eastern Kenya (46 per cent), and southern Somalia (56 per cent) seems to form a distinct unit (*Q. q. intermedia*), as suggested by Ward (1971), except that the Kenya percentage is lower than the other two (Fig.

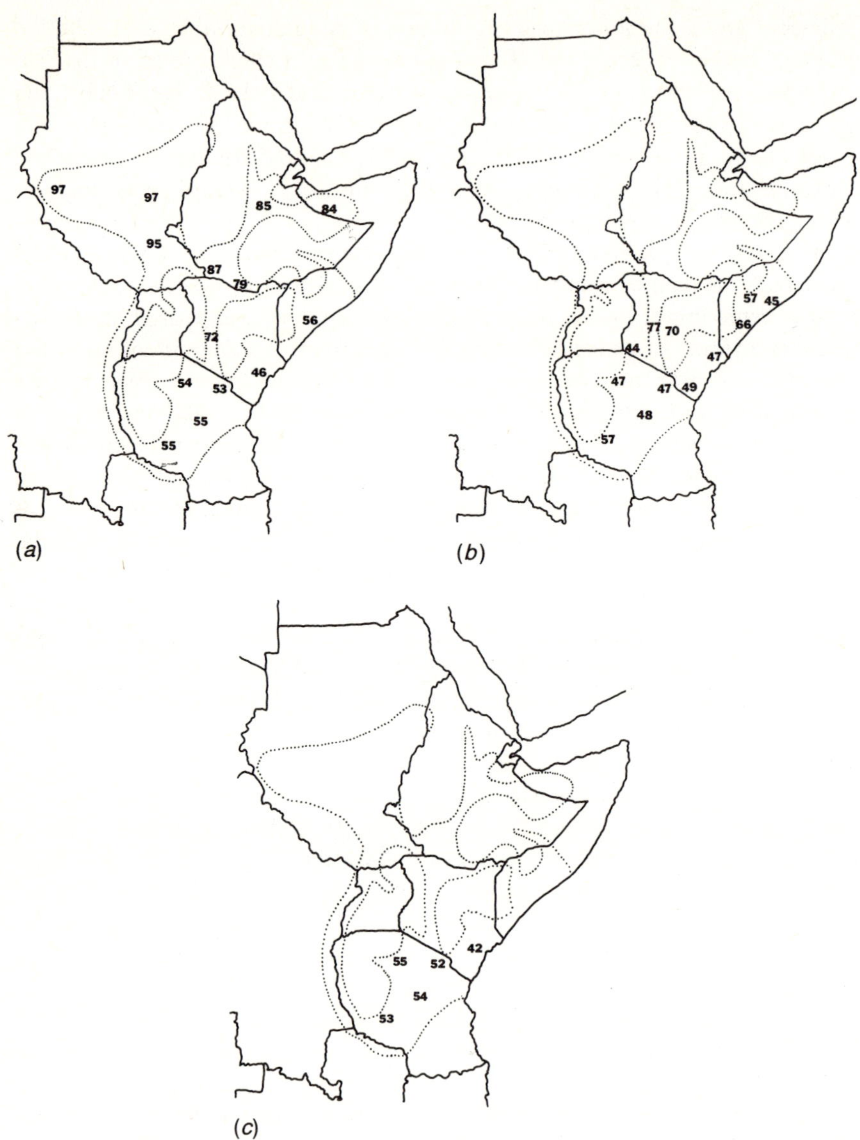

Fig. 10.8*a*. The distribution of male facial mask scores (per cent 1 and 2, no frontal band) from samples pooled over all years ($n = 26\,123$) in eastern Africa. (See Chapter 4 for facial mask score technique.)

Fig. 10.8*b*. The distribution of male facial mask scores (per cent 1 and 2, no frontal band) from samples from 1984 ($n = 5401$) in eastern Africa. (See Chapter 4 for facial mask score technique.)

Fig. 10.8*c*. The distribution of male facial mask scores (per cent 1 and 2, no frontal band) from samples from 1985 ($n = 3111$) in eastern Africa. (See Chapter 4 for facial mask score technique.)

10.8*a*). In early 1984, a drought affected the region and was particularly severe in Kenya. Figure 10.8*b* shows that in that year, mask index percentages were lower than usual in Tanzania, implying more intermixing with birds from Kenya. At the same time they were higher than usual in Somalia, indicating possible intermixing also with birds from Ethiopia. In 1985 (Fig. 10.8*c*), the pattern of a distinct difference between Kenya and Tanzania was re-established. The *Q. q. intermedia* population may, therefore, be split into sub-populations in Tanzania, Kenya, and Somalia with climatic conditions increasing or decreasing the amount of mixing.

Another population may also have an influence on the region. This is *Q. q. centralis* which is said to inhabit a restricted area of grassland associated with the Rift Valley lakes of Mobutu, Idi Amin, Kivu, and Tanganyika (Fig. 10.4; Ward 1971). Very little is known about quelea in this area. However, they may intermix with the populations in Sudan and Tanzania as indicated in Figs. 10.8*a* and 10.8*b*.

The above information indicates that the subspecies designations *Q. q. intermedia* and *Q. q. aethiopica* (Ward 1971) are unapplicable to the more complex and dynamic situation existing within the region and that, as such, use of subspecies designations should be discontinued. Clines in geographical character variation, as occur for quelea in eastern Africa, also occur in other passerines; for example, Red-winged Blackbirds *Agelaius phoeniceus* (James 1983; James *et al.* 1984; Power 1970) and House Sparrows *Passer domesticus* (Johnston 1969; Johnston and Klitz 1977; Selander and Johnston 1967).

Seasonal movements

Sudan

Ward (1971) suggested that quelea across Sudan migrate along a north-south axis approximately between 13° and 4° N. The main dry-season quelea concentrations are reported between 13° and 9° N (Bruggers *et al.* 1984*a*; Ward 1971). According to Ward's model, as the rainfront progresses northward, quelea undertake an early-rains migration southward in June into the rains where fresh grass seed is available. These birds are believed to nest if conditions allow, then to make a return breeding migration in August and September, nesting in the general area where dry-season concentrations occur. Sufficient information is not available to know whether itinerant breeding does, in fact, occur. However, opportunities seem to be available as nesting colonies have been found in the south from May to July (Kibish Hills 5°05′N × 35°37′E, Bor 6°15′N × 31°50′E) as well as in central Sudan during September and October, where enormous concentrations have been reported (Figs. 10.5, 10.6, and 10.7).

Very little is known about quelea movements in Sudan, such as the size and composition of groups and the distances and directions they travel. A single ringing recovery supports a movement between south-western Ethiopia and southern Sudan. An adult ringed in May 1980 at a nesting colony along the Kibish Hills (5°05′ N × 35°37′E) was recovered 260 km to the west at Kapoeta, Sudan (4°48′N × 33°35′E) on 26 July 1982. In May 1980, large numbers of quelea were reported moving south-east along the Akobo River (7°50′N × 33°00′E), a tributary of the Sobat River (Fig. 10.4), which forms the border with Ethiopia (T. Matonovich, pers. comm.). Possibly this was an early-rains migration to nesting areas along the lower Omo River (Fig. 10.4) in south-western Ethiopia. P. Ward (unpubl. data) has suggested an extra-seasonal movement of quelea to the Red Sea coast where December–January rains may allow nesting. In June 1979, Ward collected 6-month-old quelea near Khartoum (Lenton 1980), which he believed must have originated from either the Red Sea coast, or the southern flood zone in Jonglei Province, where grass growth can be months behind the rains when the flood waters recede. Evidence for winter breeding along the Red Sea coast of Sudan has not yet been found (J. Jackson, pers. comm.; G. Lenton, pers. comm.). Hailu (1984), however, found two nesting colonies near the Red Sea coast of Ethiopia along the lower Awash River (11°40′N × 41°30′E; Fig. 10.5) in March 1983. This was apparently due to to the unseasonally good rainfall in the previous January associated with the Red Sea Convergence Zone.

Ethiopia

Quelea movements within the Ethiopian Rift Valley were investigated from 1976 to 1982 (Bruggers *et al.* 1983; Erickson 1979; Jaeger *et al.* 1979, 1986). The main dry season (Oct.–May) concentration of adult quelea seemed to be in the north, particularly in the Awash River Basin (Fig. 10.4). Evidence from marking recapture, 'fingerprint' matching, and the pattern of disappearance and reappearance of quelea supports the migration model proposed by Ward (1971): an early-rains migration into the rainfront in May and June and a return breeding migration in August and September. Quelea disappear from the Awash Basin and nearby Lake Zwai (8°00′N × 38°53′E) in May and June, shortly after onset of the rains, coincident with the appearance of nesting colonies in south-western Ethiopia and adjoining Kenya (Fig. 10.5). 'Fingerprint' evidence also supports June–July nesting to the east of the Awash Basin in the northern Ogaden where the main rains begin in April and May. Quelea return to the Awash Valley in August and September where a second wave of nesting occurs. By November these colonies have dispersed, and quelea damage to ripening sorghum begins along the base of the highlands bordering the Awash Basin (Bruggers and Jaeger 1982; Jaeger and Erickson 1980). Quelea remain in the Awash Basin until May–June when the cycle is repeated (see Jaeger *et al.* 1979 for sizes and locations of roosts).

Evidence for two successive breeding cycles by the same adults was obtained in the Ethiopian Rift (Jaeger *et al.* 1986). Quelea were mass-marked with aerially applied fluorescent particles in two separate nesting areas in south-western Ethiopia during June 1981. Marked adults from both areas were recovered up to 100 days later during August and September in nesting colonies in the Awash River Valley 500–700 km to the north of the spray sites. Nesting colonies in both the south-west and in the Awash Valley were scattered in time and space. Colonies in the Awash were distributed over more than 300 km of the valley, and were established over a 2-month period, which coincided with local differences in maturation of grasses. This wide distribution probably increases nesting success in areas of locally variable rainfall. This contrasts with a strategy of mass migration where concentrated breeding occurs where and when suitable conditions are first encountered (see Jaeger *et al.* 1986 for further explanation).

The recoveries of quelea ringed in Ethiopia between 1976 and 1982 (W. Erickson, unpubl. data) are presented in Table 10.1. One recovery further supports a seasonal movement from the southern Rift northward to the Awash River Basin (Ambo Pond to Lake Zwai). Eleven of thirteen recoveries are of adults ringed in the Awash Valley and recovered at the same or nearby sites, and in two instances at 33 and 35 months post-ringing. This suggests an affinity of quelea to familiar areas.

Somalia

Quelea in Somalia seem to be represented by two separate populations (Fig. 10.8*a*). Quelea in the north have similar mask indices to those from the nearby Awash River Basin of Ethiopia, while in the south they are similar to those from south-eastern Ethiopia and north-eastern Kenya. Ward (1971) believed that the quelea in southern Somalia had their major dry-season (June–Oct.) concentration in Tanzania. According to his model (Figs. 10.1, 10.2, and 10.3), with onset of the short rains in Tanzania in November, this concentration migrates to the north-east and into southern Somalia (Ward and Jones 1977) where the rains begin several weeks earlier and where nesting could begin. Ward (1971) believed that with favourable rainfall most quelea return to the south after a few weeks and nest in southern Kenya (Dec.–Jan.) and again in central Tanzania (Feb.–Mar.). From central Tanzania quelea turn back to the north and follow behind the passage of the long rains, nesting once or more often, and finishing this breeding migration in southern Somalia in June and July. Therefore, the same birds would have the potential to nest four or five times within a year.

Quelea movements and nesting in southern Somalia are now known to be

Table 10.1. Summary of recoveries of quelea ringed in Ethiopia (1977–1982).

Ringed				Recovered			
Location	Date	Sex	Age group	Location	Date	Distance (km)	Time (months)
Sodere[a] (8°25′ N × 39°25′ E)	12 May 77	M	Adult	Sodere[a]	11 July 77	0	2
Wonji[a] (8°30′ N × 39°15′ E)	21 May 77	Unknown	Adult	Nazareth[a] (8°35′ N × 39°15′ E)	22 Apr. 80	10	35
Melka Werer[a] (9°35′ N × 40°20′E)	10 June 77	M	Adult	Melka Werer[a]	29 Mar. 79	0	22
Sodere[a]	8 July 77	F	Adult	Lake Zwai[a] (8°05′ N × 39°00′ E)	4 Nov. 80	60	33
Tabila[a] (8°40′ N × 39°45′ E)	16 Sept. 77	Unknown	Adult	Melkassa[a]	18 Oct. 77	60	1
Melkassa[a] (8°25′ N × 39°20′ E)	13 Nov. 77	M	Adult	Melka Sadi[a] (9°25′ N × 40°20′ E)	17 Apr. 78	160	5
Melkassa[a]	13 Nov. 77	Unknown	Adult	Melkassa[a]	17 Jan. 78	0	2
Melkassa[a]	14 Nov. 77	Unknown	Adult	Melkassa[a]	18 Jan. 78	0	2
Melkassa[a]	14 Nov. 77	Unknown	Adult	Melkassa[a]	18 Jan. 78	0	2
Saluki[a] (8°40′ N × 39°25′ E)	11 Nove 77	Unknown	Adult	Saluki[a]	14 Jan. 78	0	2
Saluki[a]	11 Nov. 77	Unknown	Adult	Saluki[a]	14 Jan. 78	0	2
Kibish (5°05′ N × 35°37′ E)	28 May 80	M	Adult	Kapoeta, Sud (4°48′ N × 33°35′ E)	26 July 82	260	26
Ambo Pond (4°38′ N × 37°31′ E)	28 July 80	F	Juvenile	Lake Zwai[a]	28 Oct. 81	400	15

[a]Location is in, or in close proximity to, the Awash River Valley.

more complex than implied by Ward's model. First, quelea nesting in southern Somalia seems to be on a larger scale than that proposed by Ward (1971). Nesting has been found near the Juba and Webi Shebelli rivers (Figs. 10.4 and 10.5), where colonies are now regularly found in both May and June and November and December (Barré 1983; Bruggers 1980; Elliott 1980*a*). Nesting was also found along the Juba River in late August 1984 (M. Jaeger, pers. obs.), where M. Barré (pers. comm.) suggests that nesting at this time is a regular occurrence. Second, quelea nesting can occur simultaneously over a wide area (Figs. 10.1, 10.5, and 10.6). For instance, evidence of May–June nesting has been found in northern Tanzania, the southern Rift Valley of Kenya, south-eastern Kenya, southern Somalia, and south-eastern Ethiopia. Third, quelea have been found in southern Somalia during most of the year (Ash and Miskell 1983). Fourth, simultaneous collections have been made in southern Somalia where the 'fingerprints' do not match; for instance, quelea sampled at three different locations during the period July to September 1984 seem to be acting independently of one another (Table 10.2). 'Fingerprinting' analysis shows that quelea collected at Baidoa had not bred in 1984, while those from Jiohar bred in May–June, and those from Gilib in August–September. Furthermore, male mask indices differ between these samples ($P<0.001$) (Fig. 10.8*b*). Such a pattern of discrete variation in 'fingerprints' would seem to suggest separate movements between and cohesion within groups of quelea.

Table 10.2. 'Fingerprint' collections from southern Somalia during 1984.

Location[a]	Date collected	Nesting period	Male mask index[b] 1 and 2 (%)
Gilib (0°30′ N × 42°30′ E)	Sept.	Aug.-Sept.	66 ($n=176$)
Jiohar (2°30′ N × 45°30′ E)	July	May-June	45 ($n=264$)
Baidoa (3°00′N × 43°30′ E)	July	No nesting	57 ($n=1334$)

[a] By 1/2° grid square.
[b] Chi-square = 20.3, $P<0.001$.

Kenya

Present evidence in Kenya supports two distinct populations (Figs. 10.8*a* and 10.8*b*), one across southern and eastern Kenya associated with southern Somalia and Tanzania, and the other from the central to northern Rift Valley connecting with quelea in the Ethiopian Rift Valley; however, nothing is known about quelea movements and their affiliations in north-eastern Kenya. Nesting occurs in south-eastern Kenya in May–June, August–

September, and December–January (Allan 1983; J. Thompson and M. Jaeger, unpubl. data), similar to southern Somalia. From December 1984 to May 1985, a period of exceptionally good and prolonged rains, nesting colonies were found in Tsavo East National Park, where evidence was found of two nestings in succession by the same birds. Likewise, in this area first-year quelea hatched in May–June or August–September 1984 seemed to be breeding. It seems, therefore, that in years of favourable rainfall, quelea can remain in eastern Kenya throughout most of the year, and that this general area can be an important regional focus for reproduction.

Recoveries of quelea marked with fluorescent particles support some movement between Tanzania, eastern Kenya, and southern Somalia following the general direction of movement proposed by Ward (1971) and Ward and Jones (1977). A single adult male quelea marked on 15 June 1984 at a nesting colony on the Tanzania-Kenya border near Taveta, Kenya (3°30′S × 38°00′E),* was recovered on 28 September 1984 at a roost 450 km to the north-east on the Kenya–Somalia border near Kiunga, Kenya (1°30′S × 41°00′E). Similarly, two adult females marked at nesting colonies in the Tsavo East National Park of eastern Kenya on 19 January (2°30′S × 38°00′E) and 2 March (3°00′S × 39°00′E) 1985, respectively, were recovered from a nesting colony on 29 March 1985, 460 km to the south-west near Dodoma, Tanzania (5°30′S × 35°00′E). Meanwhile, other quelea remained in and around the Tsavo East National Park, where nesting continued until June 1985. This further indicated a fragmented nature of quelea movements in the region.

In Kenya, quelea are also found in the Rift Valley, where they are an occasional pest of ripening wheat and barley grown along the highlands forming the Rift (Allan 1983; Williams 1954). Surveys have focused in the southern portion of the Kenyan Rift, where nesting is commonly found during May and June along the Tanzanian border near Lakes Natron and Magadi (Fig. 10.4; Kitonyo 1981, 1983). It has been suggested that these quelea attack wheat to the north near Ngorengore (1°00′S × 35°30′E) in May/June and Nakuru (0°00′S × 36°00′E) during July/August. Quelea collected at Ngorengore during August 1984 had a ‘fingerprint’ matching that of quelea nesting in northern Tanzania during May and June 1984 and supporting a northward movement into Kenya. Quelea collected further to the north near Nanuyuki (0°00′N × 37°00′E) in June and at Nakuru in July 1984 had similar ‘fingerprints,’ but differed from those from Ngorengore and Tanzania. The available evidence (Figs. 10.8*a* and 10.8*b*) suggests that the quelea found in the central Rift of Kenya are more closely aligned with quelea from northern Kenya and south-western Ethiopia, in agreement with Williams (1954). There is also a possible connection with southern Sudan and eastern Uganda.

*This coordinate and all subsequent coordinates represent 1/2° grid squares.

Quelea are not commonly reported from western Kenya, except to the south near Kisumu (0°00′S × 34°30′E) on Lake Victoria where a collection was made in August 1984. The 'fingerprint' was similar to that from Ngorengore, 1984, with evidence of recently completed nesting since May and an earlier nesting during March, probably in central Tanzania. In April/May 1986, quelea were found breeding in sugar-cane near Kisumu, but no mask index data were collected (J. Gatimu, pers. obs.).

Tanzania

The gradual northward movements of quelea from breeding in central and southern Tanzania in February/March to breeding again in May/June is long established (Disney and Haylock 1956). Migration through Tanzania is fragmented and is probably spread across a wide front including western Tanzania, from where little information is available. It is not clear what occurs following the May/June nesting, and when or by what routes quelea return to breed the following spring. There may be an exodus of adults in June that has been masked by the presence in northern Tanzania of a substantial population of juveniles and some adults. If such an exodus occurs, it could involve a movement of birds across Kenya into southern Somalia where breeding occurs in June/July. The only evidence for this is the single recovery of an adult marked in June at a nesting colony on the Kenya-Tanzania border near Taveta, Kenya, and recovered in late September on the Kenya–Somalia border near Kiunga, Kenya. Many quelea probably remain in northern Tanzania through the dry season from June to October (Disney and Haylock 1956; Ward 1971).

There are two theories for the movement which occurs with the onset of the short rains in November. One is that the quelea move north-east into southern Kenya, where some nesting could occur during December and January before the birds move back to southern Tanzania in February and March to start a new breeding cycle (Disney and Haylock 1956). In support of this, quelea ringed at nesting colonies in central and northern Tanzania were recovered to the north and north-west in the Kenyan Rift and Lake Victoria area (Disney 1960). Support is also provided by the 1984 'finger-printing' described above for Kenya, and the theory in general is consistent with the evidence collected in Tanzania since 1979 (Figs. 10.6 and 10.7).

The second theory is that with the onset of the short rains, quelea depart north-eastern Tanzania for southern Somalia, whence they initiate a south-westward wave of breeding back to central and south-western Tanzania by March (Figs. 10.1, 10.2, and 10.3; Ward 1971). Ward's evidence is weak, being a failure to locate quelea anywhere in Kenya or Tanzania in any numbers in late November and early December in 1969, and a single observation of southward-flying flocks in eastern Kenya in December. Given the vastness of the area, it would be easy to miss quelea concentrations. Subsequent work by FAO projects has shown that breeding regularly occurs in southern Somalia in November/December, but no connection between this and the Tanzania population has been established other than that the mean mask indices for the two areas are similar. The evidence suggests that shorter, overlapping migrations may be occurring along this route, as implied in earlier discussions on quelea movements in southern Somalia and eastern Kenya (Fig. 10.7). The timing and routes of these migrations, together with the occurrence of nesting, are probably subject to considerable year-to-year variation depending on the pattern of rainfall.

Present evidence also tends to refute a southward wave of breeding through Tanzania from December to February following the short rains as proposed by Ward (1971). First, nesting colonies have rarely been found in northern Tanzania during this time. Second, the 'fingerprints' from collections made at nesting colonies in central Tanzania during February and March suggest that these birds have not bred within the previous 6 months (Luder and Elliott 1984). The nature of the southward movement of quelea back to south-western and central Tanzania is unknown.

Quelea movements in Tanzania may be further complicated by a regular influx of quelea from northern Zambia and/or western Tanzania (Ward 1966, 1971; Figs. 10.1, 10.5, and 10.8*a*). For instance, extensive nesting by quelea was reported in the Ruhkwa Valley of south-western Tanzania (Fig. 10.4) from March to May 1956 (Vesey-FitzGerald 1958). This implies an influx of quelea from the Rift Valley of western Tanzania, Burundi, Rwanda, and western Uganda. Unfortunately, very little is known about the quelea from this area.

Group cohesion

Group cohesion may be another important characteristic of quelea movements in addition to their being fragmented and changeable depending on the rainfall pattern. There seems to be a tendency for the same birds to remain together when migrating or when otherwise moving from one roosting area to another. This is of potential importance to a strategy of selective control, if it allows the prediction of where quelea that damage cereals come from. Evidence for group cohesion was mentioned earlier in regard to the variation between 'fingerprints' from three different localities in southern Somalia in 1984. In addition, the grouped pattern of recoveries of ringed or fluorescent masked quelea supports group cohesion (Jaeger *et al.* 1986). For example, from one marking site in Kenya in 1984, all 36 recoveries were made in the same area, 100 km away and up to 128 days post-marking. In another instance, an estimated 900 000 adults and 2 million juveniles were marked in four neighbouring nesting colonies near Makuyuni, Tanzania, during June 1984. By early July, all of these colonies were deserted and quelea seemed to have departed the area. Subsequent collections from 11 sites in Tanzania ($n = 3966$ birds), 6 sites across Kenya ($n = 3131$ birds), and 3 sites in southern Somalia ($n = 1493$ birds) between July and October failed to find a single marked bird. One explanation is that the sampling was not sufficiently widespread and that the marked birds remained grouped and were missed.

Implications for control

Much remains to be learned about the seasonal distribution and movements of quelea in eastern Africa as they relate to cereal damage. What we do understand about these movements suggests that the regional picture is complex. The following findings support this complexity:

(1) Quelea are widely distributed within the region.
(2) Based on the composition of male mask types, three or more populations are present with widespread intermixing among them.
(3) Nesting is widely scattered in time and space.
(4) Multiple nesting by the same birds can occur within a year.
(5) Migrations within a population are fragmented in time and space.
(6) Quelea may be long-distance migrants in some situations and have much shorter movements in others.

This complexity suggests a more localized approach to studying quelea distribution and movements as opposed to a regional population approach. Susceptible cereals, such as dryland sorghum and millet, are not uniformly distributed within the region, but rather are concentrated in areas such as the Singida region of Tanzania, the Gedaref area of Sudan, the Awash Basin of Ethiopia, and the Hargeisa area of northern Somalia. It is relative to those areas that quelea movements and distribution should be studied. The resulting information would allow determination of the most effective control strategy for local conditions. In some cases, it would be better to destroy nesting colonies before they disperse to cereal areas (Jaeger and Erickson 1980), while in other cases it would be more effective to leave them and wait to destroy the roosts that may develop in close proximity to cropping areas (Elliott 1983*a*). Selective control based on understanding the major concentrations of susceptible cereals and quelea movements in relation to them, proved to be effective in Ethiopia (Bruggers and Jaeger 1982; Jaeger and Erickson 1980).

11

Distribution, populations, and migration patterns of quelea in southern Africa

PETER J. JONES

Introduction

The interpretation of evidence on the migration patterns of quelea within southern Africa is made difficult by the possible presence of two distinct subspecies. In addition to *Quelea quelea lathamii*, which hitherto has been the only race accepted as occurring in the region, subspecies status has been claimed for *Quelea quelea spoliator* breeding in the south-eastern part of the *Q. q. lathamii* range (Clancey 1960, 1968, 1973). Both subspecies occur during the non-breeding season (April to late October/November) throughout the *Q. q. lathamii* breeding range (Fig. 11.1). *Quelea quelea spoliator* is not mentioned in the standard works for Zambia and Malawi (Benson and Benson 1977; Benson *et al.* 1973), but is accepted by Irwin (1981) for Zimbabwe.

Quelea quelea spoliator was recognized from 18 specimens collected in the high interior of Natal. The specimens were in non-breeding plumage that had darker grey-brown upper parts compared to the warm buffy-brown of true *Q. q. lathamii* (Clancey 1960). The new race was rejected by Lourens (1961) and Ward (1966) as being simply one colour variant of a variable population and unsatisfactorily distinguishable from it. Lourens (1961) claimed, without presenting any data, that the offspring of both forms could be light- or dark-backed. Similarly, there is no information on how variable the accepted quelea subspecies are in the darkness of their dorsal plumage. Subspeciation in quelea is recognized morphometrically mainly by the characteristic frequencies of different plumage morphs of males in breeding dress (Ward 1966). If *Q. q. lathamii* and *Q. q. spoliator* are reproductively isolated to the extent implied by subspecific recognition, it seems reasonable to expect the proportions of plumage morphs of breeding males in both populations to show significant differences to the degree shown by other subspecies. Unfortunately, adequate morphometric data are not available for *Q. q. spoliator* to test this hypothesis.

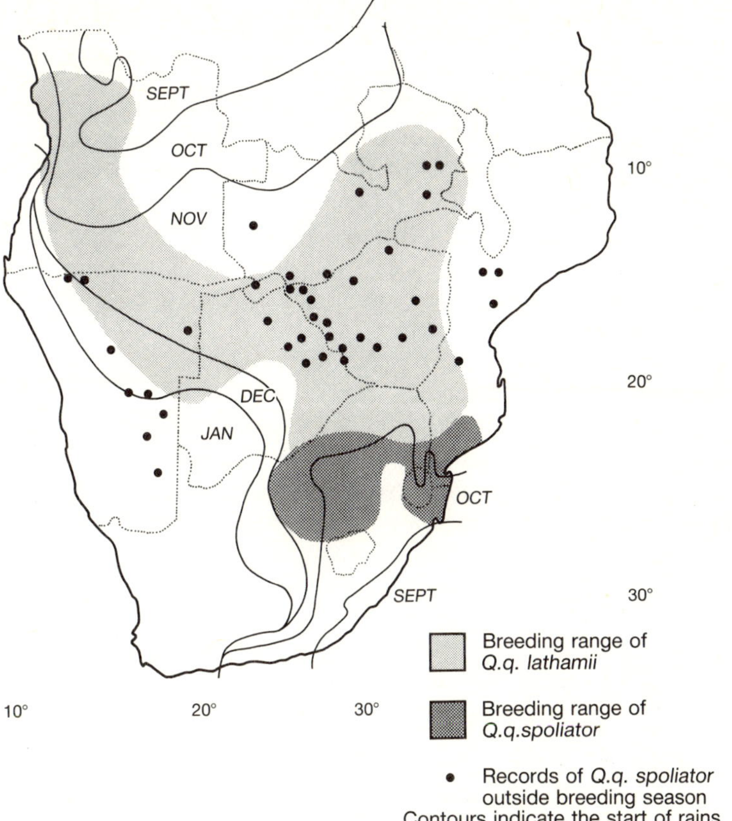

Fig. 11.1. Distribution of *Q. q. spoliator* in relation to the timing of the wet season (sources: Clancey 1973; Magor and Ward 1972; Thompson 1965).

The early-rains migration

A small area of the east coast of South Africa receives the first rainfall in September. The rains spread north and west during October and by November have spread over most of the subcontinent; only the south-western edge of the quelea breeding range receives its first rain as late as December (Fig. 11.1). The region of southern Africa receiving early rain varies in extent from year to year, but in general it corresponds to much of the suggested breeding range of *Q. q. spoliator* (Clancey 1973). Quelea are present in this region throughout the dry season and move further south as the dry season progresses to reach their southernmost latitude of 31°S on the central plateau of South Africa during August and September (Lourens

1963). This distributional limit is set by unsuitable vegetation to the south and south-east.

With the beginning of the rains in September and October and the widespread germination of grass seeds, most of the quelea population must move. They do not have the option of flying to areas of earlier rainfall where fresh seed may be available (Ward 1971), but they must move in the only directions possible, north and west, to areas where it has not yet rained and where they can still find dry seed. A hurried northward movement takes place with flocks seldom remaining in any area for more than 2 weeks, except where winter wheat is available (Lourens 1963). On arrival, they must join the quelea (presumably mostly true *Q. q. lathamii*) that have spent the whole dry season there. When it rains in November all birds, including the new arrivals, must move once again. However, some birds remain in South Africa from September to November (Kieser and Kieser 1978), probably feeding on dry grass seed that did not germinate in areas unaffected by rain.

The large area receiving rain in November includes the major zones where quelea concentrate in the dry season, such as the Zambezi, Limpopo, Sabi

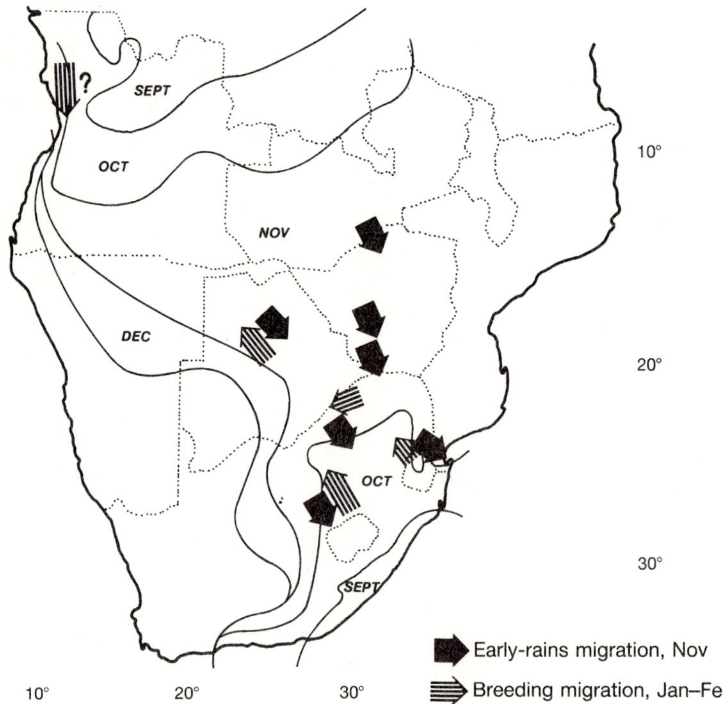

Fig. 11.2. Timing and probable direction of quelea migrations observed at localities mentioned in the text. The contours indicate the beginning of the rains.

and Luangwa valleys, and the vast area of the northern Kalahari around the Okavango Swamps in north-western Botswana. Pre-migratory fattening, followed by sudden large-scale departures of quelea from these regions with the first rains in November, has been reported from Matabeleland and Matopos in Zimbabwe (Plowes 1953, 1955), from the Zambezi escarpment in Zambia (Tree 1962), and from the Okavango (Jones 1972; Ward and Jones 1977) (Fig. 11.2), with later departures at the end of December from the western Kalahari (P. Jones, pers. obs.). By this time there are a few desert regions to the south-west where it will not yet have rained, but these are unsuitable for quelea. The safe option is for the birds to converge on areas where the rain fell in September and October and where fresh seed would be available. Such conditions could be found in two places: to the west in Angola, and to the south-east in southern Mozambique and South Africa. There are no data available from Angola, but there is evidence from the south-east, where massive increases in quelea numbers occur during November in the suggested early-rains quarters (Fig. 11.3). Quelea become much more common around Johannesburg in November and December (R. A. Reed's data in Rowan 1964, re-analysed by Ward 1971), and there are

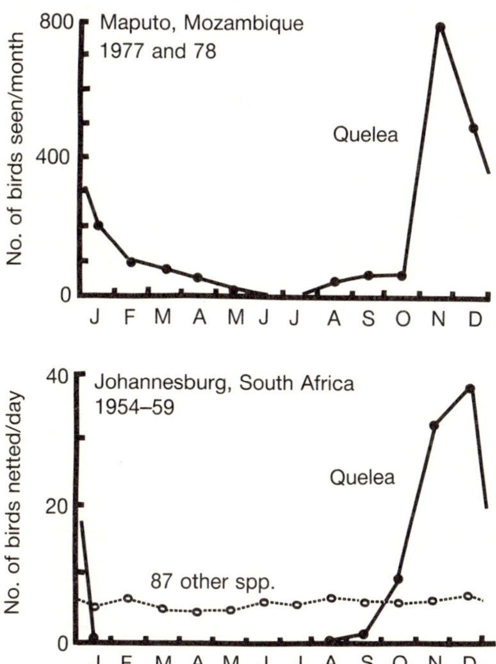

Fig. 11.3. Evidence of influxes of quelea into the early-rains quarters in November and December (sources: A. Vittery, unpubl. data; Ward 1971).

similar influxes at this time in southern Mozambique (Lamm 1955; da Rosa Pinto and Lamm 1960; A. Vittery, unpubl. data) and in the Orange Free State (Maclean 1957). Breeding begins soon after.

The breeding migration and itinerant breeding

As expected in southern Africa, the earliest breeding each season, in December and January, takes place in the south-eastern part of the breeding range of both *Q. q. lathamii* and *Q. q. spoliator*, in the area that must have served as the main early-rains quarters (Fig. 11.4). It is possible that some quelea use parts of Angola as an early-rains refuge. Some breeding is reported to occur in Angola as early as the beginning of January (Traylor 1963), although most breeding takes place between February and May (da Rosa Pinto 1960 and unpubl. data). Over the rest of their range, quelea do not begin breeding until February or March, long after they must first have come into breeding condition at the start of the rains. There is, therefore,

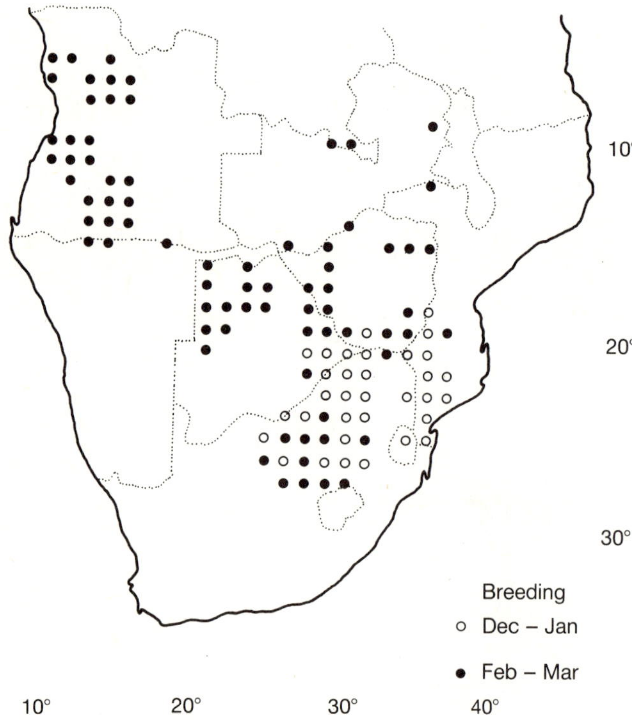

Fig. 11.4. The month of earliest recorded quelea breeding in each 1° square (compiled by J. Magor from unpublished records and Government reports).

ample time for birds that have already bred once in South Africa and Mozambique (or perhaps Angola) in December and January to make a breeding migration, as suggested by Ward (1971), to catch up with the advance of the rainfront (Fig. 11.2).

Apart from a large southerly migration witnessed near Luanda, Angola in February 1960 (da Rosa Pinto 1960), which may have been a breeding migration out of the so-far-unidentified Angolan early-rains quarters, quelea movements at this time of year are best known from the south-eastern part of the range. Quelea have been reported to leave southern Mozambique in January (Lamm 1955) and Swaziland between early January and early February in different years (K. B. Armstrong, unpubl. data). It has often been claimed that quelea migrate along river valleys wherever possible; an aerial survey in January 1957 recorded a massive movement of birds upstream along the Limpopo at 23°S into the main breeding areas of the Limpopo basin, where 2 weeks earlier no quelea had been present (Lourens 1963). Both the timing and direction of movement suggest that these birds were on their breeding migration from early-rains quarters in southern Mozambique. The previous year, quelea flocks had first reappeared in the central Transvaal on 14 January and had begun breeding immediately. Such January breeding is probably by birds that had performed the early-rains migration about 2 months earlier, but perhaps had not bred in their early-rains refuge. By late February, almost all quelea must have had the opportunity to breed once. One of the few eyewitness accounts of the breeding migration (Naude 1955*a*) described successive flocks of quelea flying low in a north-westerly direction along the Rhenoster River in the northern Orange Free State on 20 February 1955. They had presumably just left an early-rains refuge in Natal to the south-east, and over the next few days a number of new breeding colonies were located along about 60 km of the line of flight. These birds may have been breeding for the first time that season, but taking into account the time of year, they probably had bred once already.

Abandonment of fledged young by their parents in order to continue the breeding migration appears to be sudden. In the Rhenoster valley one day in March 1956, the entire adult population, estimated at 150 000 birds, left their breeding site within half an hour of sunrise. Although they were followed for 15 km downstream (NW), none was seen to rest or feed and they did not return. The young left in the same direction 2 days later in small flocks throughout the morning (Lourens 1963).

The destination of these birds was unknown. Although breeding continues within South Africa until May, many quelea leave South Africa altogether for their second breeding attempt. Large-scale departures occur between early January and mid-February, with sudden departures of large concentrations from the Limpopo valley reported in the second week of February

(Naude 1955*b*). Quelea nesting in the Okavango region in February and March were thought to be making their second breeding attempt of the season (Jones and Ward 1976). Certainly, quelea had not been present there until February, and there would have been enough time for them to have bred successfully once already within the early-rains region.

Both in South Africa and northern Botswana, laying can continue into late April (Lourens 1963; P. Jones, pers. obs.). At such a late date these birds could be making a third breeding attempt, although circumstances would have to be unusually good to stimulate breeding at a time when the post-nuptial moult would normally begin. Quelea laying near Kimberley, Cape Province, at the end of April had interrupted their moult in order to do so (R. Liversidge, unpubl. data), but there seem to be no recorded instances of this occurring elsewhere in southern Africa (Chapter 13).

Post-breeding dispersal

The end of the breeding season varies greatly, depending on local conditions and the readiness of the birds to make a further breeding attempt. An absolute limit to the breeding season is normally set by the onset of the ensuing post-nuptial moult, which has begun throughout the population by the end of May (P. Jones, pers. obs.). Some adults that finish breeding early must, therefore, disperse into the dry-season feeding areas well before moult begins, while others exceptionally delay moult until any late breeding attempt is completed.

Invasions of quelea into southern Mozambique between April and July after a period of absence since January (Lamm 1955) must represent post-breeding dispersal, though whether these are of juveniles or adults is not known. Similarly, in April at Heilbron in the northern Orange Free State, successive flocks of quelea were observed flying south-east throughout the day to return to their dry-season feeding area some 60 km further along the flight line, from where they had been absent throughout the breeding season (Naude 1955*a*). On the Zambian plateau, which is largely *Brachystegia* woodland unsuitable for breeding, quelea do not reappear until the dry season, although they may have been breeding not far away at lower altitudes in the main river valleys (P. Jones, pers. obs.)

Young quelea frequently remain in the vicinity of the colonies for some time after their parents have departed. However, sometimes young leave the area altogether, as in the Rhenoster valley in 1956 (Lourens 1963), and may fly long distances in the same direction as the adults continuing the breeding migration. Some juveniles caught in early March in north-west Botswana were much older than any locally born young and could only have come from colonies several hundreds of kilometres away (P. Jones, pers. obs.).

Plate 1. The quelea has a very wide distribution in Africa and affects cereal crops in more than 25 countries (photo: FAO).

Plate 2. In Nigeria, local people benefit by collecting guano from enormous quelea roosts and using it on their farms as fertilizer (photo: E. Dorow).

Plate 3. Feeding flocks can rapidly decimate small rice paddies (photo: R. Bruggers).

Plate 4. Surveys to monitor quelea populations in the remoter parts of its habitat require considerable logistic organization (photo: J. Jackson).

(a)

(b)

Plate 5. (a) The spray tanks of a DLCO-EA Beaver can be filled with fluorescent marker particles mixed with diesel and boiled linseed oil and sprayed over quelea roosts of colonies (photo: M. Jaeger). (b) The dust-sized particles stick especially to the underside of the quelea's wings and fluoresce under UV light. Tens of thousands, sometimes millions, of birds can be marked in one evening's spray operation (photo: B. Johns).

(a)

(b)

Plate 6. (a) Radio-transmitters weighing only 0.9–1.1 g can be glued to the tails of quelea to track their daily movement patterns (photo: M. Jaeger). (b) Quelea, equipped with 1.1 g transmitters, were tracked in Ethiopia to determine colony formation and feeding movements (photo: M. Jaeger).

Plate 7. Statistically reliable methods are now available for quantifying bird damage over large crop areas such as the Ethiopian Rift Valley (photo: M. Jaeger).

Plate 8. Quelea form pre-breeding flocks from which birds that are physiologically ready segregate to nest when they find optimum habitat conditions (photo: B. Davidson).

Plate 9. The quelea's nest is loosely woven and made in only about 2 days by the male. The clutch size is normally 3 (photo: M.-T. Elliott).

Plate 10. An adult male quelea about to regurgitate food for its chick (photo: B. Davidson).

Plate 11. Dense colonies are established in reeds in river beds such as the Niger River delta in Mali (photo: R. Bruggers).

(a)

(b)

Plate 12. (a) Sometimes hundreds of Marabou Storks *Leptotilos crumeniferus* gather at breeding colonies to eat the chicks as they emerge from the nest; (b) Kori Bustards *Ardeotis kori* pick up those that fall on the ground (photos: B. Davidson).

Quelea movements shown by ringing

Between 1951 and 1960, the Department of Agriculture of South Africa sponsored the ringing of nearly 60 000 quelea to determine their migration patterns. Ringing was first started on a small scale in Zimbabwe, but the majority of the birds were ringed in the Transvaal between 1954 and 1960. Since then ringing has been carried out on a much smaller scale, with an average of about 2000 ringed annually. A total of 92 794 quelea ringed by the end of the 1978/1979 season had yielded 214 recoveries (0.23 per cent) (see also Jones 1980). The data used here are derived from annual ringing reports (Ashton 1957; Elliott and Jarvis 1970, 1972–1973; McLachlan 1961–1967, 1969) and a corrected computer printout of recoveries kindly supplied by the South African Bird Ringing Unit.

The long-distance recoveries that are of most interest are plotted separately for birds ringed within the early-rains quarters (Fig. 11.5) and for areas of later rainfall outside South Africa (Fig. 11.6). The recovery patterns partly reflect the geographical distribution of bird ringers and quelea control operations, but also lend support to Ward's (1971) model of quelea migration patterns in southern Africa. Birds ringed within the early-rains

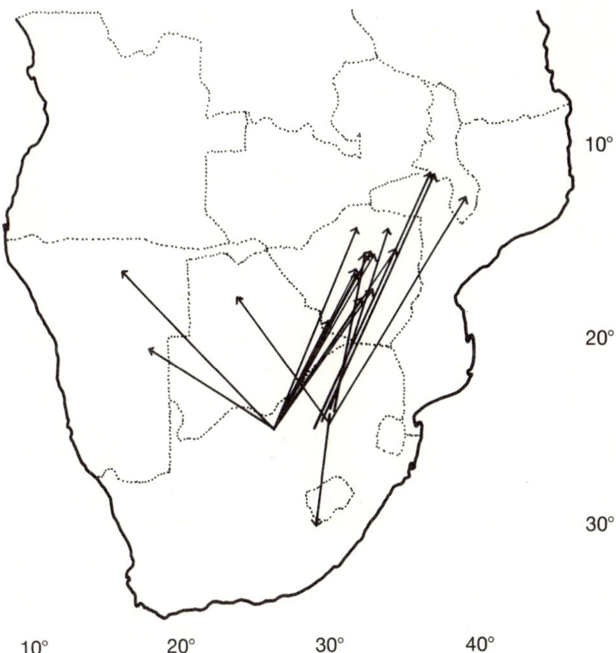

Fig. 11.5. Recoveries of quelea at more than 500 km from ringing sites within the early-rains quarters (data courtesy of the South African Bird Ringing Unit).

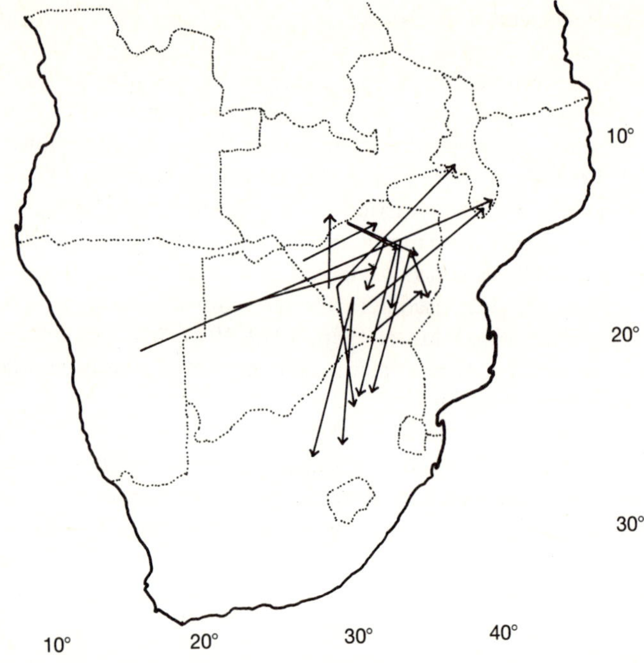

Fig. 11.6. Recoveries over 200 km away from the ringing sites of quelea ringed outside the early-rains quarters (data courtesy of the South African Bird Ringing Unit).

quarters in South Africa (mainly between August and January) have been recovered throughout southern Africa, as would be expected. For birds ringed outside South Africa, the majority again moved in the predicted directions, on a track towards or away from the early-rains quarters.*

Some data suggest that juvenile quelea might breed far from their natal area. One juvenile was ringed in Namibia and recovered in Malawi after 11 months, and another was ringed in Botswana and recovered in Zimbabwe after 20 months, i.e. after presumably having performed one and two early-rains migrations, respectively. It is, therefore, unlikely that they flew directly east-northeast to their recovery points, but passed at least once through the early-rains quarters. In contrast, no adults moved in an unexpected direction;

*Unfortunately, these data do not help resolve the *Q. q. spoliator* issue discussed earlier. Birds ringed in South Africa between October and April and recovered outside the breeding season in Malawi, Zimbabwe, and Zambia were claimed to be *Q. q. spoliator* (Clancey 1973), but no notes were made of their plumage characteristics. It is not possible to distinguish subspecies on the basis of migration movements alone, since both would be expected to follow the same routes. There are no recovery data to show whether the offspring of probable *Q. q. spoliator* parents in South Africa would breed with *Q. q. lathamii* elsewhere or vice versa.

all six birds ringed as adults outside the early-rains quarters and recovered after at least one return migration were either recovered where they were ringed (four birds), or along the expected heading route between the ringing locality and South Africa. This difference is statistically significant ($P = 0.036$, one-tailed Exact Test) but the sample sizes are regrettably small, because so few quelea were properly aged when ringed. This is either because adults and juveniles were not reliably identified early in the year or because they could not be distinguished after the post-juvenile moult. However, the results suggest that there may be a real difference in migratory behaviour; first-year birds may not necessarily return to breed in their natal area, whereas older birds that have already bred once return along the same heading route to breed in succeeding years. Unfortunately, the recoveries of adults and young ringed in South Africa (the great majority of cases) do not help us to distinguish between the migratory behaviour of first-time and experienced breeders. First-year birds born in South Africa and adults that may or may not have been born there, are expected to return annually to the early-rains quarters. Both groups, therefore, yielded recoveries locally within the early-rains quarters as well as along all bearings radiating from it.

The ringing recoveries reveal another important aspect of quelea behaviour: quelea flocks may be surprisingly cohesive. Four adult birds ringed at the same time and location in mid-September at Ventersdorp were recovered together in the same breeding colony 240 km to the north-east in early January. In the same colony there were two other adults that had been ringed together in late August the previous year 110 km to the south. Even more remarkable was a double recovery in Malawi of two birds ringed together in the Transvaal 18 months earlier (McLachlan 1962). Jaeger *et al.* (1986) found similar cohesiveness of fluorescent-particle-marked birds in Ethiopia.

The likelihood of reproductive isolation between *Q. q. lathamii* and *Q. q. spoliator*

There is no information on the relative proportions of *Q. q. spoliator* and *Q. q. lathamii* among the birds that achieve full breeding condition early in the rains. There is considerable variation in the state of advancement of the prenuptial moult and gonad development of birds undergoing pre-migratory fattening (Chapter 9 and Fig. 9.5). The reason for this is not clear, but presumably the birds that are the most sexually advanced breed first. Because events in the nesting cycle are so highly synchronized within a quelea colony (Ward 1971), a difference of less than a fortnight between two individuals in the attainment of full breeding capability may be sufficient to prevent them from breeding in the same colony. If the earliest birds to come into breeding condition were predominantly *Q. q. spoliator*-type individuals, this may provide a possible isolating mechanism. However, it is not known if *Q. q. spoliator*-type individuals do achieve breeding condition earlier than the *Q. q. lathamii*-type, nor, even if they did, whether the two forms would remain isolated throughout the breeding season. Many *Q. q. lathamii* will rapidly come into breeding condition after arrival in South Africa, and before the end of the breeding season of *Q. q. spoliator* in March or April (Clancey 1973). Reproductive isolation could then be maintained only if the two forms continued to breed in discrete colonies. In reality, South African colonies contain both light- and dark-backed forms (Lourens 1961), as do the colonies in north-western Botswana (P. Jones, pers. obs.), but quantitative data are lacking.

It seems improbable that genetic isolation between *Q. q. lathamii* and *Q. q. spoliator* would be maintained under these conditions alone. Other discrete quelea populations, some of which are formally accorded subspecific status (Ward 1966), are kept apart by their migration patterns (Ward 1971; Chapter 9 and Fig. 9.6). The migration patterns of *Q. q. lathamii* and *Q. q. spoliator* seem likely to bring them together at all times of the year and, most importantly, when they are breeding. At present, there is insufficient information to say whether partial isolation may occur. However, it is of great interest that the ringing results show that the same individuals may remain together in a flock for long periods, even when migrating, and that they breed in the same colonies. If this cohesive behaviour is usual, it could offer a mechanism whereby different genotypes could persist at different frequencies in sympatric groups.

The influence of man on migration patterns

It has been suggested that the migratory behaviour of quelea in South Africa has changed since the development of intensive agriculture during this century (Lourens 1963). During the first half of the century, quelea occurred in small flocks foraging over all of the central plateau during the dry season, but moved north with the rains to breed. Apart from two isolated instances of breeding in the eastern Cape (James 1928), there were no breeding records south of the Vaal River. After 1954, however, increasing numbers of colonies were reported on the central plateau, with breeding occurring later in the season further south to about 30°S (Lourens 1963). Whether this indicated a real change in the breeding distribution, or whether it reflected better surveying and reporting of colonies with the introduction of intensive quelea control during the same period, is uncertain.

What is certain, however, is that nowadays the breeding and migration patterns of quelea in South Africa are considerably disrupted by quelea control operations. It has been claimed that migration can now occur only into South Africa (early-rains migration), with most birds being destroyed while breeding and before they can leave on the return breeding migration. While vast numbers are undoubtedly killed, many others survive to breed outside South Africa. The many ringing returns of birds that have lived through several years of intensive control, when they must have migrated annually into and out of South Africa unscathed, are evidence of this (Jones 1980; Chapter 15).

The intensive growing of small-grain crops, especially winter wheat in South Africa, Zimbabwe, and Zambia, may have modified migration patterns to a small extent by inducing quelea to remain longer in areas where there would otherwise be little food. Elsewhere, cultivation accounts for little of the land surface or food available to quelea. Of greater importance must be the provision of boreholes and dams in otherwise waterless regions that the quelea could not previously exploit; there will also have been habitat changes brought about by severe overgrazing by domestic stock. The effects of these changes on quelea migration patterns in southern Africa remain a matter for conjecture.

12

Distribution, populations, and migration patterns of quelea in western Africa

STANISLAW MANIKOWSKI,
LOUIS BORTOLI, and
ALIOUNE N'DIAYE

Introduction

Western African farmers have protected their cereal crops against granivorous birds for years (Mallamaire 1959*b*; Serrurier 1966). Apparently, birds are most numerous and their damage is most severe after floods. In 1935, the great flood of the Senegal River was followed by 3 years of heavy bird damage; the flood of 1950 was followed by a proliferation of birds (Anon. 1954). Most of the damage was caused by quelea. This species was perceived as an important pest of cereals in Mali (Bocquet and Roy 1953) and Senegal (Morel 1968).

Rainfall and vegetation

Quelea habitat in western Africa extends from Senegal in the west to Chad in the east (Magor and Ward 1972). The presence of quelea is limited by the subdesert steppes which lie north of 20°N and by wooded savannahs south of about 10°N (nomenclature, UNESCO 1959). Quelea distribution can be further subdivided into the Sahelian zone in the north, characterized by wooded acacia steppes and a mean annual rainfall between 100 and 600 mm (Stroosnijder and van Hempst 1982), and the Sudanese zone in the south, characterized by a mean annual rainfall up to 1100 mm per year (Boudet 1975). This layered and parallel structure of climate and vegetation is disrupted by three zones where flooding occurs. Zone 1 comprises the Senegal River Basin in Senegal and western Mali, with average floods of 9 m at Kayes and 2.8 m at Richard Toll. The floods begin in June–July, reach their maximum in September–October, and end during the last week of

December. Zone 2 is the interior delta of the Niger River in Mali, with average floods of 5.5–6.2 m from July/August to March that reach their maximum in December (ORSTOM 1970). Zone 3 is Lake Chad with its surface varying from 10 000 to 25 000 km^2, mainly according to the amount of water received from the Chari River. The lowest water level in Lake Chad is in July and the highest is in December (Bouchardeau and Lefevre 1965). These three zones support greater concentrations of quelea than areas depending only on annual rains, because the borders of the flood zones are covered by annual grasses such as *Echinochloa* spp., *Oryza* spp., *Panicum* spp. and *Pennisetum* spp. (Rattray 1960), all eaten by quelea (Ward 1965*a*), and the abundance of vegetation is largely independent of capricious Sahelian rainfall (Gillet 1974).

There is one dry and one wet season. In the north, the rains start in June and end in September; in the south they last from April to October (Table 12.1). The beginning of regular rain stimulates the germination of annual grasses, first in the south and later in the north. For example, in Mali the vegetation starts growing at 13°N slightly before 15 July and at 16°30′N just before 15 August.

The vegetation of the Sahel is composed of annual grasses and thorny acacia bushes that cover some 5 per cent of the surface, except in depressions where acacia may form dense forests (Cisse and Breman 1982). At the 300-mm isohyet, annual grasses produce 28.8 kg of seeds per hectare by the end of the rainy season (Bille 1976). In areas of higher rainfall and in depressions, the annual production of wild grain may reach 200–500 kg/ha (Gaston and Lamarque 1976). Seeds fall onto the soil surface during the dry season, where many are gradually consumed by insects, birds, and rodents. The annual consumption of seeds is estimated at 30 per cent of the initial production (Bille 1976).

In June and July the seeds germinate. However, due to the irregularity of rainfall and the heterogeneity in germination time of seeds of the same species, not all of them germinate during any one rainy season (Breman *et al.* 1982). In Senegal at the 300-mm isohyet, which is close to the northern limit of quelea distribution, the amount of seeds germinating during the wet season has been evaluated at only 10 per cent of total seeds that survived the dry season (Bille 1976). Shrubs, trees, and perennial grasses provide the main vegetation cover in the Sudanese zone. They become less dense toward the south of the zone (Boudet 1975).

Distribution and migration patterns

Information on quelea distribution and migration was obtained from data accumulated since 1953 by organizations that conducted surveys, control,

Table 12.1. Mean rainfall (mm) by month and year (over 30 years of records) and the characteristics of the rainy season near the northern and southern limit of quelea distribution in western Africa (source: FAO 1984*a*).

	Mean rainfall (mm)												
	Month								Total for year	Start pre-rainy season	Start rainy season	Rainy season duration (days)	End growing season
Location	Apr.	May	June	July	Aug.	Sept.	Oct.	Nov.					
Northern limit													
Kiffa	1	4	25	91	121	86	17	2	352	17 July			24 Sept.
Aioun-El-Altrouss	1	2	19	97	115	56	15	2	316	15 July			5 Sept.
Tombouctou	1	3	19	65	95	37	5	0	225	2 Aug.			25 Aug.
Gao	1	6	27	75	110	30	5	0	261	1 Aug.			29 Aug.
Menaka	0	8	22	73	112	43	4	0	263	1 Aug.			29 Aug.
Tahoua	2	18	45	117	144	66	14	0	407	12 July			2 Sept.
N'Guigmi	0	7	8	57	141	22	0	0	235	30 July			31 Aug.
Bol	0	6	9	72	189	48	7	0	331	20 July			6 Sept.
Southern limit													
St. Louis	0	1	7	44	161	97	29	2	346	27 July	5 Aug.	27	27 Sept.
Linguere	0	4	31	101	209	136	45	4	536	8 July	29 July	52	5 Oct.
Matam	0	4	30	129	202	122	22	2	535	29 June	28 July	45	30 Sept.
Kayes	4	13	90	197	227	162	45	3	723	17 June	8 July	78	5 Oct.
Segou	9	20	94	202	239	127	27	3	724	15 June	3 July	71	29 Sept.
Yola	48	125	156	172	197	197	82	6	992	30 Apr.	1 July	123	17 Oct.
Garoua	38	122	155	178	224	214	75	1	1014	5 May	10 June	114	14 Oct.
Pala	32	94	156	230	224	210	84	4	1044	15 May	10 June	115	18 Oct.

and research; namely the Organisme de Lutte Antiaviaire (OLA), Dakar; the Organisation Commune de la Lutte Antiaviaire (OCLA), Dakar, Gao, and Zinder; the Organisation Commune de Lutte Antiacridienne et de Lutte Antiaviaire (OCLALAV), Dakar, Richard Toll, Gao, Zinder, and Garoua; the Food and Agriculture Organization of the United Nations (UNDP/FAO) Quelea Projects in Chad, Senegal, Niger, and Mali; the Plant Protection Service in Maiduguri, Nigeria; and the Deutsche Gesellschaft für Technische Zusammenarbeit (GTZ), Niamey, Niger. Since 1970, the UNDP/FAO Project teams have augmented these data with many unpublished observations of feeding flocks and day roosts in the region. The teams have also extensively sampled the quelea populations in the three important flood zones mentioned above.

In Nigeria, almost all intra-African migrants move northward in June–July at the beginning of the rainy season and southward after the end of the rains (Elgood *et al.* 1973). The northward migration to the Sahel and sub-Saharan regions is related to the rapid development of the vegetation and to the appearance of many insects and other animals with the beginning of the rains (Salvan 1967).

The generally accepted pattern of quelea migration in western Africa was first proposed for the quelea population in the Nigerian sector of the Lake Chad Basin (Ward 1965*a*). According to this proposal, quelea are the only species that fly south across the advancing rains in an 'early rains migration'. This migration permits quelea to arrive in the southern part of their range at a time when abundant fresh seed, induced by the earlier rains, will be available. The alternative to the early-rains migration is for birds to exploit local areas to the north that have received less rain and contain ungerminated seeds, or move further north to areas where the rains have not yet fallen. They can also begin feeding on insects, mainly termites that swarm abundantly at the beginning of the rains. However, these alternatives are temporary, and according to Ward (1965*a*, 1971), quelea must eventually move south.

In Nigeria, quelea are said to fly 300–600 km south in June–July and stay in the Benoue River valley where fresh seed is available. After about 6 weeks, they migrate northwards and establish breeding colonies in the first suitable habitat, and, thereafter, in sites progressively further to the north as the habitat develops. Ward (1971) suggested that this pattern of north-south migration would apply to all western African quelea populations such as those of the Lake Chad population in Chad and Cameroon and for those of the interior delta of the Niger River. It was presumed that in Senegal and western Mali the migration would be north-west/south-east in line with the Senegal River valley. Ward's (1971) hypothesis was put forward despite earlier studies indicating that in some parts of its range, quelea was a nomadic or sedentary species (Morel and Bourlière 1955; Dekeyser 1958; Salvan 1969).

New data on the displacement of quelea in western Africa indicate that their migration patterns in Mali and Senegal do not follow Ward's (1971) hypothesis. The only other possibilities seem to be a migration northwards in June–July and southwards in August–September as suggested by Elgood *et al.* (1973) or a nomadic dispersal of quelea in June and their reconcentration for breeding in August–September as proposed by Dekeyser (1958) and Morel and Bourlière (1955). We will examine the applicability of these proposals for each of the three important quelea population zones.

The Lake Chad area

One of the well-documented events in the quelea's life is their abrupt disappearance following the first rains of the wet season (COPR 1977; Dekeyser 1958; GTZ 1982; Manikowski 1981). In the Lake Chad area at this period some quelea accumulate fat (Ward and Jones 1977). The evident conclusion is that birds with fat migrate. In Nigeria, Ward and Jones (1977) reported that they migrate about 300 km south to the valley of the upper Benoue River.

In June, quelea were recorded arriving in the Yagoua and Bongor area at 10°25′N, south of the great dry-season concentrations, half way in their movement to the south (Elliott 1979). From mid-July to mid-August, the GTZ team working in Nigeria recorded the presence of small roosts in the Benoue River valley (GTZ 1982). In two other studies by Ward (COPR 1977) and Jackson (1973) of quelea distribution in the Lake Chad area, large concentrations of quelea disappeared in June and July from the plains to the south of Lake Chad and reappeared in the proximity of their southern distribution limit at about 9°N. They also reported the presence of quelea in the north between 13°N and 13°30′N. The latter pattern also was noted at the beginning of the rains by Salvan (1969); he observed quelea in Abeche, Chad, at 13°50′N in May and June. Newby (1980) reported the presence of several large flocks (with up to 1500 individuals) in July in Derba, Chad, at 14°47′N. He noted that 'occasionally, after the beginning of the wet season birds arrive in small parties from the south.' Some quelea stay throughout June, July, and August in the plains south of Lake Chad. At the beginning of the rainy season, they profit from the presence of drinking water in areas far from permanent rivers (Manikowski 1975; GTZ 1982; Ward 1965*a*); they also eat insects, mainly termites (Ward 1965*a*), and ungerminated seeds.

Interior delta of the Niger River

During the dry season, quelea are seen throughout the delta, but concen-

Table 12.2. Average and standard error (SE) of the latitudinal position by month of quelea roosts to the nearest 0.1° N. Student t-test values and probability (P) levels are given where there is a significant difference between months.

	Month											
	Jan.	Feb.	Mar.	Apr.	May	June	July	Aug.	Sept.	Oct.	Nov.	Dec.
Lake Chad Basin (Chad and Cameroon)												
No. of roosts	34	57	34	24	43	33	30	13	–	5	5	26
Av. latitudinal position ± SE	11.6 ± 0.2	10.7 ± 0.2	11.3 ± 0.3	12.4 ± 0.2	12.7 ± 0.2	12.8 ± 0.2	13.2 ± 0.2	13.1 ± 0.3		14.1 ± 0.7	11.7 ± 0.6	12.1 ± 0.3
t		2.95		2.93							3.96	
P		0.01		0.01							0.01	
Zinder (Niger)												
No. of roosts	2	–	–	–	28	49	29	18	5	7	14	3
Av. latitudinal position ± SE	13.3 ± 0.2				13.8 ± 0.5	13.8 ± 0.5	13.6 ± 0.7	14.0 ± 0.3	12.9 ± 0.3	13.4 ± 0.2	11.9 ± 0.3	12.4 ± 0.7
t			2.96								3.01	
P			0.01								0.01	
Niamey (Niger)												
No. of roosts	2	–	–	4	17	22	5	4	8	5	23	5
Av. latitudinal position ± SE	14.6 ± 0.0			14.4 ± 0.1	14.4 ± 0.4	14.4 ± 0.1	15.1 ± 0.2	14.9 ± 0.1	15.4 ± 0.2	15.2 ± 0.1	14.5 ± 0.1	14.0 ± 0.2
t							2.58				3.33	2.40
P							0.02				0.01	0.01
Interior Niger Delta (Mali)												
No. of roosts	–	9	11	73	122	69	46	14	21	53	88	20
Av. latitudinal position ± SE		15.8 ± 0.1	15.8 ± 0.1	15.7 ± 0.0	15.3 ± 0.1	15.9 ± 0.1	15.9 ± 0.1	15.6 ± 0.2	15.9 ± 0.1	15.2 ± 0.1	15.0 ± 0.1	15.1 ± 0.2
t						3.19				3.37		
P						0.01				0.01		

Table 12.3. Average and standard error (SE) of the geographical position by month of quelea roosts in the Senegal River Valley in Senegal and western Mali to the nearest 0.1° N. Student *t*-test values and probability (*P*) levels are given where there is a significant difference between months.

	Month											
	Jan.	Feb.	Mar.	Apr.	May	June	July	Aug.	Sept.	Oct.	Nov.	Dec.
Position on north–south axis												
No. of roosts	21	13	10	12	13	7	–	–	–	8	16	11
Av. geographical position ± SE	16.2 ± 0.1	16.4 ± 0.3	14.7 ± 1.6	16.1 ± 0.2	16.2 ± 0.2	16.4 ± 0.2	–	–	–	16.1 ± 0.2	16.3 ± 0.2	15.0 ± 0.2
t												
P												
Position on west–east axis												
Av. geographical position ± SE	11.9 ± 0.6	13.9 ± 0.3	13.8 ± 1.5	13.4 ± 0.6	12.1 ± 1.0	14.4 ± 0.7	–	–	–	13.7 ± 0.9	13.3 ± 0.5	12.4 ± 0.8
t	2.40											
P	0.05											

Table 12.4. Average and standard error (SE) of the latitudinal position by month of quelea nesting colonies to the nearest 0.1° N. Student *t*-test values and probability (*P*) levels are given where there is a significant difference between months.

	Month				
	Aug.	Sept.	Oct.	Nov.	Dec.
Lake Chad basin (Chad and Cameroon)					
No. of colonies	49	172	42	3	–
Av. latitudinal position ± SE	12.2 ± 0.1	12.3 ± 0.06	12.6 ± 0.2	10.4 ± 0.2	
t				3.48	
P				0.01	
Zinder (Niger)					
No. of colonies	–	4	13	–	–
Av. latitudinal position ± SE		12.5 ± 0.57	13.3 ± 0.27		
t					
P					
Niamey (Niger)					
No. of colonies	2	39	9	–	–
Av. latitudinal position ± SE	15.8 ± 0.1	15.6 ± 0.04	14.9 ± 0.2		
t			6.71		
P			0.001		
Interior delta of the Niger River (Mali)					
No. of colonies	–	77	36	11	3
Av. latitudinal position ± SE		15.3 ± 0.1	15.0 ± 0.1	15.0 ± 0.2	16.1 ± 0.1
t			3.08		2.24
P			0.01		0.05

Taxonomy

Ward (1966) described the quelea races for western Africa as *Q. q. quelea* (from Senegal and neighbouring countries) and a hybrid swarm in the Lake Chad area where *Q. q. quelea* overlaps with *Q. q. aethiopica* (first described from Sudan). He also pointed out that the white-faced morph is present throughout the quelea range in Africa as 5–25 per cent of the total male population. The white-faced morph can be subdivided into classes according to the combination of white and brownish colouration. S. Manikowski (pers. obs.) found that in Mali among 181 males examined, 65 per cent had

Table 12.5. Average and standard error (SE) of the geographical position by month of quelea nesting colonies in the Senegal River Valley in Senegal and western Mali to the nearest 0.1° N. Student *t*-test values and probability (*P*) levels are given where there is a significant difference between months.

	Month			
	Aug.	Sept.	Oct.	Nov.
North–south axis				
No. of colonies	3	46	44	5
Av. geographical position ± SE	16.8 ± 0.0	15.7 ± 0.2	15.7 ± 0.4	16.4 ± 0.4
t				
P				
West–east axis				
Av. geographical position ± SE	15.2 ± 0.0	12.9 ± 0.3	15.7 ± 0.2	14.2 ± 1.1
t			6.84	
P			0.001	

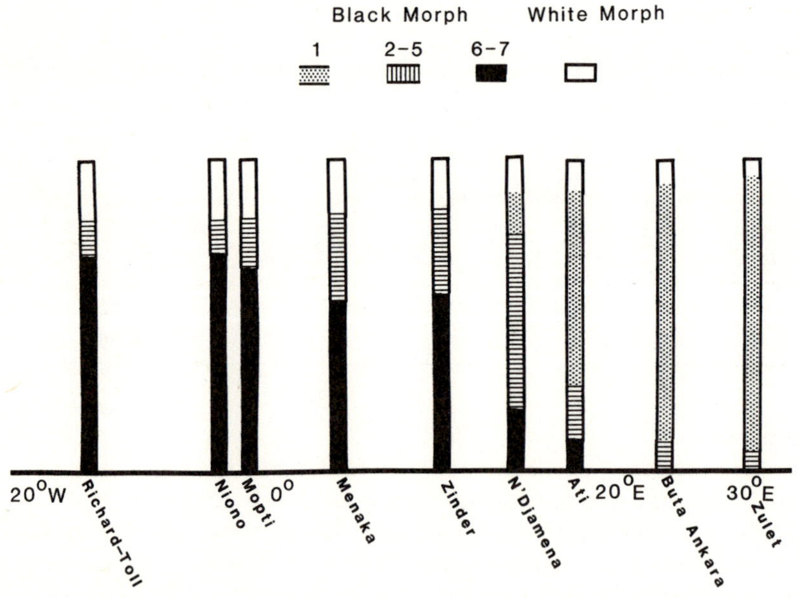

Fig. 12.2. Percentage of black morphs with indices 1, 2–5, 6–7, and white morphs of quelea sampled in Richard Toll, Niono, Mopti, Menaka, Zinder, N'Djamena, Ati, Buta Ankara, and Zulet. Distances between bars are proportional to the distances between sampling areas; abscissa in degrees.

Table 12.6. Number of male quelea in breeding plumage belonging to different categories of morph and mask indices in western Africa.

Location	Latitude	Longitude	Black morph indices			White morph	Total birds	Source
			1	2–5	6,7			
Richard Toll	16°25′ N	15°40′ W	0	89	559	162	810	Manikowski (1981)
Niono	14°15′ N	05°59′ W	0	76	514	131	721	Manikowski (1981)
Mopti	14°37′ N	04°59′ W	0	110	519	140	769	Manikowski (1981)
Menaka	15°52′ N	02°13′ E	0	97	217	74	388	Bortoli and Jackson (1972)
Zinder	13°47′ N	08°59′ E	0	76	176	52	304	Bortoli and Jackson (1972)
N'Djamena	12°08′ N	15°02′ E	85	365	139	78	667	Manikoswki (1980)
Ati	13°13′ N	18°18′ E	214	94	11	14	333	Manikowski (1980)
Buta Ankara	10°40′ N	24°10′ E	248	28	0	24	300	Bortoli and Jackson (1972)
Zulet	12°35′ N	39°45′ E	174	9	0	10	193	Bortoli and Jackson (1972)

brownish foreheads and white cheeks, 28 per cent had white foreheads and white cheeks, 5 per cent had brownish foreheads and brownish cheeks, and 2 per cent had white foreheads and brownish cheeks.

The characteristic that distinguishes *Q. q. quelea* from *Q. q. aethiopica* is the presence of a large, black, frontal band in *Q. q. quelea* (mask index 6 and 7; Fig. 12.2) and little or no black on the forehead of *Q. q. aethiopica* (mask index 1 and 2; Fig. 12.2). In the Lake Chad population both forms exist, as well as many intermediate ones with frontal bands of different widths. Independent of mask index, differences among males are also reflected in the colouration of other parts of the head, throat, and underparts, which may be of different shades of buff and in some individuals are suffused with varying amounts of pink.

Little is known about other morphological differences related to the polymorphism of quelea. Ward (1966) presented some data on the decrease in size (wing, tail, and bill lengths) in birds from Sudan westwards to Senegal. There may also be differences in size among males from the same area. In a sample of 94 of the black morph quelea, S. Manikowski (unpubl. data) found wings and tarsi (in millimetres) were significantly longer for quelea with frontal bands than for those without [wings: $\bar{x}=68.2\pm0.2$ SE; tarsi $\bar{x}=19.3\pm0.2$ SE for birds with frontal bands, and $\bar{x}=67.2\pm0.5$ SE, $\bar{x}=18.8\pm0.1$ SE for birds without frontal bands]. For the white morph males, 45 individuals with white foreheads had wings significantly ($t=2.37$, $P<0.05$) shorter ($\bar{x}=67.0\pm0.4$ SE) than 26 individuals with a brownish forehead ($\bar{x}=68.4\pm0.2$ SE).

Samples of males taken since 1972 (Table 12.6; Fig. 12.2) suggest a racial distribution of quelea that is more complicated than previously thought. A

Table 12.7. The number of black and white morphs in samples of male quelea in breeding plumage in Senegal and Mali from 1955 to 1981.

	Number of birds		
	Dark morph	White morph	Source
Senegal River Valley	86	21	Dekeyser (1958)
	79	21	Ward (1966)
	160	54	Bortoli and Jackson (1972)
	645	162	Manikowski (1981)
Chi-square = 4.98 (not significant)			
Interior delta of the Niger River	112	24	Dekeyser (1958)
	590	131	Manikowski (1981)
Chi-square = 0.27 (not significant)			

greater degree of polymorphism occurs across its entire distribution in western Africa than reported by Ward (1966). Figure 12.2 shows that in each sample at least three categories of mask types were present. In particular, Ward's (1966) description of *Q. q. quelea* and *Q. q. aethiopica* are misleading since even at the extremities of the range, a significant proportion (31 per cent in Senegal, 10 per cent in Sudan) of the male population does not have the typical colouration. Although the proportions among morphs and mask indices vary geographically, they appear to be stable over time within the same locality. Data from Senegal and Mali since 1955 indicate no significant changes (Table 12.7). We, therefore, suggest that either the races have to be redefined to account for polymorphic variation, or the racial distinction at least between *Q. q. quelea* and *Q. q. aethiopica* has to be abandoned.

13

Factors determining the breeding season and clutch size

PETER J. JONES

Introduction

Even into the early years of this century, so little was known about the breeding biology of quelea that they were widely believed to be brood parasites, with the Red Bishop *Euplectes orix* the most probable host (Roberts 1909; Taylor 1906). The first eggs and nests were described by James (1928), but it was not until the 1950s that biologists gave serious attention to quelea breeding ecology. Since then the quelea has become one of the best studied tropical savannah birds, largely because their vast breeding colonies are a natural focus for the efforts of quelea control teams, but also because knowledge of their breeding biology has given insights into other crucial aspects of their ecology, particularly their migration patterns and population dynamics.

Food supply and the breeding season

The seasonal timing of events in the annual cycle of a species is as much a product of natural selection as are its behavioural and morphological characteristics. One of the most important factors determining the timing of the breeding season is likely to be the availability of the additional high-quality food that parents feed to their young. The greater the extent to which the availability of this food shows a sharp seasonal peak, the greater will be the selection pressure restricting the period when breeding is attempted. Parents that breed earlier or later in the year will rear fewer young and contribute less to the gene pool of future generations than those breeding at the optimum time. Different lines of evidence suggest that in any one locality the period of peak food availability for nesting quelea may be very short.

Quelea breed during the wet season, but only after the rains are well advanced, when the new green seed of annual grasses is becoming plentiful. At the same time, easily caught larval and nymphal insects are becoming sufficiently well-grown and abundant to provide an important protein supplement to the diet of adults and nestlings. A large quelea breeding colony may contain several million nestlings that for their first 5 days of life are fed on a mixture of insects and green grass seeds. The ready availability of insects is especially important, for in the period before its body feathers develop, the nestling does not thermoregulate and can take maximum advantage of a high protein food for efficient conversion into body growth (Ward 1965*b*). During the next 10 days, there is a gradual changeover to a diet consisting almost entirely of green grass seeds. At no time other than the wet season could a large quelea colony find enough easily gathered protein food to satisfy this initial growth requirement of its vast numbers of young, but even in the rains neither type of food is available to quelea for long.

The insects taken are mainly Harvester Termites *Hodotermes mossambicus*, grasshopper nymphs, and lepidopteran caterpillars (Disney and Marshall 1956; Jones and Ward 1976; Morel *et al.* 1957; Ward 1965*b*). The termites, the only species to harvest grass above ground in daylight, are especially important in some areas because they remain available for long periods during the rains, but the bulk of the insect food is often provided by grasshopper nymphs and caterpillars. No sooner have these grown to a size at which they are worthwhile for a quelea to take than they metamorphose. Adult grasshoppers are almost impossible for quelea to catch, and caterpillars pupate below ground where they are unobtainable. The availability of grass seeds is similarly ephemeral: they are not worth eating before they have swelled to the 'milky' or 'doughy' stage, but they then very quickly mature and fall from the seed head. Although great quantities of seed are produced, it is not necessarily readily available; quelea evidently have difficulty finding seeds on the ground among the tangle of collapsing mature stems, for they feed preferentially along paths and tyre tracks or on burnt and trampled areas where seeds are exposed. Thus both insects and seeds, although present in quantity during the rains, escape quelea predation quite suddenly, as continuing events in their own life histories render them unavailable. In addition, not only do quelea compete with other species of birds and rodents for part of their food, but they must also significantly deplete it themselves in the immediate vicinity of a large colony. One rough calculation suggests that a quelea colony of 50 ha may remove 50 per cent of the insects available within a 10-km radius during the rearing period (Morel 1968).

It is almost certainly because of the restricted period of food availability that quelea colonies in one locality tend to be more or less synchronous and only rarely are colonies established in the same place several weeks later. By this time, optimal conditions for breeding are more likely to be found elsewhere. It is probably also for these reasons that the breeding schedule of

quelea is so abbreviated. The complex nest is built in only 4 days, and the incubation period of 10 days (Ward 1965*b*) is among the shortest known for any bird (cf. Nice 1953). The young leave the nest at 11–13 days old although they cannot yet fly, and the period of post-fledging care is also remarkably short; within a week or so the fledglings are abandoned by their parents, some of whom may be breeding again a few days later, perhaps hundreds of kilometres away, under optimal conditions once more (Chapter 9; Jaeger *et al.* 1986). Finally, it is probably because mature grass seed on the ground is available only with difficulty that newly independent juveniles take more readily accessible cultivated cereals, which are not a normally preferred food (Chapter 17; Ward 1965*c*).

If the plentiful supply of insect and seed food, essential for successful rearing of young, is available in any one locality for only a short time, adult quelea must ensure that they begin the physiological preparation for laying well beforehand. They must, therefore, use environmental cues, the so-called proximate factors that trigger them to nest and lay in the right place at the right time.

Proximate factors controlling breeding

Photoperiod and rainfall

Most work on identifying the proximate factors controlling the breeding season in temperate regions has highlighted the prime importance of the increasing photoperiod in spring for bringing birds into breeding condition, with temperature and food availability assuming only secondary importance (for review see Murton and Westwood 1977). Physiologists have been curious to know whether the reproductive cycles of tropical birds were also modifiable by photoperiod changes, although in the wild at locations near to the equator they would not be exposed to significant variation in day length. An early experiment in western Africa showed that male quelea in eclipse plumage given an additional 5 h light after dusk underwent much more rapid maturation of the testes and development of nuptial plumage than control birds maintained on natural day length (Morel and Bourlière 1956). A similar experiment in eastern Africa gave much less convincing results (Marshall and Disney 1956); although it was claimed from more detailed later work that more rapid testis maturation occurred under a 17-h photoperiod, this is not demonstrated by the data (cf. Tables 1 and 2 of Disney *et al.* 1961). However, a further experiment by Lofts (1962) showed that gonad growth was indeed more rapid under a 17-h than a 12-h photoperiod and that when day length was reduced to 8 h, maturation did not occur at all.

Although these experiments showed somewhat surprisingly that the reproductive cycles of quelea could be manipulated by artificial photoperiods quite different from those they would ever experience living near the equator, they also showed that testis growth, nevertheless, took place quite normally and spontaneously under a natural 12-h photoperiod. Natural changes in day length were, therefore, discounted as a proximate factor controlling the breeding seasons of quelea in the wild (Disney *et al.* 1961; Disney and Marshall 1956; Lofts 1962; Marshall and Disney 1957), even in those populations living at the latitudinal extremes of the quelea's distribution, for example at 16°N in Senegal, where there is a 1.8-h annual variation in daylight and at 25°S in South Africa where the annual variation is 2.5 h (Lourens 1960; Morel and Bourlière 1956).

Instead, it was suggested that the proximate factors stimulating breeding were most likely to be found in the greatly changed environmental conditions following rainfall, since there was no evidence that quelea ever bred except after substantial rain (Disney 1957; Disney and Marshall 1956; Marshall and Disney 1957). Marshall and Disney (1957) provided separate groups of caged quelea with different combinations of dry or green grass, dry or green grass seed and millet, insects (Harvester Termites, grasshoppers, and fly larvae), and artificial rainfall. Male quelea could not weave nests with dry grass, and even with fresh grass they could not complete nests acceptable to females unless they had access to long stems, leaf blades, and seeding heads with which to build. Eggs were laid first by females provided with green seeds alone, a month after both sexes had completed their pre-nuptial moult. These females laid again 2 months later, quickly followed by those that had also fed on insects and had experienced rainfall. It was concluded that neither rainfall nor the seasonal appearance of protein food was a necessary stimulus to reproduction; rather, the availability of seeding grasses was sufficient to enable males to complete nests acceptable to females. In turn, the sight of green grass and the presence of vigorously building males was enough to induce females to lay. As these authors pointed out, seeding grasses are invariably associated with a high insect abundance, and at no time other than the wet season is there enough protein available to feed rapidly growing young.

The experiment is not easy to interpret. The cage containing females that laid when denied insect food was apparently invaded by insects that may have been available as food (Disney 1957). More puzzling is the fact that quelea given only dry grass and dry seed eventually bred anyway, 5 months after first assuming full nuptial plumage.

The results of the later photoperiod experiments suggested that gonad maturation occurs spontaneously under the light regimes occurring naturally, leading Disney *et al.* (1959, 1961) to suppose that the precise timing of breeding might be achieved by the interaction of an environmental stimulus with an endogenous sexual rhythm. The presence of green grass would stimulate only those birds that had already spontaneously begun spermatogenesis. Such an endogenous rhythm was evident when male quelea were kept on an unvarying light schedule of 12 h light and 12 h dark (Lofts 1964). For 30 months, birds showed a clear 12-month cycle of testis enlargement for about 7 months, followed by spontaneous regression and a refractory period of 6 weeks, during which the gonads could not be restimulated, but which ended with spontaneous testis regeneration (Fig. 13.1). Light periods longer

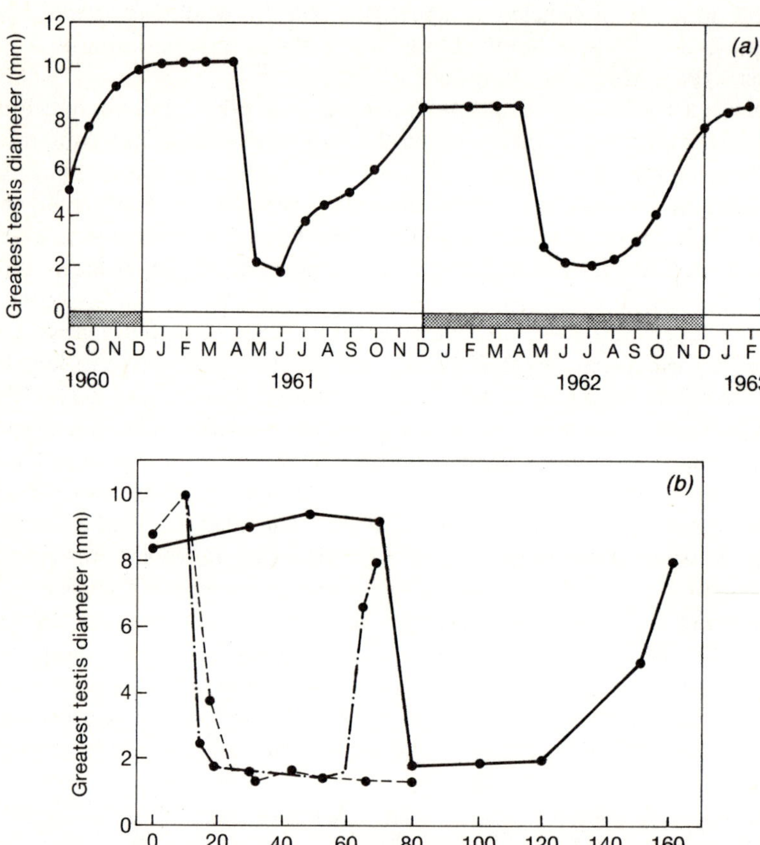

Fig. 13.1. (*a*) Seasonal variations in testis size of *Quelea quelea* on a constant light-dark (LD) 12:12 h photoperiod show an autonomous periodicity (source: Lofts 1964). (*b*) Experimental manipulation of the testicular cycle of *Quelea quelea*. The solid line indicates birds held on LD 12:12 h as in (*a*). The broken dotted line (· — · —) indicates birds exposed to LD 17:7 h, in which the duration of peak breeding condition was greatly shortened, the refractory state was entered sooner and subsequent testis growth was much more rapid. The dashed line (– – –) indicates birds exposed to LD 8:16 h, in which subsequent testis growth was permanently suppressed (source: Lofts 1962).

than 12 h speeded up testis growth and greatly reduced the time for which testes remained active, but the length of the refractory period was unchanged (Lofts 1962, correcting Wolfson and Winchester 1959). Some argument has centred on whether Lofts' experiment demonstrated the existence of a true autonomous rhythm of reproductive activity, because such a rhythm can only be demonstrated unequivocally in constant darkness (Murton and Westwood 1977). Nevertheless, whatever its physiological origin, the rhythm would not be disrupted by the light regimes encountered in the wild.

The adaptive value of the refractory period in temperate birds, where it is usually much longer, is said to be that it prevents breeding at unseasonal times (Lofts and Murton 1968). The much shorter refractory period of quelea is typical of birds living where suitable conditions for breeding may be erratic, ensuring that the recovery of breeding potential is rapid and enabling birds to take advantage of any unseasonal rainfall that may occur (Lofts 1962). Murton and Westwood (1977) argued that quelea are probably in a state of near readiness to breed almost throughout the year, requiring only the presence of suitable releasers such as green grass and/or a suitable food supply for breeding to occur.

For two main reasons the interaction between endogenous rhythm and environmental cues cannot be as simple as that. First, the 6-week refractory period is clearly associated with the onset of the post-nuptial moult into eclipse plumage, which begins late in the wet season as the last broods are being reared. In Morel and Bourlière's (1956) photoperiod experiment, quelea subjected to a 17-h light period showed a greatly advanced post-nuptial moult. Not only had these birds attained breeding condition much more rapidly than normal, but they must also have entered the refractory phase much more rapidly under the longer light regime, as discovered in later experiments (Lofts 1962) and, therefore, moulted early. The refractory period perhaps signals the end of the breeding migration (Chapter 9) and ensures that the moult can get underway at a still favourable time of the year, the late wet/early dry season, without being interrupted by a further breeding attempt that would be completed too late in the year to be successful. However, this does not explain why an inability to breed does not develop earlier in the season after the first breeding attempt, but allows itinerant breeding to occur (Chapter 9); nor does it explain why the post-nuptial moult is sometimes interrupted to allow a further late breeding attempt.

Quelea breeding in South Africa in late May 1967 had renewed up to three primaries by the time the young had left the nests (R. Liversidge, unpubl. data); their post-nuptial moult must have begun before the colony was founded and was presumably interrupted during breeding. In this instance, moult was interrupted at an early stage; it is not known if breeding could still occur once the birds were well moulted into eclipse plumage. The quelea that Lofts (1964) used to demonstrate the existence of the endogenous rhythm did not help settle the problem; these males moulted directly from one nuptial plumage to the next, without any intervening eclipse plumage and were, therefore, much better prepared to breed again when suitable conditions arose. This implies that the normal sequence of plumage change is not controlled by light, nor is there a simple alternation of eclipse and nuptial dress. If, as Murton and Westwood (1977) suggested, the nutritional state of the birds was involved, a factor that was not controlled in Lofts' experiment, the existence of a semi-autonomous rhythm in wild birds becomes much more debatable.

The second line of evidence arguing against a simple interaction of endogenous rhythm and environmental cue is what actually happens at the start of the rainy season in the wild. Neither rainfall nor green grass can be sufficient on their own to trigger a breeding attempt, for it is evident that in the wild they elicit a quite different response. Their initial effect is to stimulate fattening, and a migration away from the region (Chapter 9), followed by breeding sometime later in response to a possibly quite different set of releasers.

Obviously, a breeding attempt cannot take place unless both sexes are ready to breed. Although the photoperiodic responses of male birds of many species have been investigated in great detail, much less attention has been given to females, apparently because their ovarian responses are much less clearcut. Female quelea responded to long day lengths in Morel and Bourlière's (1956) experiment in much the same way as the males, showing rapid assumption of the yellow bill colouration and follicular development to about 3 mm in diam. None of Marshall and Disney's (1956) females responded to long day lengths, and although females in their later experiments developed yellow beaks, their follicles developed only to a maximum of 1.1 mm in diameter (Disney *et al.* 1961). Clearly, no females were stimulated by long day lengths into the phase of rapid yolk deposition leading to ovulation; this has generally also been the finding in studies of other species (Lofts and Murton 1968). It has been suggested that female quelea require the additional visual stimulus of building and courting males in order to begin rapid yolk deposition (Disney and Marshall 1957), but it is now known that yolk deposition often begins before arrival at a nesting site, and even while a female is still feeding the young of an earlier brood (Jones and Ward 1976; Ward 1971; P. Jones, unpubl. data). Yet, some further proximate factor must be necessary to carry the female forward into full egg production. Because this is a nutritionally demanding process (the normal clutch weighs about 30 per cent of the female's body weight and is produced in a week), the other stimulus is likely to be the female's food supply and her nutritional balance.

Nutrition

Work on temperate birds has underlined the fact that whatever their dependence on day length to trigger the initial physiological preparations for a breeding attempt (perhaps particularly important in species where the male must establish and defend a territory well in advance of nesting), the female is finally dependent on the availability of sufficient food to form her eggs. This dependence may be so critical in some species that she may not always be able to lay sufficiently early to ensure that her young hatch in time for the peak in food availability (Perrins 1970), or she can do so only if assisted by courtship feeding from the male (Royama 1966).

Among equatorial forest birds that experience almost unvarying daily photoperiods, uniformly high temperatures and humidity, and high rainfall at any time of the year, breeding is nonetheless highly seasonal. There is some evidence that seasonal changes in the nutritional state of the female, particularly changes in her protein balance, dictate at what times of the year egg formation is possible and at what times it is not (Fogden 1972; Ward 1969). For example, Yellow-vented Bulbuls *Pycnonotus goiavier* in good condition, such as those about to breed, had higher lean dry weights of their flight muscles than those in poor condition, during periods when insect food was scarce (Ward 1969). Seasonal variations in protein balance occur among birds of the tropical savannahs, and a site where additional lean material is stored has been identified in quelea. Electron microscopy of the pectoralis major flight muscle of quelea has shown that under starvation the space between adjacent myofibrils (the contractile elements of the muscle fibres) decreased by about 60 per cent compared with well-fed individuals, without any change in the myofibrils themselves. The main ultrastructural change in the muscles of starved quelea was, therefore, a marked reduction in the amount of sarcoplasm, which alone was probably sufficient to account for the observed gross loss of muscle tissue (Kendall *et al.* 1973).

The nutritional state of the female, especially the amount of this labile reserve protein stored in the muscle, will in turn be dependent on a range of other factors. Foremost is the availability of protein-rich food, such as insects, above the amount required for normal daily body maintenance. However, the maintenance requirement is itself determined by ambient temperature and thermoregulatory costs; food dispersion, as it determines daily foraging costs; day length, as it alters the relationship between the time available for foraging and the length of the night when feeding is not possible; and other nutritionally demanding processes such as moult and migration.

Quelea are most adversely affected by these factors during the dry season. In particular, insect food of the types easily taken by quelea (caterpillars and grasshopper nymphs) is not available, and only dry seed with a low protein content can be found. Even this food becomes increasingly patchily distributed, obliging quelea to incur increasing foraging costs as they fly greater distances between night roosts and feeding grounds (Ward and Zahavi 1973). Nights may be longer relative to the daylight hours available for feeding and they may be cold, reaching freezing temperatures in parts of the quelea's dry-season range. In addition, not only have many birds left the previous breeding attempt in poor condition after rearing young (Jones and Ward 1976), but all are obliged to undergo a complete moult (Ward 1973*b*). Moult is a protein-demanding process that in most passerines is metabolically incompatible with breeding (Payne 1972). Moult in quelea normally takes place on a low-protein diet of dry seeds alone; there would be little spare protein available for a female to store for egg formation, regardless of all the other demands on her daily food intake at this time of the year.

Whatever the adaptive value of the 6-week refractory period, it is clearly not long enough to prevent quelea from breeding throughout the 6 months of the dry season. Inadequate nutrition probably prevents them from doing so. We may then ask if the alleviation of these nutritional constraints, either by suspending or completing the post-nuptial moult or by migrating to a more favourable physical environment where rain has fallen and insects are available, is sufficient to permit breeding, or whether other environmental cues are still required.

The results of Disney and Marshall's (1956) experiments are inconclusive. Their experimental quelea were undoubtedly in a more favourable environment than the local wild population, because all caged females assumed the yellow breeding colouration of the beak before the wild birds (Marshall and Disney 1957). Green grass seeds, with which these birds were provided, normally contain higher protein levels than dry seeds (W. Meinzingen, unpubl. data), so that even without insect food these birds could have attained a nutritional status sufficiently high to enable them to breed. The fact that even the quelea maintained on dry seed alone eventually managed to lay is not incompatible with these ideas; relieved of the nutritional burdens of moult and foraging costs, and given warm shelter and plentiful food, it would not be surprising if they could slowly acquire sufficient levels of reserve protein to stimulate a breeding attempt in the absence of any of the other environmental stimuli normally operating in the wild.

Evidence in the field of the role of nutrition in quelea reproduction was obtained primarily at a colony in north-western Botswana that was most probably a second breeding attempt for the majority of the birds present (Jones and Ward 1976). Yolk development had already started in these birds as they arrived at the nesting site in early February, and laying began 4 days later. The females that were sampled as they arrived had considerably abraded plumage indicative of an earlier breeding attempt and had probably only recently abandoned young from their first broods. Evidently these birds had completed a leg of the breeding migration of up to 600 km (though possibly not non-stop) from the likely area of these first colonies in the Limpopo River valley. Far from being in poor condition after the breeding attempt and a long migration, these females were in peak condition and just beginning the phase of rapid yolk deposition in the first follicle of the three-egg clutch.

The fat reserves of these females were 20–50 per cent above average dusk values (between 1.1 and 1.4 g, compared with non-breeding values of just under 1.0 g), but were not as high as those present before migration (Chapter 9). The most striking indication of their peak condition was the lean dry weights of the major flight muscles, the pectoralis major and supracoracoideus. These were higher than at any other time of the year (0.9–1.3 g, compared with non-breeding values of 0.7–1.0 g) and higher even than the normally heavier flight muscles of males. This peak represents an increase of 80 per cent in reserve protein over normal and was associated with a high percentage of insects in their diet (15–20 per cent by dry weight compared with 10 per cent dry weight of the males' diet; Fig. 13.2). Nevertheless, despite abundant protein-rich food, during the 5 days from the start of rapid yolk deposition to the laying of the first egg a rapid and severe decline in muscle protein and total body fat occurred (Fig. 13.3), which in some females reached starvation levels (cf. Pope and Ward 1972). In a later study in Nigeria, females deteriorated in condition even more rapidly during yolk formation and many were found dead under the nesting bushes after laying their first egg (Jones and Ward 1979). The survivors were by this time under

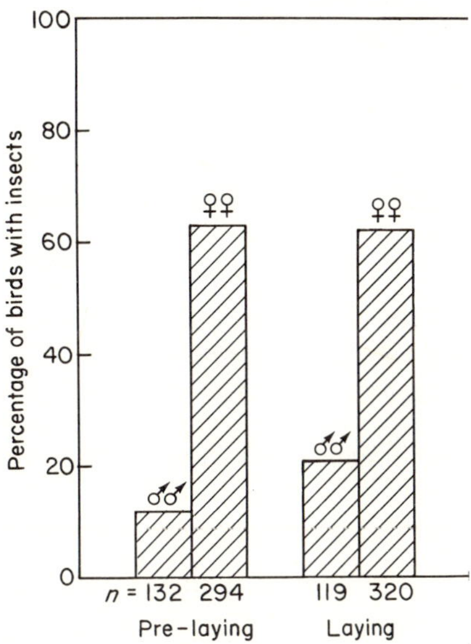

Fig. 13.2. Percentage of quelea sampled during the periods of yolk formation and laying that had insects among their crop contents. Sample sizes of males 132 and 119; of females 294 and 320 (source: Jones and Ward 1979).

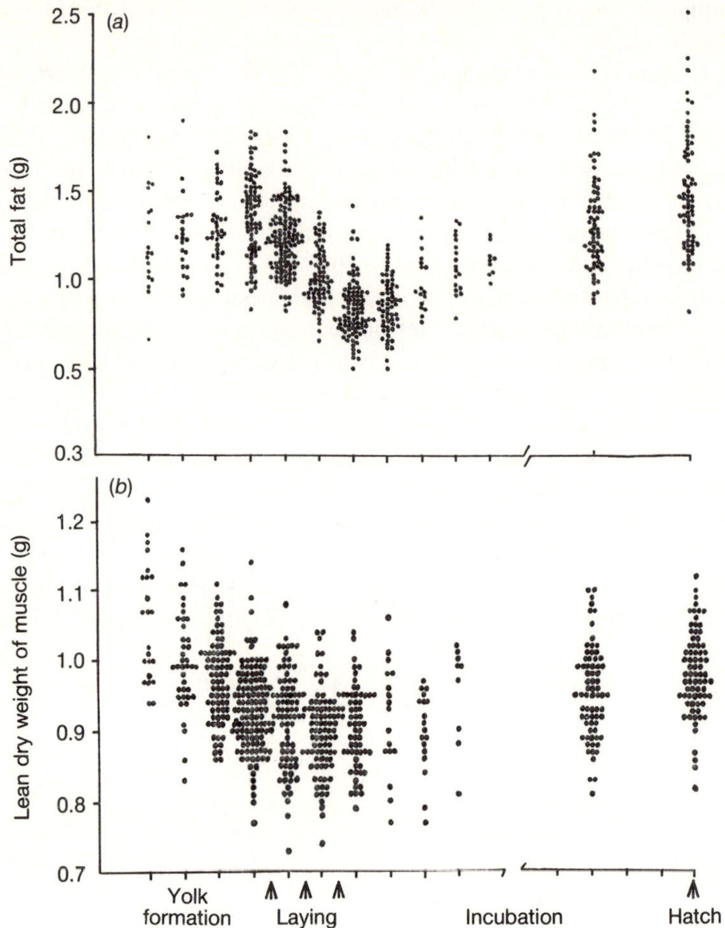

Fig. 13.3. Changes in dusk levels of total body lipid (*a*) and lean dry weight of the flight muscles (*b*) of female quelea during yolk formation, egg-laying and incubation. Until the end of laying only birds producing the normal three-egg clutch are included; thereafter all females are shown. The base line at 0.3 g in (*a*) approximates to zero reserve fat; that at 0.7 g in (*b*) approximates to zero reserve protein (source: Jones and Ward 1976).

such severe nutritional stress that none attempted to incubate their eggs and all abandoned the colony.

The loss of condition may affect other physiological processes. Some female quelea in the Nigerian colony suffered such an adverse protein balance that red blood cell production ceased. It is probable that this commonly occurs even among females that go on to rear their young successfully (Jones 1983). The packed cell volume of the blood (haematocrit) declined from normal levels of 50–60 per cent to anaemic values of under 30 per cent, with females in poorest condition showing the lowest values (Fig. 13.4). Males did not show abnormally low haematocrits, even when in equally poor condition otherwise, strongly suggesting that it is the special demands of egg production that causes so severe a deterioration in females.

Recovery occurs gradually during the incubation period, when generally both sexes are free to forage for much of the day, leaving the eggs to incubate by solar heat (Ward 1965*b*). By the time the eggs hatch, muscle protein levels have returned to normal non-breeding values, fat reserves have increased to higher levels than before breeding, and the recovery from anaemia has been speeded by the enlargement of the thymus gland as an erythropoietic organ (Kendall 1980; Kendall and Ward 1974; Ward and Kendall 1975).

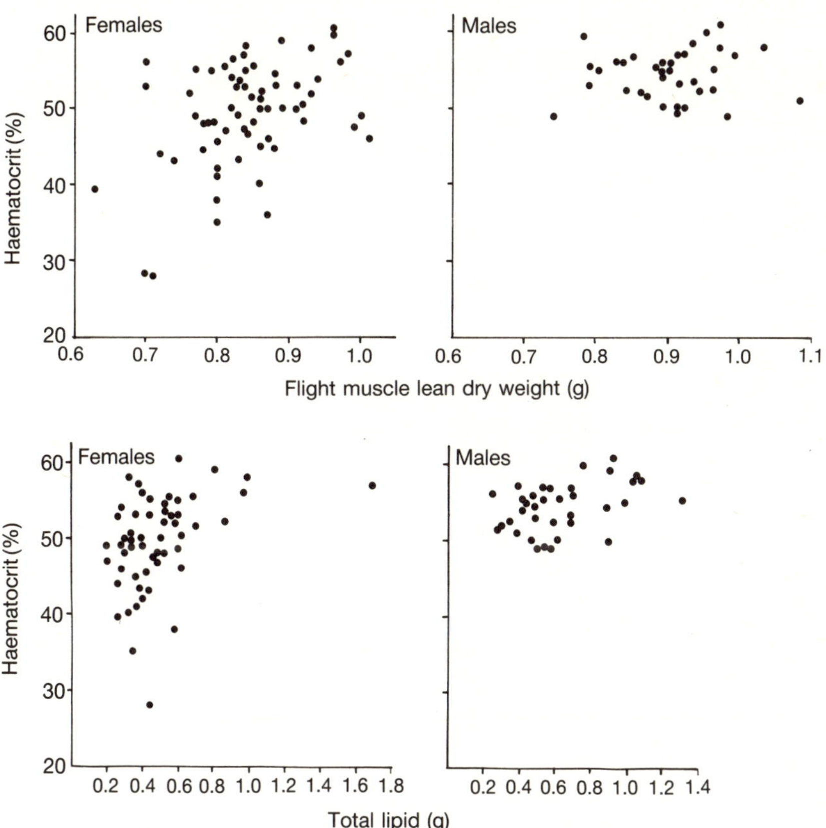

Fig. 13.4. Relationship between haematocrit per cent (packed red blood cell volume) and reserves of protein and fat in female and male quelea during laying (source: Jones 1983).

The initial stages of egg production undoubtedly cause a great drain on the nutritional reserves of laying females, even when abundant high-quality food is available. The implications of this are of fundamental importance; unless the initial high level of reserve protein can be achieved before yolk formation, the production of a normal clutch would not be possible. Indeed, the attainment of a threshold level of stored labile protein could on its own be the proximate factor controlling the onset of breeding (Jones and Ward 1976), an explanation proposed earlier for equatorial forest birds (Fogden 1972; Ward 1969), although disputed by Murton and Westwood (1977).

Nutrition and the control of clutch size

Further evidence of the nutritional limitations on laying quelea came from a study of the factors determining clutch size in the same Botswana colony. Three-egg clutches make up 50–85 per cent of quelea clutches throughout their range in Africa (summarized in Ward 1965b). Four-egg clutches normally comprise only 2–16 per cent of the total; yet, 52 per cent of females in the Botswana colony actually began yolk deposition in four follicles, only to resorb the fourth yolk at about the time the first egg was laid (Fig. 13.5). A simple model developed by King (1973) may help us to understand why this occurs. In birds such as quelea, where a succession of yolks begin development at daily intervals, the daily metabolic output required for the formation

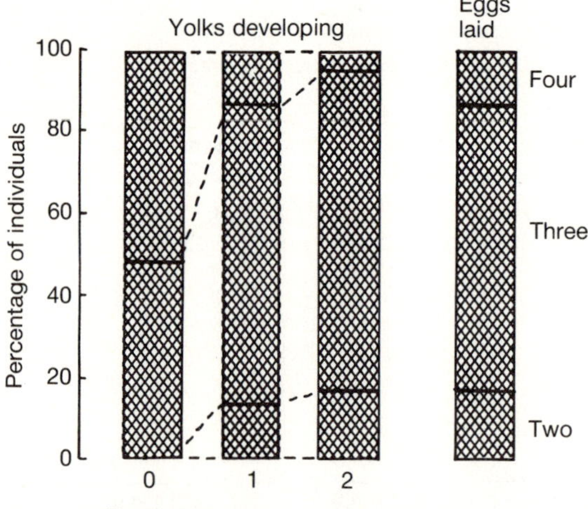

Fig. 13.5. The percentages of female quelea producing two, three, or four yolks on 3 successive days in their laying cycle, compared with the percentages of nests containing two, three, or four eggs observed in the colony during incubation. Sample sizes were 122, 90, 111, and 98, respectively (source: Jones and Ward 1976).

of the clutch increases to a peak over a period that is 1 day shorter than the time a single follicle takes to develop. This peak investment must, thereafter, be sustained daily for as long as new follicles begin enlargement. In quelea, where yolk deposition takes 4 days, peak investment is, therefore, reached on the third day (Fig. 13.6). It is apparent that quelea cannot sustain this level of investment. Their protein reserves start to decline as soon as yolk deposition begins and reach their lowest level shortly after the fourth yolk starts to enlarge. Although the fourth yolk begins development, its growth cannot be sustained, and it is soon resorbed, apparently in response to the reserves having reached some lower threshold limit. Perhaps a female whose reserves fell below this limit might die, or be too weak to rear her brood.

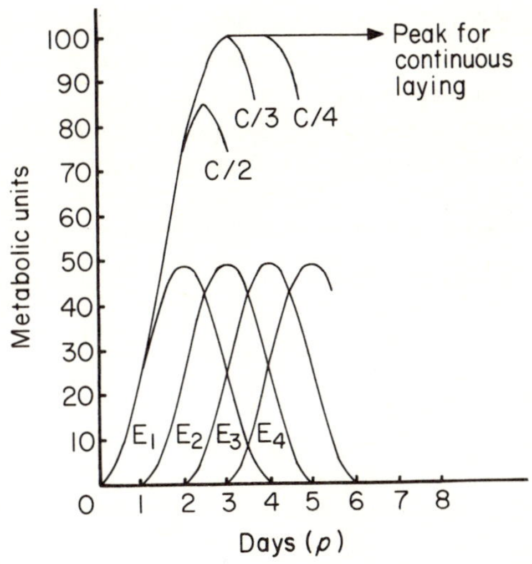

Fig. 13.6. Model of daily metabolic output required to produce the yolks of individual eggs (E_1–E_4) and complete clutches (C/2–C/4), where the total investment in a single follicle (area under each curve E) is an arbitrary 100 units. In quelea, the period of yolk formation, *p*, is 4 days with a 1-day laying interval; peak investment is reached on day $p-1=3$ and is equivalent to the total investment in a single follicle (source: King 1973).

The amount of material removed from the muscle during yolk formation is about the same as the total protein content of the yolks of a normal three-egg clutch. However, it is not known whether this material is translocated directly to the developing yolks, or whether it is used to maintain essential body processes while the daily protein intake in the food is diverted to the ovary. Whatever the function of the protein store, a high level is not a prerequisite for a large clutch. The initial level achieved before yolk development seems to be the same for a two-, three-, or four-egg clutch (Fig. 13.7) and, therefore, cannot determine clutch size. Similarly, the lowest level reached at the end of laying does not reflect the number of eggs laid. Instead, the rate at which the reserve is lost correlates best with clutch size; the protein reserve of females laying only two eggs is depleted most rapidly, while it is depleted most slowly by those laying four eggs. Therefore, the rate of depletion seems to act as the proximate control on clutch size. In other words, the greater the proportion of the total yolk material that a female can supply from her daily food intake, the smaller the amount of reserve withdrawn daily to supplement it; the reserve, therefore, declines more slowly and will last for more days, permitting a larger clutch to be laid (Fig. 13.8). This simple model points to a single high threshold level of reserve muscle protein which, once exceeded, both triggers yolk deposition and contributes to it. The rate at which it decreases, and hence the final clutch size, may then depend on the availability of protein-rich food and, perhaps, the foraging efficiency of the individual female.

The nutritional limit on egg production provides a proximate mechanism whereby clutch size could be adapted to the maximum number of young that

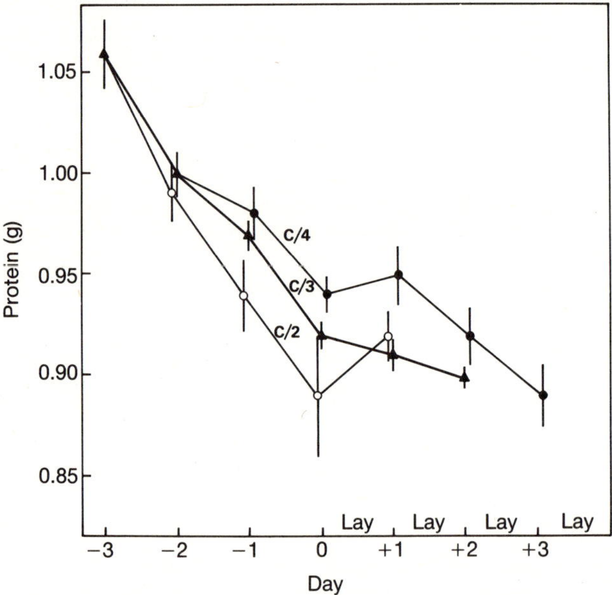

Fig. 13.7. Changes in the lean dry weight of the flight muscles during the laying cycle of female quelea producing two, three, or four eggs. Birds producing only two eggs could be distinguished from the remainder only 2 days (Day − 2) before the first egg was ovulated (Day 0), while birds laying three could be distinguished on the following day (source: Jones and Ward 1976).

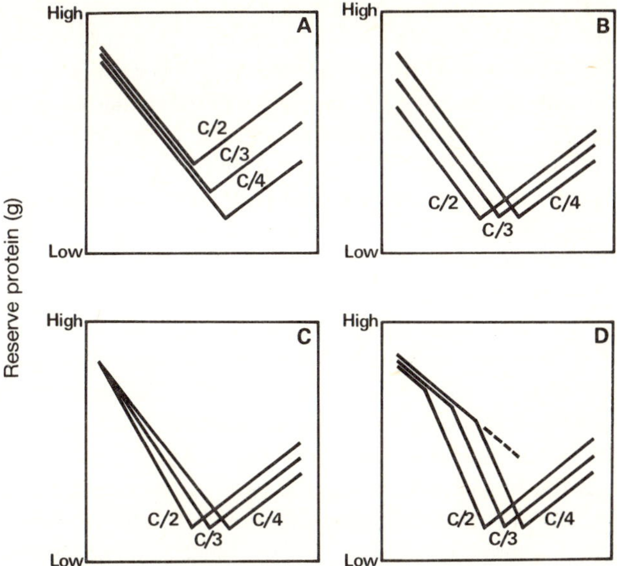

Fig. 13.8. Four possible models suggesting how the decline in protein reserves of a quelea during laying might determine the number of eggs laid. Neither Model A, where the protein reserve at the end of laying reflects the number laid, nor Model B, where the initial level predetermines the clutch size, are thought to apply (see Fig. 13.7). Models C and D are similar and both match the real situation. In each, yolk development starts from a single threshold and ceases at the same level, whatever the final clutch. The final clutch size is determined by the rate at which the reserve declines, though the patterns of decline may differ (source: Jones and Ward 1976).

the parents might successfully rear. In quelea, the most common clutch size of three eggs is also the maximum brood size that the parents can normally feed (Ward 1965*b*). If the food supply is poor or a female inefficient, she might be able to produce only two eggs. Nevertheless, if this is predictive of the circumstances still prevailing a fortnight later, as is quite possible, when the young require large amounts of insects that are in short supply, a small brood may be the most productive. Conversely, an efficient female with a plentiful supply of insects may lay a large clutch that she can easily rear.

Unfortunately, the existing data are barely adequate to test these ideas but do provide some support. Morel *et al.* (1957) suggested that in years when the rains start very late in the season, are less abundant than usual, and cause a reduction in the amount of food available, both clutch size and fledging success are lower than in years with good rain. In the Senegal valley in 1956 (a poor year), the frequencies of four- and five-egg clutches decreased by 14 and 44 per cent, respectively, compared with 1955 (a good year). Of these larger-than-average clutches, 97 per cent failed to fledge all their young, compared with 39 per cent in the good year. Clutches of two eggs increased 1.8-fold, and the frequency of broods with only one survivor more than doubled in the poor year. At the abandoned Nigerian colony in 1973, a year of extreme drought in the Sahel, 70 per cent of the females found dead were about to lay a clutch of only two eggs (Jones and Ward 1979). This colony was beyond the northern limit of successful breeding that year, but even in successful colonies further south the predominant clutch was only two eggs (W. Meinzingen, pers. comm.).

Whereas the attainment of a high threshold protein level is necessary before breeding can begin, a high fat level does not seem to be required. Although the lipid reserve is slightly higher than normal at the start of yolk development, it does not decline until the first yolk has been ovulated. It is then depleted rapidly, and the rate at which it declines is once more correlated with the clutch size (Fig. 13.9). Its primary function seems to be to supplement the daily energy intake in the food during a critical period, i.e. the days when the female is carrying the eggs of the clutch in turn in the oviduct, where they acquire their albumen and shell. The amount of protein constituting the albumen of each egg is four times that incorporated in the

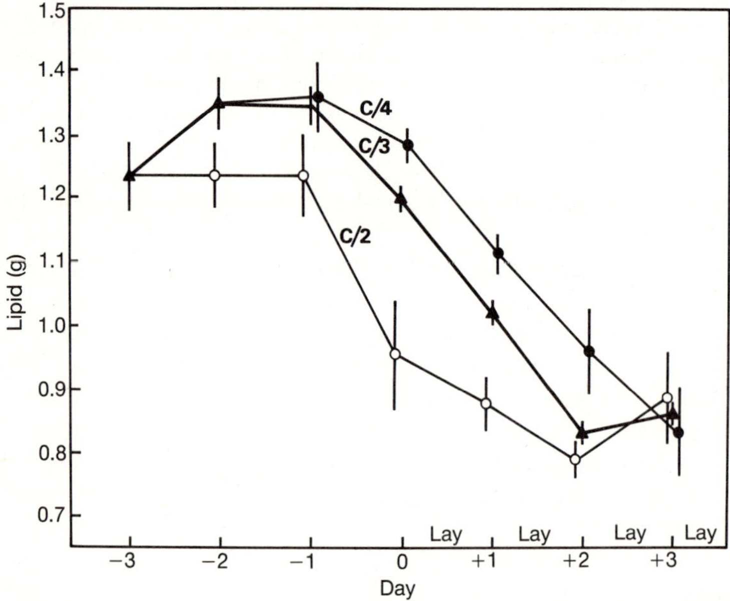

Fig. 13.9. Changes in the total lipid reserves (at dusk) during the laying cycle of female quelea producing two, three, or four eggs. Birds producing only two eggs could be distinguished from the remainder only 2 days (Day − 2) before the first was ovulated (Day 0), while birds laying three could be distinguished on the following day (source: Jones and Ward 1976).

yolk, while the three eggshells of a clutch together contain about as much calcium as the bird's entire skeleton (Jones 1976; Jones and Ward 1976). Neither of these components seems to be stored in the body, and it is known that females must forage specifically for calcareous material for eggshell formation on these days (Fig. 13.10). Presumably they must also forage for additional insect food to produce the large albumen component, which cannot come from the muscle store as this is by now totally depleted. Both of these requirements may be difficult to obtain and must be satisfied during foraging time that might otherwise be spent maintaining the bird's daily energy balance and depositing sufficient fat for overnight survival. The function of the fat store depleted during this period is presumably to subsidize the shortfall in energy intake that inevitably occurs on these days (Fogden and Fogden 1979; Jones and Ward 1979). The subsequent correlation with clutch size results from differences in food availability and/or the female's foraging efficiency.

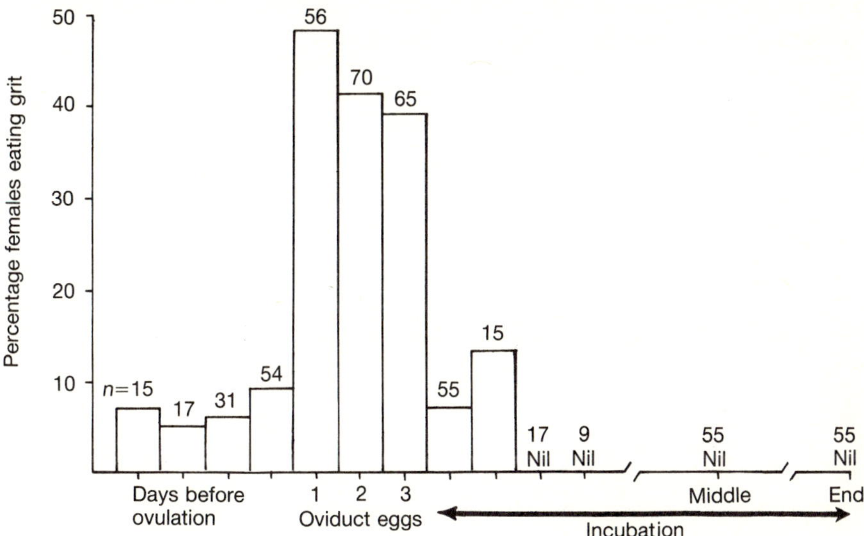

Fig. 13.10. Incidence of female quelea eating calcareous grit during the breeding cycle. Data are given for birds laying the normal clutch of three eggs. Males were never recorded eating such material (source: Jones 1976).

The role of the male

The photoperiod and cage breeding experiments described earlier, while showing that male quelea probably did not respond to changes in day length in the wild, did not properly identify the environmental factors to which they do respond. During the early rains, both male and female quelea begin the pre-nuptial moult involving the contour feathers of the head and chest. This is most obvious in the males, which acquire bright breeding plumage, and it is accompanied by development of the testes (Fig. 13.11). It is often observed that there is considerable asynchrony in the timing of these changes, with males in eclipse plumage being found in company with males in full breeding colours (Chapter 9 and Fig. 9.5). The factors stimulating this change are unknown, but the great variation in its onset make it unlikely that any widely operating environmental factors are involved, such as rainfall or the sight of green grass. It is probable that the early acquisition of breeding dress in the wild is dependent on nutritional status, since testis growth is correlated with the lean dry weight of the flight muscles (Fig. 13.12). Insects become available in small numbers in the late dry season as the humidity rises; once the rains break, the process of pre-migratory fattening allows some individuals to improve their protein reserves (Ward and Jones 1977). It is likely that not all quelea will benefit equally from improvements in feeding conditions, especially if these occur patchily at first, so that asynchrony between individuals might easily arise in this way.

The favourable nutritional balance that allows the start of the pre-nuptial moult may not be sufficient to permit a breeding attempt. While females

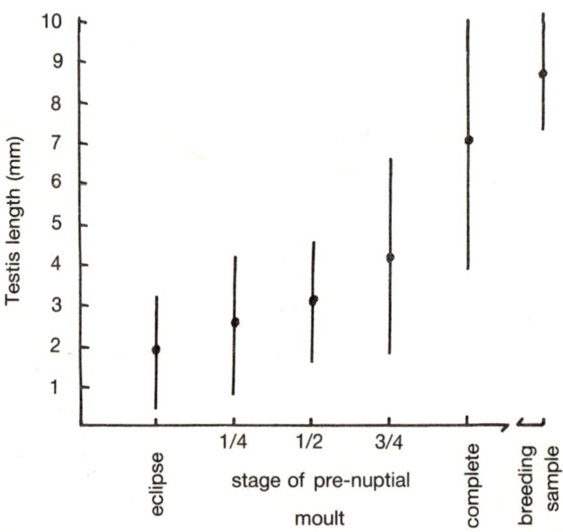

Fig. 13.11. Relationship between stage of pre-nuptial moult and testis growth in a sample of male quelea collected within 2 weeks of the start of the rains, compared with a later sample of breeding males. Means are given with 95% confidence limits of the individual measurements. Sample sizes are 94, 109, 36, 29, 55, and 100, respectively.

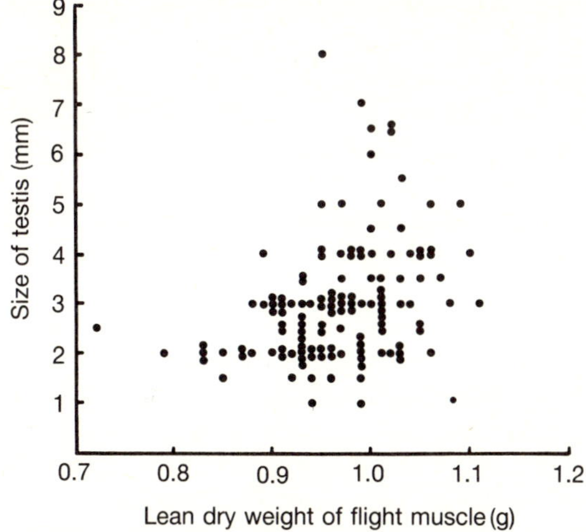

Fig. 13.12. Relationship between testis growth and reserve protein among male quelea during pre-migratory fattening in Botswana ($r = 0.351$, $P < 0.001$).

suffer great nutritional stress during laying, males likewise must experience a considerable loss of feeding time during the hectic period of nest building and nest-site defence, when any protracted absence for feeding inevitably means the theft by rivals of nesting material already collected.

The courtship period, when there is intense competition for receptive females that are usually fewer than the number of displaying males, involves many hours of continuous display, again without the opportunity to feed. It is not surprising, therefore, that males deplete fat and protein reserves during this period and, like females, accumulate reserves beforehand in anticipation of this depletion. However, while the initial fat reserves of males and females are similar and not particularly high, the protein reserve of the males is only 20 per cent of that accumulated by females (Jones and Ward 1976), and the depletion of protein in males by the end of courtship is not nearly as drastic as in females. Of the dead birds found in the Nigerian colony abandoned at the start of laying, only 5 per cent were males (Jones and Ward 1979).

Males are apparently not as limited nutritionally as females. To breed, females must acquire an initial protein reserve five times higher than a male and continue to forage for insects throughout the laying period. Males can much more easily achieve the nutritional status required to start breeding, and although they lose condition, they can subsist on a largely seed diet thereafter (Fig. 13.2; Jones and Ward 1976, 1979). This may account for the frequency with which groups of male quelea initiate colonies by constructing nests to the ring stage, only to abandon them after a day or so. Presumably females, although present in the area, have been unable to acquire the higher threshold level of reserve protein necessary for laying. This probably also explains the much greater ease with which males may be brought into full breeding condition in cage experiments. Females do not respond so readily.

Courtship feeding does not occur in quelea, so the male cannot supplement his mate's diet to enable her to lay sooner or alleviate the depletion of her body reserves that laying entails. Pairing occurs so late during egg development that courtship feeding probably would not help. Although it has been suggested that visual stimulation by vigorously displaying males serves to accelerate the female's sexual development (Marshall and Disney 1957), there is no evidence from the wild that the onset of breeding condition in the female is in any way dependent on the male. As pointed out above, females must arrive at the colony unmated, many of them with eggs already having begun development without social stimulation. Female quelea choose a mate and copulate only just before ovulation of the first egg. The very high degree of synchronization in laying within a colony probably occurs because only those females that already have rapidly developing yolks are attracted to displaying males; others try elsewhere when they are ready (Ward 1972*b*; R. Bruggers and M. Jaeger, unpubl. data). In rare instances, females that are ready to lay may arrive at a colony in greater numbers than there are males building nests. These females have no option but to lay in the few nests sufficiently complete to hold eggs. Up to 35 eggs have been found in the same nest, with many more on the ground beneath (Jones and Ward 1979; Morel and Bourlière 1955).

It seems unlikely in these circumstances that the physiological state of the males determines the readiness of females to breed, or even that social stimulation can accelerate the rate of yolk deposition in an already actively developing clutch. Yolk deposition is probably already occurring at its maximum rate set by the urgency of the nesting schedule and the limitations of the female's nutrition. Although the male must display vigorously and advertise a properly built nest to attract a mate, the female is by this time already committed to laying somewhere. If the breeding site is too poor, the attempt ends in failure.

Whether or not the male has been physiologically capable of breeding for weeks previously, either as a result of an endogenous gonadal rhythm (Lofts 1964), or because he was more easily able to exceed the necessary nutritional threshold, the precise timing of laying depends on the ability of the female alone to acquire a level of reserve protein that is higher than at any other time of year. This mechanism, which needs no environmental cues or releasers, ensures that quelea will breed only during the latter half of the local wet season and only in localities where they can find the large amounts of insect food that their young will demand.

Itinerant breeding and the end of the breeding season

It is probably normal in most parts of Africa for quelea to breed more than once in a season, although usually in a different place. Their ability to breed itinerantly depends on their capacity to maintain or regain the necessary reserves after they have been depleted by the previous breeding attempt. The recovery of protein and fat reserves during incubation prepares the parents for the difficult rearing period, when the demands of the brood may not allow them to satisfy their own metabolic demands every day. By the time the clutch hatches, the female's fat reserves at dusk may reach levels more typical of pre-migratory fattening (up to 2.5 g; Chapter 9), which she may then use to meet any shortfall in her daily energy balance while catching insects for her

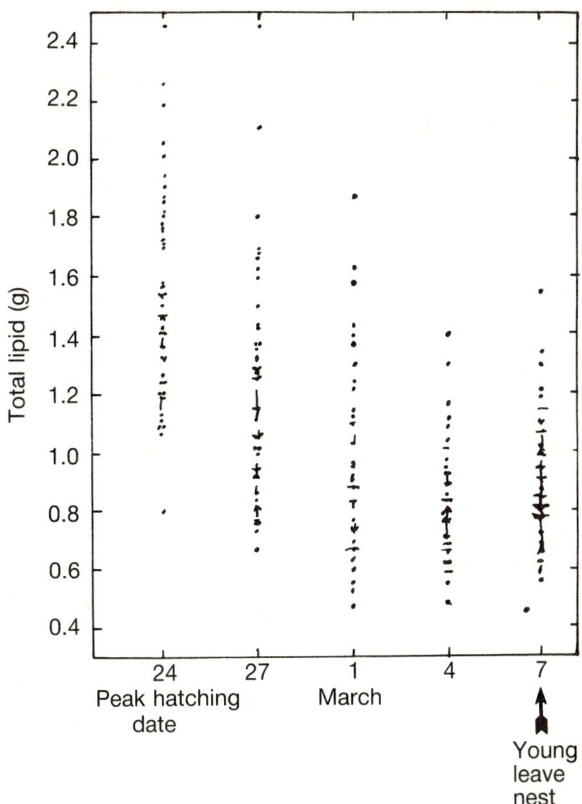

Fig. 13.13. Decline in the total lipid reserves (at dusk) of females during the rearing period. The base line at 0.3 g approximates to zero fat reserve (source: Jones and Ward 1976).

brood. The fat reserve is drawn upon by most females, for by the time the young leave the nest many of the females are in poor condition and some may once more reach starvation levels (Fig. 13.13). The protein reserves show similar, though not so dramatic changes. Females that remain in good condition during the rearing period are those that begin development of the next clutch immediately; up to 20 per cent may begin yolk deposition while still feeding well-grown young that are about to leave the nest. In most of these females the first follicle grows to about 4.0–5.0 mm in diameter, with the second showing some yolk deposition before they leave the colony (Ward 1971; P. Jones, pers. obs.). It is not known whether yolk development can be arrested at this stage. If not, these birds must urgently seek a new colony where they can mate and lay immediately on arrival. Other birds that complete the breeding attempt below peak condition, as is probably usually the case, take much longer to replenish their reserves to the threshold level necessary for laying.

The termination of breeding may be under equally simple nutritional control, but evidence is still lacking. It may be that the inability to regain condition after breeding is sufficient for birds to enter a refractory period with its accompanying gonadal changes. This, in turn, allows the post-nuptial moult to supervene. However, some quelea populations regularly encounter conditions once more favourable for breeding only many weeks after the last attempt and even after the post-nuptial moult has begun. In Ethiopia, about 2–3 months elapsed before quelea bred again up to 700 km from the original nesting area. During this period the post-nuptial moult had begun but had been interrupted to permit breeding (Jaeger *et al.* 1986). Such moult interruption is commonplace among eastern African quelea populations, but rare in southern Africa and has not been recorded in western Africa (C. Elliott, pers. comm.; P. Jones, pers. obs.). The opportunity for interrupted moult to occur and permit further itinerant breeding seems to depend on the length of the wet season within the region occupied by a particular quelea population. Once the wet season finally ends, quelea cannot regain the body reserves necessary for breeding, and the post-nuptial moult is completed into eclipse plumage.

14

Formation, sizes, and groupings of quelea nesting colonies

MICHAEL M. JAEGER,
RICHARD L. BRUGGERS, and
WILLIAM A. ERICKSON

Introduction

Red-billed Quelea nest in large, dense, synchronized colonies that can contain more nesting birds than colonies of any other living land bird (Lack 1968). These colonies are typically established in semi-arid savannah that has been temporarily transformed by rainfall into a luxuriant habitat with a short-lived superabundance of grass seed and insects. Quelea colonies are remarkable for the high degree of synchrony between the tens of thousands of nests, with almost all the nests being within 2 or 3 days of the same age as each other. They are also remarkable for their rapid turnover, the time from nest-building to abandonment by young being about 42 days. Records of incubation lasting 9–10 days are the shortest known for any bird (Lack 1966).

Lack (1966) found the reported nesting success of quelea (83 per cent of the eggs laid giving rise to young which left the nest) as extremely high for a small passerine. He attributed it to the comparative safety of the thorny vegetation usually used for breeding and the probable adaptive value of synchronization in 'swamping' predators. Ward (1972*b*) postulated that synchronization develops due to the simultaneous detachment of individuals in breeding condition from an overall large population consisting of individuals in all stages of breeding preparedness. Colonies may also serve as information centres on the location of patchily distributed food and water (Ward and Zahavi 1973) or as a mechanism for group cohesion and information maintenance during seasonal movements and subsequent breeding (Jaeger *et al.* 1986).

Breeding colonies are the principal targets of control operations and also have a general ornithological interest as uniquely large gatherings of land birds; yet, little has been published on colony formation, size, and groupings.

Such information is basic for developing local strategies of selective control (Chapter 22). In this chapter, we describe the areas and nest densities of ten colonies measured in Ethiopia from 1978 to 1981 and two colonies from Kenya in 1985. In addition, spatial and temporal relationships among groups of colonies in the two areas are described and the biological implications are discussed.

Measuring, sampling, and monitoring techniques

Colonies were located, mapped, and measured using techniques described by Elliott (1981*b*), Bruggers *et al.* (1981*a*), and in Chapter 4. The total number of nests in a colony was estimated either by simple random sampling of quadrats (eight colonies), stratified random sampling (one colony), or by random transects (three colonies). Quadrat size varied from 5 × 5 m in extremely dense thorny vegetation of *Acacia nilotica* to 50 × 50 m in relatively scattered *A. mellifera* thorn-bush. The number of quadrats sampled ranged from 31 (Weyto-B, Ethiopia), through 16 (Abidir, Ethiopia) and 10 (Galana-1, Kenya), to between 3 and 5 in remaining colonies.

The number of nests in each colony and the variance of the estimates for each sampling method were determined according to Seber (1973) (Table 14.1). In eleven of twelve colonies measured, productivity was determined by sampling between 40 and 1000 nests across all sampling quadrats to obtain the percentage of occupied nests. The number of nestlings per nest was determined in three colonies. Synchronization of nesting activities was estimated by ageing the oldest nestling according to Ward (1973*b*). Populations were sampled during different reproductive stages for routine data dissection as described by Ward (1973*b*).

Table 14.1. Estimating the number of nests in a colony and the sample variance (source: Seber 1973).

	Sampling method[a]	
Estimate	Random transect/random quadrat	Stratified
Number of nests	$\hat{N}=\bar{x}S$	$\hat{N}=\Sigma N_i$
Variance	$\sigma_N^2=S^2\frac{v}{s}\left(1-\frac{s}{S}\right)$	$\sigma_N^2=T_i^2\frac{v_i}{s_i}\left(1-\frac{s_i}{T_i}\right)$

[a]$\bar{x}$ = average number of nests in each quadrat sampled; S = the total number of quadrats in a colony; s = number of quadrats sampled; v = variance of quadrat counts ($x_i s$); T_i = number of plots in the ith stratum.

Rainfall and colony formation

In Chapter 13 it was pointed out that quelea breed only during the rainy season and then only after the rains are well advanced. Rain is a critical factor in producing conditions suitable for breeding. In Ethiopia, the main rains in the southern Rift Valley occur seasonally from March through September but are most intense between July and September, and in the Middle Awash Valley they occur mainly between July and August (Jaeger *et al.* 1979). In 1981, colonies in the southern Rift Valley were installed from mid-April to the end of May. The area is drought prone, but in 1981 rainfall was plentiful (Erickson and Damena 1982).

At Galana, Kenya, the two colonies were formed near the end of a prolonged nesting season that began in the general area of Tsavo East National Park in mid-November 1984 and ended in May 1985. The peak time for colony installation in the area typically is in late December/early January, which coincides with the normal rainfall pattern. In 1984/1985, the rains beginning in October were above average and in February were three times the normal average. The latter caused a burst of annual grass seed and insect production that allowed the Galana-1 colony to be formed unseasonably late.

The conditions triggering quelea breeding are not fully understood. Disney and Marshall (1956) suggested that the onset of the rains or the greening of

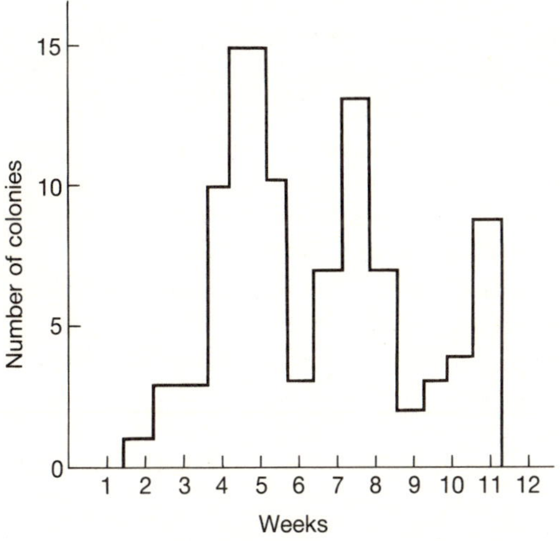

Fig. 14.1. The relationship between the number of quelea colonies installed and the elapsed number of weeks between the calculated onset of the rains and installation (source: Luder 1985*b*).

vegetation is responsible for gonadal incrudescence. Quelea need green grass leaves to construct their nests and ripening seeds and insects to feed young (Ward 1965*b*). In the Lake Chad Basin, Gaston (1973) concluded that a total of 200–450 mm of rain was necessary for the development of the annual grasses and subsequently the seeds, which form a crucial component of the food needed to support a colony. In Tanzania, Luder (1985*b*) determined that the installation dates of 72 per cent of 90 colonies were between 4–9 weeks after rainfall started (35 mm of rain falling in a 3-day period; Fig. 14.1). In this situation, colony installation peaked in the fifth and eighth weeks; a peak at ll weeks presumably was produced by prolonged rainfall such as resulted in the Galana-l colony in Kenya.

Nesting colony descriptions

Vegetation

The 12 colonies described here were situated in thorn-bushes or thorn-trees, most of which were 2–4 m high. Some, such as *A. nilotica*, reached 7–9 m (Table 14.2). Seven colonies were in *A. mellifera*, which is common nesting vegetation for quelea in eastern Africa (Disney and Haylock 1956; Disney and Marshall 1956); four were in other acacia species, and one was in *Terminalia brevipes*. Quelea seem to prefer the hooked thorns that make *A. mellifera* nearly impenetrable to most avian and mammalian predators (Fig. 14.2). Nests always were built within the protective tangle of thorns and from l m above ground to within l m of the tops of bushes.

In eastern Africa, quelea nesting in reed-thickets along river edges have been recorded regularly only in eastern and southern Tanzania (Kilosa and the Ruaha River at Mbarali, respectively) and in Cattail *Typha* spp. at both Lake Zwai and along the Awash River at Melkassa and Abidir (Jaeger *et al.* 1986). Colonies in tall grass such as *Sorghum macrochoeta* have only been recorded in the extreme south of Tanzania (Vesey-FitzGerald 1958). Breeding in sugar-cane has only been recorded since 1984 (near Kisumu, Kenya, C. Elliott, pers. comm.; at Joihar, NW of Mogadishu in 1987, J.-U. Heckel, pers. comm.). In western Africa, thorn-bush is also a typical nesting habitat, although colonies are regularly found in marshes of the Niger River Delta (Plate 11) and sugar-cane fields in Senegal (Morel *et al.* 1957; Ward 1965*b*).

Colony area

Quelea nesting colonies vary tremendously in size. Those reported in eastern Africa between 1978 to 1984 ($n = 266$) ranged from an estimated 0.25 to 114 ha (Table 14.3). Nearly 40 per cent of these colonies were less than 10 ha,

Table 14.2. Support vegetation in quelea nesting colonies described in Ethiopia and Kenya between 1978 and 1985.

Colony	Area (ha)	Nesting vegetation			Average no. active nests/tree or bush	Sample method[a]
		Species	Type	Average no./ha		
Ethiopia						
Weyto-B	14.0	*Acacia nubica*	Bushes, 2–4 m	NR[b]	–	RT
Weyto-D	8.1	*A. reficiens/A. nubica*	Bushes, 2–4 m	NR	–	RT
Abidir	8.0	*A. nilotica*	Trees, 7–9 m	NR	–	RT
Nura Hera	12.0	*A. mellifera*	Bushes, 2–4 m	NR	–	RQ
Awore Melka	7.0	*A. nilotica*	Trees, 7–9 m; bushes, 2.0–2.5 m	725	127	SRQ
Mulu Marsh	14.1	*Terminalia brevipes*	Trees, 3–5 m	980	48	RQ
Issa Plain	77.5	*A. mellifera*	Bushes, 2–4 m	64	40	RQ
Denku	12.1	*A. mellifera*	Bushes, 2–4 m	214	70	RQ
Omar	41.1	*A. mellifera/A. reficiens*	Bushes, 2–4 m	74	59	RQ
Kenchero	9.1	*A. mellifera/A. reficiens*	Bushes, 3–5 m	180	198	RQ
Kenya						
Galana-1	40.0	*A. senegal/A. reficiens/A. mellifera*	Bushes, 2–4 m	143	18	RQ
Galana-2	10.0	*A. senegal/A. reficiens/A. mellifera*	Bushes, 2–4 m	97	24	RQ

[a] RT, random transect; RQ, random quadrat; SRQ, stratified random quadrat.
[b] NR, not reported.

Fig. 14.2. Dense colonies are established in stands of acacia trees (photo: FAO).

Table 14.3. General description of nesting colonies of quelea in Africa (sources: FAO 1979*a*, 1980*b*, 1984*b*).

Country	Year	No. colonies	$\bar{x}$ colony area (ha) (range)	$\bar{x}$ estimated population/ colony (in millions)	Support vegetation
Tanzania	1978	30	24	–	*Acacia* spp.
	1979	33	25 (3–100)	–	*Acacia* spp.
	1980	30	23 (5–114)	–	*Acacia* spp.
	1981	27	17	–	*Acacia* spp.
	1982	67	11 (0.5–60)	–	*Acacia* spp.
Ethiopia	1979	4	39 (3–80)	–	*Acacia* spp., *Typha* spp.
	1980	4	21 (10–40)	–	*Acacia* spp., *Typha* spp.
	1981	21	14 (0.25–60)	–	*Acacia* spp., *Typha* spp.
Somalia	1978–79	4	48 (12–78)	1.2	*Acacia* spp.
	1981	8	36 (10–70)	3.0	*Acacia* spp.
	1982	35	35 (10–100)	–	*Acacia* spp.
Kenya	1980	1	9	3.0	*Acacia* spp.
	1981	2	15 (10–20)	2.4	*Acacia* spp.

and about 80 per cent less than 40 ha. Our 12 colonies averaged 21 ha (7.0–77.5 ha; Table 14.2), the same average size as 7 colonies Thiollay (1978*a*) measured in Chad and Cameroon. Similarly, Ward (1965*b*) noted that colonies in Nigeria usually ranged from 15 to 30 ha. With the exception of Mali, the average size of 44 other colonies measured in Senegal, Mauritania, and Chad was similar (17–19 ha; Table 14.4).

Table 14.4. General description of nesting colonies of quelea in western Africa (source: unpubl. progress and trip reports of UNDP/FAO project RAF 73/055).

		Colonies		
Country	Year	No.	Area $\bar{x}$ ha ± SD (range)	Support vegetation
Senegal	1967–77	18	17 ± 17 (1–70)	*Acacia* spp., *Typha* spp., sugar-cane
Mali	1973–77	23	63 ± 144 (1–600)	*A. senegal* *A. pennata* *A. nilotica* *A. sieberiana*
Mauritania	1967–77	16	16 ± 23 (1–80)	NR[a]
Chad	1975–78	10	19 ± 12 (3–36)	*A. nilotica*

[a] NR, not reported.

Nest density

Nest density varies widely both within and among colonies and depends mainly on vegetation species and its height and density. For example, colonies established in *A. mellifera* bushes are often large in area but contain a low nest density because bush clumps are usually scattered. Table 14.2 depicts the difference in *A. mellifera* density between the Issa Plain colony, with an average of 64 bushes/ha, and the Awore Melka colony that was continuous bush and averaged 725 trees/ha. The nest density at Awore Melka was estimated at 92 210 nests/ha compared with 2587 nests/ha on the Issa Plain (Table 14.5). Bushes in the Issa Plain colony were relatively small (2–4 m high); the mean number of nests per bush was 40 with a maximum of 161. At Kenchero, the larger bushes (3–5 m high) averaged 198 nests/bush with a maximum of 898 nests in a single bush; the estimated nest density was 35 720 nests/ha. Overall, in our 12 colonies, the average nest density among colonies was 24 954 nests/ha (Table 14.4), and the average tree density among the eight colonies where counts were made was 310 trees/ha (Table 14.2).

Table 14.5. Description of the area, number of nests, nest density, and number of breeding birds in quelea colonies sampled in Ethiopia and Kenya between 1978 and 1985.

Colony	Area (ha)	Estimated total nests	Coefficient of variation (%)	Active nests (%)	Active nests/ha	Average no. nestlings/ nest	Estimated no. nesting adults[a]	Estimated no. young
Ethiopia								
Weyto-B	14.0	195 104	13	98	13 657	2.8	382 396	535 365
Weyto-D	8.1	121 759	13	95	14 280	2.7	231 336	312 312
Abidir	8.0	541 848	42	95	64 344	NR[b]	1 029 504	1 420 725
Nura Hera	12.0	71 680	51	87	5 197	NR[b]	124 728	172 118
Awore Melka	7.0	777 675	20	83	92 210	NR[b]	1 290 940	1 781 498
Mulu Marsh	14.1	783 396	22	85	47 226	NR[b]	1 331 773	1 837 847
Issa Plain	77.5	217 930	23	NR[c]	2 587	NR[b]	400 985	553 368
Denku	12.1	191 974	18	94	14 914	2.9	360 919	523 321
Omar	41.0	186 435	50	96	4 365	2.8	357 930	501 137
Kenchero	9.1	342 160	43	95	35 720	2.6	650 104	845 135
Kenya								
Galana-1	40.0	114 912	17	90	2 586	NR[b]	206 880	285 441
Galana-2	10.0	25 928	19	91	2 359	NR[b]	47 180	65 121
$\bar{x}$ (±1 SE)	21.1 (6.16)	297 567 (75 762)		92 (1.48)	24 954 (8 419)	2.76 (0.05)	534 556 (128 312)	736 116 (176 513)

[a]Assume one male and female adult per nest.
[b]Assume 2.76 young per nest.
[c]Assume 92% occupancy.

Nest densities reported elsewhere in Africa indicated high variability among colonies. In Somalia, for instance, estimates ranged from 2892 to 8210 nests/ha among four thorn-bush colonies (J. S. Ash, in Bruggers 1980). Morel (1968) reported an average of 12 400 nests/ha in a colony in Senegal, and Thiollay (1978*a*) estimated 13 500 nests/ha among seven colonies in *A. nilotica.* In colonies located in wild sorghum and *Echinochloa* sp., Vesey-FitzGerald (1958) estimated about 5.5 nests/m^2 or 55 000 nests/ha. The highest estimate of nest density was 141 000 nests/ha in a small colony in *Typha* spp. (Fuggles-Couchman 1952 in Vesey-FitzGerald 1958), which is comparable to the highest density found in Ethiopia of 145 200 nests/ha estimated for a discrete 2.5-ha portion of the Awore Melka colony (Table 14.2).

Colony productivity

Our 12 colonies averaged 297 567 nests (Table 14.5), close to the average of 285 700 found by Thiollay (1978*a*) in Chad and Cameroon. Active nests averaged 92 per cent (range 83–98 per cent). Nesting adults ranged from 47 180 in Galana-2 to 1 331 773 at Mulu Marsh with an overall average of 534 556. Nest activity of 84–99 per cent was found in colonies in Somalia (J. S. Ash, in Bruggers 1980).

Mean clutch sizes of quelea reportedly range from 2.0 to 3.8 eggs per nest with clutches of 2.8–2.9 most common (Jarvis and Vernon, in press-*b*; Jones and Ward 1976; Thiollay 1978*a*; Ward 1965*b*). However, egg and nestling mortality occur (Chapter 16). In southern Africa, Jarvis and Vernon (in press-*b*) reported mean clutch sizes of 2.8 eggs in two colonies but only 2.1-2.4 chicks were successfully raised by each pair of adults. Three colonies in Senegal produced 2.1–2.7 young per nest (Morel *et al.* 1957). The estimated numbers of young produced in our colonies (Table 14.5) ranged from 65 121 at Galana-2 to 1 837 847 at Mulu Marsh, with an average of 736 116 per colony. Mean brood size was 2.8 young per nest. Combined numbers of adults and young averaged 1.3 million birds per colony.

Groups of colonies

Characteristics

Nesting colonies frequently occur in groups of two or more active colonies (Jaeger *et al.* 1986). The two Galana colonies in Kenya, for example, were 10 km apart, and a third colony was only 2 km from Galana-1 and 9 km from Galana-2. The Galana-1 colony was actually two adjoining colonies at different stages of development. Egg-laying was synchronous within each of these colonies but not between them. The Galana-2 colony was the most advanced, egg-laying having occurred about 1 week prior to that in the third colony, 2 weeks prior to that in the older section of Galana-1, and 3 weeks prior to that in the younger section of Galana-1. Therefore, at least four active nesting colonies occurred simultaneously within a 5-km radius.

A group of active nesting colonies might be considered a 'supercolony'. To determine whether or not there is meaning to this concept, it is necessary to know if such groups occur regularly, if they occur because of lack of suitable nesting vegetation or due to clumped food resources, and if they are comprised of a characteristic number of colonies. Multiple colony groups have been found throughout eastern Africa with group sizes of two or three colonies most common. In the Ethiopian Rift Valley, for instance, 23 nesting colonies were found in 1981 (Jaeger *et al.* 1986). Based on spatial and temporal patterns, all colonies appeared to be part of groups: five groups of two colonies, three groups of three, and one of four. Multiple colony groups have previously been recognized only as reflecting the patchy distribution of the local nesting vegetation. However, multiple colony groups also occur where nesting vegetation is continuous or in large patches, such as in the Tuli Block area of Botswana (R. Bruggers, pers. obs.). In addition, the staggered timing of colony establishment within a group, such as at Galana, suggests that the distribution of nesting vegetation is not the cause for separate colonies. On a larger scale, the distribution of nesting habitat is often patchy, and nesting colony groups can themselves be clustered. Four clusters of colonies were found, for example, in the Middle Awash River Valley of Ethiopia in 1981.

Formation of multiple colony groups

Present evidence suggests that multiple colony groups form as a succession of individual colonies from a large pre-nesting aggregation. Enormous flocks have been observed swarming in the general area when nesting colonies are being established (Ward 1972*b*). From one such pre-nesting swarm of birds located in the Weyto River Delta of southern Ethiopia in 1981 two adult male quelea were captured and outfitted with miniature radio transmitters (Bruggers *et al.* 1983). Four days later, these birds were found at a newly established colony where nest building had just begun (colony B; Tables 14.6 and 14.7). About 200 m away was a second colony (A), only 0.4 ha in size, which had been established approximately 35 days earlier. From each colony, two more adult males were instrumented with radio transmitters. Six days after colony B was established, three of the instrumented quelea were tracked to a third colony (C), 30 km to the north of B, where nest building had just begun. Colony C was smaller than colony B, and it appeared to have been formed by birds that had departed colony B 2 or 3 days earlier, leaving

Table 14.6. Listing of numbers of adult quelea (M = male; F = female) that were radio-equipped and tracked during various stages of the nesting colony cycle in the Sagon–Weyto River Delta, Ethiopia, 1981. The date and location of attachment for each bird is indicated.

Colony		Radio-equipped quelea					
Stage	Age (days)	M	Radio no.	F	Radio no.	Date instrumented	Location[a]
Pre-nesting	–	2	51, 55	0	–	17 May	Feeding flocks
Nest-building/egg-laying	1–4	2	53, 57	0	–	22 May	Colony B
	1–4	1	58	0	–	31 May	Colony C
Incubating	5–14	0	–	0	–	–	–
Care nestlings	15–24	4	60, 61, 62, 63	2	59, 64	3 June	Colony B
Care fledglings	25–43	4	65, 69, 73, 74	2	68, 72	10 June	Colony B
Adults abandoning	43	2	54, 56	0	–	22 May	Colony A

[a]Tracking dates: Helicopter — 17, 18, 21, 22, 23, 25, 26, 28, 30 and 31 May; 3, 5, 8, 10, 11, 12, 15, 19, 20, 22, 24, 25, 27, and 28 June; ground — 22, 26, 28, and 31 May; 3, 5, 8, 10, 11, 12, 19, 20, 24, 25, 27, and 28 June.

Table 14.7. Description and chronology of four quelea nesting colonies found in the same area in the Sagon–Weyto River Delta, Ethiopia, 1981.

Colony	Location	Area (ha)	Estimated no. completed nests/colony	% nests occupied	Average no. eggs/nest	Estimated population ($\bar{x}$ 1000)		Approximate date	
						Adults	Young	Nest construction	Colony dispersal
A	5°01′ N × 36°58′ E	0.4	20 000	40–50[a]	Not determined	1–2	3–5	13–15 Apr.	1–2 June
B	5°01′ N × 36°57′ E	14.0	196 000	98	2.82	464	560	19–21 May	28–30 June
C	5°16′ N × 37°03′ E	2.1	1 652	0	0.00[b]	50–80	0	27–29 May	28–30 May
D	4°59′ N × 36°58′ E	8.1	120 176	95	2.70	240	247	30 May–2 June	10–12 July

[a] Estimated from the presence of droppings in nests.
[b] Colony abandoned before laying completed.

behind many unfinished or unoccupied nests. Colony C was completely abandoned after 3 days, possibly due to the disappearance of nearby water. The three instrumented quelea were then tracked to a fourth colony (D), 5 km from colonies A and B, where nest building was underway. Colony D later appeared to be two adjoining colonies, 3 days out of phase with one another. These findings suggest that colonies B and D were formed successively over a 10-day period by splitting of the original pre-nesting aggregation. Both colonies were successfully completed.

Unoccupied nests along the periphery are a common feature of quelea colonies. Ward (1965*b*) assumed that these unused nests resulted from colony contraction as its centre becomes defined, and birds abandon peripheral sites for more central ones. We found that the quelea that left colony B were not ready to breed, suggesting that multiple colony formation results from a stepwise sorting of birds in similar reproductive condition into individually synchronous colonies.

Reproductive synchrony within quelea colonies, or sections of colonies,

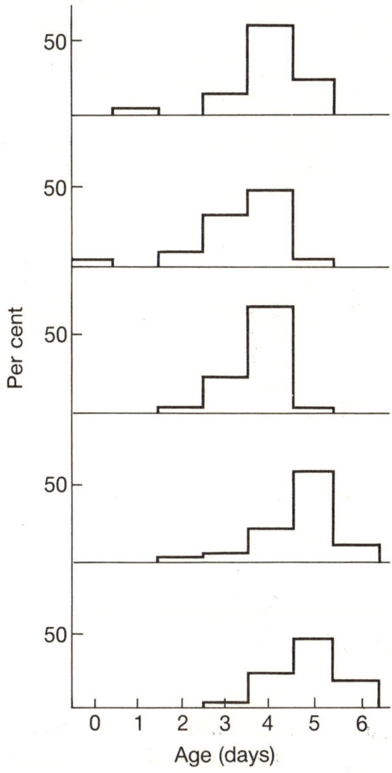

Fig. 14.3. Age of largest nestling from 100 nests in each of five randomly selected blocks of colony B in the southern Rift Valley, Ethiopia, 10 June 1981.

has been well documented. The synchrony of nestling development in colony B is illustrated in Fig. 14.3. Although the average age of the largest nestling from each of the five different areas on 10 June was significantly different ($P<0.05$), the age range was only 0.8 day ($\bar{x}=4.1$–4.9 days of age) across areas. Adult females sampled from both the pre-nesting aggregation and from colony B on the day preceding onset of egg-laying and prior to the formation of colony C were more asynchronous in their follicle development (Fig. 14.4). From those collected at colony B, approximately 60 per cent had follicles of 4–9 mm in length, indicating that they might soon lay eggs. A similar pattern was found in a sample of pre-nesting females collected at about the same time along the Omo River. Ward (1965*b*) also noted a lack of synchronization of pre-nuptial moult and gonad size in a pre-nesting aggregation of quelea.

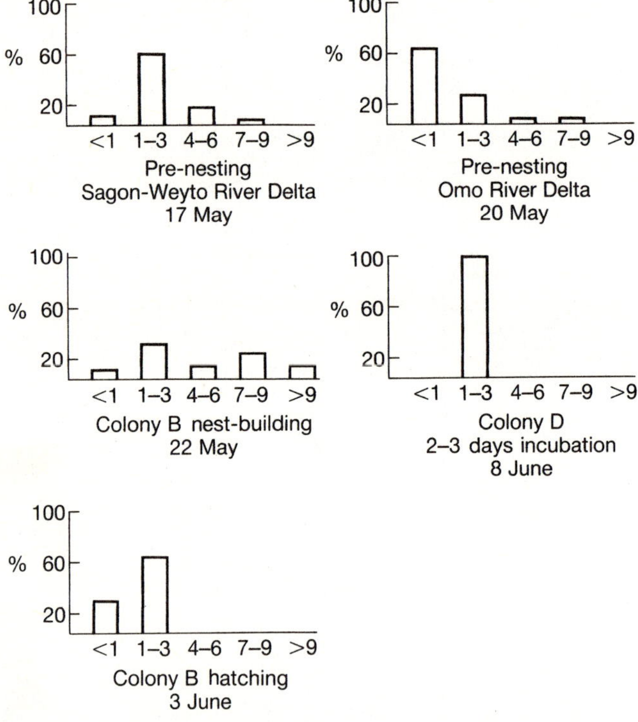

Fig. 14.4. Percentage quelea females with largest follicles in various size classes between 1 and 19 mm in length. Birds were collected in pre-nesting flocks or from inside colonies during the different developmental stages in southern Rift Valley, Ethiopia, 1981 ($n=50$–159 birds/collection).

Roosts associated with nesting colonies

Non-breeding quelea

Non-breeding adults will join a night roost occurring at a colony during its formative stage, as we described for colony B; but, presumably, these non-breeding birds depart to roost elsewhere as egg-laying and incubation begin. Their departure may be due to a general breakdown in roosting behaviour resulting from breeding birds becoming more attentive to their clutches. The behaviour of post-breeding adults towards nearby nesting colonies can vary. For example, radio-equipped adults from colony A did not immediately depart the area, but instead moved and roosted with other birds near the three other colonies. One was found at a night roost near colony C the day before that colony was discovered, and then for a short time in colonies C and D. In contrast, of nine adult quelea that were instrumented during the post-breeding stage in Galana-2 colony at the time of its completion, none were later found at the two younger colonies nearby.

Juveniles often remain in the vicinity of nesting colonies for days and sometimes weeks after adults depart. This segregation from adults can be maintained for months following breeding (Jaeger *et al.* 1979). Because of this tendency for post-breeding segregation, we assume that juveniles do not generally attach themselves to other nearby colonies. Juveniles from colony A were not found in later samples from either colonies B or D, although some were collected from the short-lived colony C. One to two months after breeding in this general area of southern Ethiopia, the adults migrated north into the Awash River Valley to breed a second time (Jaeger *et al.* 1986). Only a small percentage of the quelea collected in the Awash colonies were juveniles that had been produced earlier in the south. Many of these juveniles had begun moulting directly into adult breeding plumage and had interrupted the post-juvenile moult of the primaries. Adults that were nesting for the second time had also interrupted the post-breeding moult of the primaries. Juvenile males had enlarging testes but juvenile females, in contrast to adults, had small follicles. J. Thompson and M. Jaeger (unpubl. data) have found evidence in Kenya for breeding by quelea that have not yet completed the post-juvenile moult.

Late-colony roosts

Roosting and flocking behaviour by adults reappears in the late stages of breeding, a week or two before adults depart a colony. This is most pronounced with males and becomes apparent when young begin flying. At this time, night roosts of adults are commonly formed away from the colony. During the late fledgling stage of colony B, four radio-equipped adults were found using a night roosting site about 1 km from the colony. Late-colony roosts of adults are also found within colonies. The Galana-1 colony had a very large roost within it that appeared to include adults from the older colony nearby.

Biological implications

Synchrony is the most consistent feature of the colonies discussed in this chapter. The potential adaptive advantages of synchrony are generally attributed to minimizing losses to predators (Lack 1968) or improving communal food finding by young (Ward and Zahavi 1973). In addition to these functions, synchrony may be important for maintaining group cohesiveness for successive breeding. Results from marking studies in Ethiopia suggested that post-nesting group cohesion of adults exists and that these groups migrate independently of each other (Jaeger *et al.* 1986). P. Jones (pers. comm.) reported that quelea ringed in South Africa were captured together after a period of 18 months.

The biological implications of multiple colony groups also are intrigueing. These groups may result from the strong tendency for reproductive synchrony of birds within individual colonies. The finding of a supercolony whose components are out of reproductive synchrony with one another, further suggests that the role of synchrony may be more complex than simply minimizing the time spent nesting in an area to reduce the potential impact by predators (Jaeger *et al.* 1986). If, for instance, all quelea in an area nest in the same colony, it may increase the risk of overall failure. This could easily have happened if all birds in the Weyto–Sagon River Delta in Ethiopia in 1981 had tried to nest near the location where colony C was attempted and later abandoned due to drying up of the standing water, desiccation of grasses, and subsequent lack of insects. Therefore, one potential advantage of birds nesting in multiple colony groups dispersed in time and space may be to increase the chances of some of the colonies in the area being successful. This, in fact, was the situation in the Weyto–Sagon area.

Management implications

Surveys for quelea nesting colonies can be done more efficiently with an understanding of the vegetation and topography that characterize nesting sites. Nesting vegetation will vary with habitat; for instance, where marshes predominate or dense acacia is absent, colonies are commonly in *Typha* spp. Throughout much of their range, however, quelea nest in semi-arid savannah where colonies usually occur in discrete patches of *Acacia* sp. thorn-bush along seasonal drainages and often at the base of a slope where grass growth is relatively pronounced (Chapter 4). These patches of thorn-bush are easy to recognize from the air by an experienced observer. In addition, nesting colonies tend to re-occur in the same areas each year, if food conditions are favourable (Jaeger *et al.* 1986). Therefore, it is useful to keep accurate records of the locations of nesting colonies on maps or aerial photographs.

Nesting colonies frequently occur in groups of two or three, so that where one is found, survey efforts should be intensified as others probably also occur. Radio-telemetry can be useful for locating colonies in rough terrain (Chapter 6; Bruggers *et al.* 1983) or sufficiently early during their establishment to permit control to be conducted prior to young fledging (after which non-target predatory birds become more numerous [Bruggers *et al.*, in press]) and adults begin departing the colony.

Nesting colonies show great variability in size, nest density, and bird numbers. Therefore, not all colonies are of equal value to control. Often the success of a control campaign is based on the number of colonies sprayed and the volume of avicide used, measures which probably bear little relationship to the numbers of birds sprayed or the amount of crop saved. As an example, the largest colonies measured here (Table 14.4) in terms of area were among the smallest in numbers of nesting adults, and vice versa ($r=0.25$, not significant). A better indicator of the number of nesting adults was nest density/ha ($r=0.91$, $P<0.001$). Generally, aerial spraying is more effective on the smaller and more densely populated colonies. Nesting quelea scatter away from the colony after only two or three passes of the spray aircraft so that additional passes required of larger areas have less and less effect. Large colonies often require two or more nights of spraying for a satisfactory result. Aerial spraying then becomes less efficient as colonies get larger in area and lower in bird density. Aerial spraying of colonies is more effective if directed against the moving mass of birds as opposed to spraying colony swaths.

15

Quelea population dynamics

PETER J. JONES

Introduction

The longstanding and widespread interest in bird protection in Europe and North America has encouraged a popular belief that birds are especially vulnerable to the adverse effects of man's influence on nature. An increasing number of species are under threat of extinction, and some notable extinctions of once numerous species have already occurred, such as the Passenger Pigeon *Ectopistes migratorius* of North America. It seems to have been a similar belief among colonial governments in Africa that quelea, despite their vast numbers, would be equally vulnerable if sufficient effort were devoted to their eradication. Their habit of sometimes aggregating in huge flocks that roost communally and nest colonially, often in traditional sites, made them an easy target for control. Modern technology and lethal chemicals could achieve spectacular kills, and it seemed obvious that mass extermination should soon reduce quelea populations to levels where they could no longer cause serious agricultural losses.

It should have been no surprise that mass eradication failed to control quelea numbers. Early research workers repeatedly emphasized the species' great reproductive potential as, for example, when extravagant attempts were made to calculate the damage to crops that would result if quelea were allowed to breed unchecked (Grosmaire 1955). They seemingly failed to recognize that if quelea were endowed with such a high potential rate of increase, they would soon make good any losses due to mortality imposed by man, so that population reduction would be unlikely to offer any long-term relief from damage. In undisturbed populations, high reproductive rates are generally compensated by high mortality rates. If a species is to be harvested for man's benefit, this life history specification is a useful one; a large 'doomed excess' permits a high rate of culling without endangering the population (Chapter 23). If like quelea, the species is a pest, it is not enough simply to cull the excess. If a worthwhile reduction in population is to be achieved and maintained, artificial mortality must be imposed at a rate that is significantly greater than that which occurs naturally; otherwise a compensatory reduction in natural mortality might redress the balance. Mortality must, therefore, be imposed at a rate that cannot be compensated for by the reproductive output of the survivors.

The failure to control quelea populations was attributed initially to inadequate technology, although this was largely rectified by the replacement of ground-based explosive techniques with sophisticated aerial spraying. Then, following the realization that quelea are long-distance migrants on a large scale, failure was attributed to the awesome logistical (and political) difficulties faced by control teams in achieving sufficient coverage. Only recently have people begun to appreciate that the dynamics of quelea populations may be sufficiently robust to withstand the levels of control that most countries can afford.

In order to understand the demography of a population and to make predictions about its increase or decrease over time, we need to measure the probabilities of survival and rates of reproduction of individuals at each age, and to summarize these values in the form of a life table that can be subjected to standard mathematical analyses (Caughley 1977; Ricklefs 1973). Despite appeals at the international quelea meetings of the 1950s and 1960s for research on quelea population dynamics, no serious attempt was made to assemble the data already available (e.g. on reproductive rates), nor to collect the information that was lacking (e.g. on mortality rates). Hence, no attempt was made to model the dynamics of quelea populations, or to predict their demographic responses to the policies of total eradication that had already been put into effect. This chapter presents the still rather meagre information on quelea reproductive rates and mortality and attempts to assess the level of control over quelea populations that may realistically be attained.

Population statistics for quelea

For a complete description of the demography of a quelea population, information is required on sex ratio, the number of eggs laid per clutch, the number of clutches per year, the number of young successfully reared to fledging per clutch, the proportion of fledged young surviving to the age of first reproduction, and the proportion of adults surviving through each age interval thereafter.

Sex ratio

Usually, only the dynamics of the female segment of a population are considered, because it is much easier to determine the number of daughters to which a female has given birth than the number of sons fathered by the male.

In monogamous species where both sexes care for the young and the sex ratio is 1:1 at all ages, the dynamics of the female segment are the same as those of the males. In promiscuous and polygamous species it would be almost impossible to determine the fecundity of males; likewise, if the sex ratio is greatly skewed at any age, the analysis is more complex.

Quelea are monogamous, both sexes care for the young and the sex ratio at hatching is 1:1 (Morel and Bourlière 1955; Ward 1965*d*). However, after fledging the sex ratio may become locally biased for reasons that remain unclear. In Nigeria and Senegal, males were slightly more numerous in samples of recently fledged young and markedly predominated in samples taken toward the end of the dry season (Morel and Bourlière 1955; Ward 1965*d*). This was not the case in samples from the Lake Chad region, which showed a slight bias toward females (Manikowski 1980). In Botswana, a sample of moulting juveniles showed a significant bias toward females but by the end of the dry season, the sexes of birds undergoing premigratory fattening showed no significant difference from 1:1 (P. Jones, unpubl. data).

Ward (1965*d*) suggested that food scarcity during the dry season may cause higher mortality among females, but this explanation evidently does not always apply, nor does it account for the return to unity in the sex ratio of adult quelea during and immediately following breeding. It is more likely that sexual segregation during the dry season gives rise to biases in individual samples. A study in Ethiopia has shown that a much wider dispersal of females after breeding leaves behind local concentrations of quelea that are significantly male biased, while the overall sex ratio remains close to unity (Jaeger *et al.* 1979). More females than males die during breeding, but the numbers of birds involved seem to be very small (Jones and Ward 1979; Thiollay 1978*a*).

In species where female survivorship is lower than that of males, there may be an imbalance at breeding, with a 'floating' population of non-breeding males. Even if this occurs in quelea, their numbers would be impossible to determine. Potential irregularities in life tables caused by sex-related differences in survival are taken care of by ignoring males. The quantitative implications for the life table of unbalanced sex ratios are unknown, and for the present analysis we must assume that any imbalance is not serious.

Clutch size

Although the breeding range of *Quelea quelea* extends over a vast area of Africa between 17°N and 32°S and from 17°W to 47°E (Magor and Ward 1972), the most common clutch size everywhere is three eggs (Ward 1965b). Very occasionally clutches of four eggs may comprise up to 70 per cent of the total, presumably when breeding conditions are particularly good; when conditions are unusually poor, clutches of two may predominate (Jones and Ward 1976, 1979). In most colonies the mean clutch size is 2.7–3.1 eggs and for the present purposes it is convenient to take a mean of 2.8 eggs, so that the value used in the analyses will be 1.4 daughter eggs per female per clutch.

For almost all species studied that lay a variable clutch, it has been shown that females lay fewer eggs in their first breeding season than in later years (Lack 1966). Ideally, age-specific fecundity rates should be determined for all ages, but this has not been done for quelea and would require close monitoring of a marked study population over several years, clearly an impracticable task.

Number of broods per season

All females probably make at least one breeding attempt a year. Among populations where a strategy of itinerant breeding would permit more than one brood to be raised successfully, a variable proportion might make two or three attempts, depending on annual variations in local conditions (Chapter 10; Ward 1971). The proportion of successful attempts in any one year is unknown, however, and in some regions quelea control operations must considerably disrupt breeding.

Number of young reared to fledging

In undisturbed colonies there are always nests, mostly on the periphery, that are not completed. These may comprise up to 10 per cent of the total and as much as 14 per cent of completed nests may remain empty (Jackson and Park 1973; Thiollay 1978*a*). Clutches comprising only a single egg (about 2 per cent of total clutches) were always deserted (Ward 1965*b*) and up to 5 per cent of all clutches may be abandoned (Thiollay 1978*a*). We may assume that the birds involved in these attempts try again elsewhere, so that we are concerned here with clutches that are incubated normally.

The different causes of losses of eggs and young clearly vary in importance from colony to colony, with predation rates varying greatly. Losses through predation during incubation account for between 1 per cent and 10 per cent of the total; up to about 14 per cent additionally fail to hatch either because they are infertile (5–10 per cent) or because of embryo death in the first few days of development (up to 4 per cent) (Thiollay 1978*a*; Ward 1965*b*). In Ward's (1965*b*) study, very few young were lost to predators. Nestling mortality was density-dependent on brood size (Fig. 15.1) and was apparently due to starvation; the last nestling to hatch was usually the one that starved during its first few days of life. The mean survival to fledging of 92 per cent was the same as that found in broods of three. Thiollay (1978*a*)

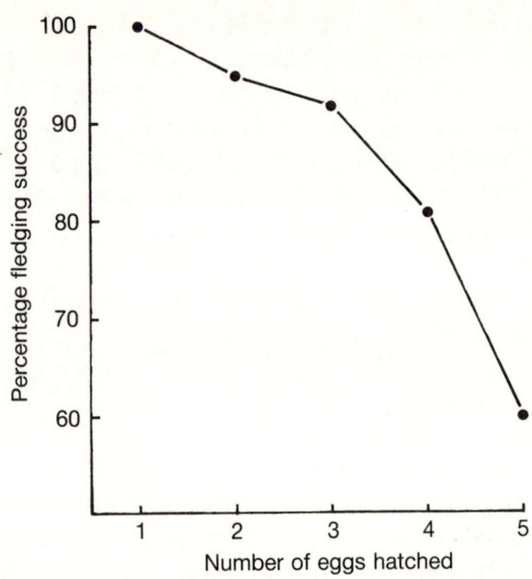

Fig. 15.1. Percentage fledging success in relation to initial brood size in a Nigerian quelea colony (source: Ward 1965*b*).

found 4.5 per cent of nestlings starved to death, mostly soon after hatching, and 11 per cent taken by predators. Sometimes predation rates can be much higher (Chapter 16; Thiollay 1975).

Estimates of average success rates are similar: at least 2.2 young fledging per pair in Nigeria (Ward 1965*b*); 2.3 per pair in Chad/Cameroon (Thiollay 1978*a*); 2.2 in Zimbabwe (C. Vernon, unpubl. data); and a range of 1.9–2.7 in Senegal (Morel and Bourlière 1956).

Survival of adults

While fecundity rates in quelea have been studied in some detail, there have been few attempts to measure survivorship. An attempt to derive the adult mortality rate from the percentage of juveniles (birds in their first year) in the population at the start of the breeding season in Chad gave two estimates in successive years of 39.2 per cent and 55.2 per cent adult survival per annum (Manikowski 1980). However, apart from errors introduced by using the degree of cranial pneumatization to age the birds (Manikowski 1980) and the likely age biases in sampling (Jaeger *et al.* 1979), ratio methods depend for their validity on the assumption that the population has assumed a stationary age distribution, which is the special case of a stable age distribution appropriate to a zero rate of increase (Caughley 1977). This was not the case; for a population's age distribution to approach a stable form, the life table schedules of fecundity and mortality must not vary for some time, yet it is clear that the two estimates, a year apart, are quite different. Manikowski's (1980) estimates should therefore be disregarded.

Much more useful figures can be derived from the results of bird-ringing programmes carried out over a number of years. Survival rates may be calculated quite simply from the age distribution of recoveries of ringed birds, provided that recoveries continue to be reported for long enough after the last bird was ringed (Haldane 1955). Such a calculation for quelea ringed in South Africa gave a mean annual survival rate for adults of 43.8 per cent (Jones 1980). Unfortunately, data derived from the considerable amount of ringing carried out in Tanzania in 1954–1959 (Disney 1960) are too disorganized to analyse this way.

The South African estimate must be treated with caution, because it was obtained during a period when quelea control was intense and millions of birds were killed annually (Fig. 15.2). The apparent decline in the breeding populations of around 20 per cent annually during 1954–1964 was possibly due, in part, to the activities of the control teams, but it is equally possible that changing environmental conditions during these years caused fewer colonies to be established in South Africa, so that the birds were not exposed to control operations. If we assume that the total southern African population was, in fact, stable during these years, but with an increasing proportion of birds breeding outside South Africa and not being counted, the survival estimate needs to be corrected upwards. With an annual population reduction in South Africa of 21 per cent ($r = -0.23$, Fig. 15.3), the corrected mean annual survival rate of adults could be as high as $43.8/79.0 = 55.4$ per cent. If

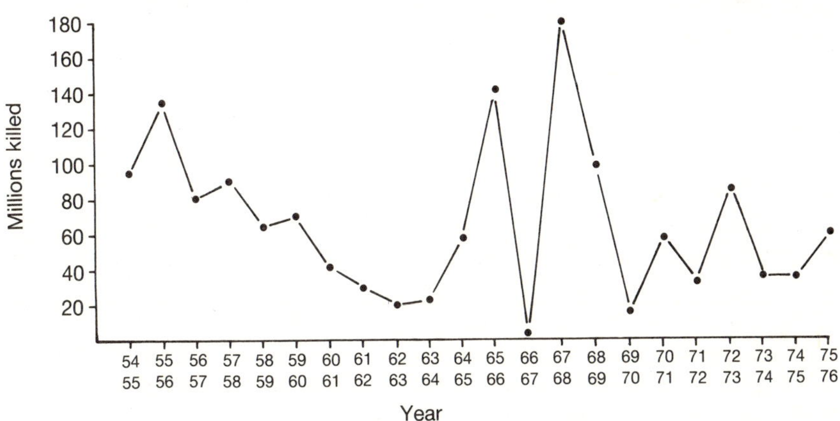

Fig. 15.2. Fluctuations in numbers of quelea killed each year in South Africa (source: Ward 1979; courtesy of the South African Department of Agricultural Technical Services).

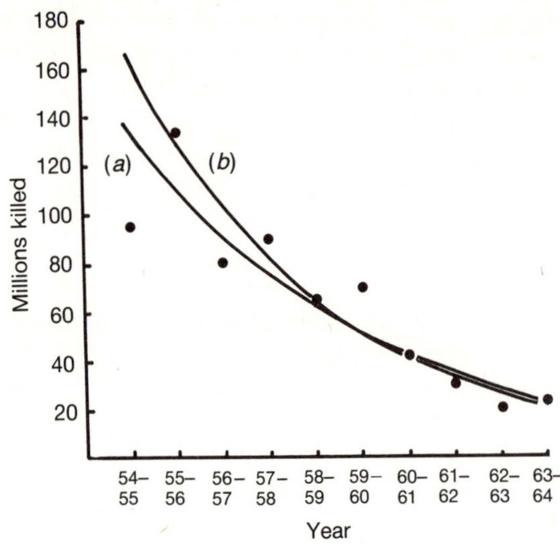

Fig. 15.3. The rate of decrease in numbers of quelea killed annually in South Africa, 1954–1964. Curve (*a*) represents an exponential rate of decrease $r = -0.20$ (a decline of 18% p.a.), and curve (*b*), which excludes the data point for 1954–1955, gives $r = -0.23$ (a decline of 21% p.a.) (source: courtesy of the South African Department of Agricultural Technical Services).

the maximum likelihood method of Seber (1970) is used, which allows both the recovery rate and the survival rate to vary annually, the South African data yield an arithmetic mean survival rate of 56.7 per cent per annum (M. Dyer, unpubl. data).

These figures give a misleading sense of precision. Survival rates probably vary greatly from year to year and from region to region, depending on local conditions of food availability and the timing of rainfall. Ward (1965*a*) supposed that quelea populations are regulated by food shortage, the effects of which become increasingly severe as the dry season advances and the standing crop of dry seed becomes scarce. Grass seed production can vary at least sevenfold between years depending on rainfall and, within any one year, its biomass is reduced to just a few per cent of the original production by the end of the dry season (Gaston and Lamarque 1976). The little remaining seed then disappears through germination as the rain begins. Food shortage may be especially acute in years when the rains begin more synchronously over large areas, so that grasses in the early-rains quarters may not have produced seed before the birds arrive there. For these reasons, large annual differences in survival may be quite usual. Nevertheless, survival estimates for quelea of around 50 per cent per annum match the few estimates available for other African savannah ploceids (Jones 1980). An exception is that Red Bishop

Euplectes orix males seem to have a higher annual survival of 60–75 per cent, which is correlated with the frequent deferment of sexual maturity in males until their second year (Rowan 1964), a common phenomenon among birds (Ricklefs 1973). There is no evidence that this occurs among male quelea. Although the analysis of ringing results did not distinguish between males and females, there is no reason to suppose that their survival rates differ greatly. It is, therefore, not unreasonable to use a figure of 50 per cent annual survival in the life table for females. This figure is assumed to remain constant with age, because as with other small passerines with similar survival rates, so few live long enough for any senescence-related reduction in survival rates to affect the life table significantly (Lack 1954).

Survival of juveniles

In some eastern African populations, where there is a long breeding season, juveniles that hatched early in the season may be able to breed for the first time in the same season at about 6 months old, as soon as they have finished the post-juvenile moult (M. Jaeger and J. Thompson, unpubl. data). Elsewhere, where breeding seasons are more restricted, quelea breed for the first time at 9–12 months old. The survival of independent juveniles until first breeding has not been measured in any population, because ringing studies have not yielded enough recoveries for analysis. However, estimates may be derived from values established for adult survival and the number of young fledging per pair each year. For example, if we assume a female rears two broods a year, the annual production of quelea fledglings to independence may be:

$$\frac{2.8\text{ eggs/clutch} \times 0.80\text{ egg success} \times 2\text{ nestings}}{2\text{ adults per pair}} = 2.24$$

fledglings per adult per year (or daughters per female per year). If the adult mortality is 50 per cent, then only 0.5 fledglings are necessary to replace the adult that dies during the year. In a stable population fledgling survival during the first year to breeding must, therefore, be 0.5/2.24 = 22 per cent. Smaller clutch sizes, lower nesting success, fewer successful nestings per year, and higher adult mortality all increase the rate of juvenile survival necessary to balance adult losses and vice versa. In general, juvenile survival in small land birds varies between one-quarter and three-quarters that of adults (Ricklefs 1973). Of course, different estimates for the survival of juvenile quelea can be obtained by substituting other values in the above calculations within the limits of variation measured in the field. Obviously in some years juvenile survival cannot be high enough to compensate for losses elsewhere and the population will decrease. In other years, all the variables will combine favourably to give an increase.

A life table for *Quelea quelea*

The basic parameters of a life table are age-specific fecundity and age-specific survivorship, from which the growth rate of a population can be determined. The effects of altering these parameters can then be assessed in terms of the population's rate of increase or decrease. This is of special interest in quelea population dynamics in permitting predictions of the impact of control measures of different intensities at different stages of the life cycle.

Age-specific fecundity (m_x) is properly the number of female eggs laid by females at different ages. Eggs, nestlings, and fledglings have different survivorships which may be entered separately in the life table, although this is cumbersome where they differ depending on the age of the parent. More conveniently, the life table simply records in the m_x column the product of the mean clutch size, nestling survival, survival of fledglings to independence, and the number of successful nestings per year for each age of parent female. As mentioned earlier, where adequate data for other species exist, they show that young females perform less well in rearing their young than older birds. Unfortunately, this relationship remains unknown for quelea and, therefore, fecundity is entered as a constant for all ages.

With all the parentally determined differences in survivorship accounted for in this way, the initial age (0) for the survivorship (l_x) column of the life table is the age at which the young become independent of parental care. The proportion of independent juveniles alive at this point is entered as unity and, subsequently, devalued by their survival rate in the year until first breeding. This proportion is then reduced yearly thereafter by the survival rate of adults, here entered as a constant for each age, until the oldest adults constitute such a small proportion of the total population that their contribution to the population's reproductive performance is negligible.

The product $l_x m_x$ gives the expectation of fecundity at each age x. The total expectation of fecundity during a female's entire life span is, therefore, $\sum l_x m_x$, usually referred to as R_0, the net reproductive rate. A value of less than 1.0 for R_0 means that each female will not replace herself during her lifetime and that the population will therefore decline, whereas a value greater than 1.0 indicates an increasing population. A life table for a stable population of quelea is constructed in Table 15.1 using values established earlier for the survival of juveniles from independence to first breeding of 0.22, annual adult survival thereafter of 0.50, and annual fecundity of 2.24 daughters per female. The value of R_0 is approximately 1.0, indicating that on average each individual replaces itself during the course of its lifetime.

The *annual* rate of population increase or decrease is not so easily derived from R_0, because the generation length must be taken into account. Generation length is the mean age at which individuals reproduce and requires knowledge both of age-specific mortality rates and age-specific variations in

Table 15.1. A life table for *Quelea quelea* using juvenile survival from independence to first breeding = 0.22, annual adult survival thereafter = 0.50, and mean annual fecundity = 2.24 daughters per female.

Age (years)	Survival l_x	Fecundity m_x	$l_x m_x$
0	1.000	0	0
1	0.220	2.24	0.4928
2	0.110	2.24	0.2464
3	0.055	2.24	0.1232
4	0.028	2.24	0.0616
5	0.014	2.24	0.0308
6	0.007	2.24	0.0154
7	0.003	2.24	0.0077
			$\Sigma l_x m_x = 0.9779 = R_0$

fecundity. An additional complexity is that in bird populations successive generations overlap, so that before a female has completed her lifespan, some of her oldest daughters will have their own offspring which in turn may have bred. To obtain annual rates of population increase or decrease in such circumstances, each row of the life table is further multiplied by the quantity e^{-rx}, where r takes the value necessary to correct the column total to unity, i.e. where $\sum l_x m_x e^{-rx} = 1$. This is widely known as Lotka's equation and its important feature is the inclusion of the quantity e^r to express the finite rate of increase of a population (though in negative form as a correction factor). Thus, when r is zero, Lotka's equation is identical to that for the net reproductive rate R_0 in a stable population ($\sum l_x m_x = 1$). Because in an expanding or declining population the value for $\sum l_x m_x$ is above or below unity, respectively, the value of r that is sufficient to correct $\sum l_x m_x e^{-rx}$ to unity also measures the rate at which the population is expanding or declining. Lotka's equation can be solved iteratively by computer to obtain an exact value for r, the exponential annual rate of increase (positive values) or decrease (negative values) of the population (for further discussion see Caughley 1977).

Figure 15.4 shows rates of population change produced by different combinations of adult and juvenile survivals at two different levels of fecundity that represent one and two successful breeding attempts per year, respectively. If populations where individuals breed twice are to remain stable at levels of adult survival between 40–60 per cent per annum, juvenile survival need be only 15–30 per cent . If only one breeding attempt is

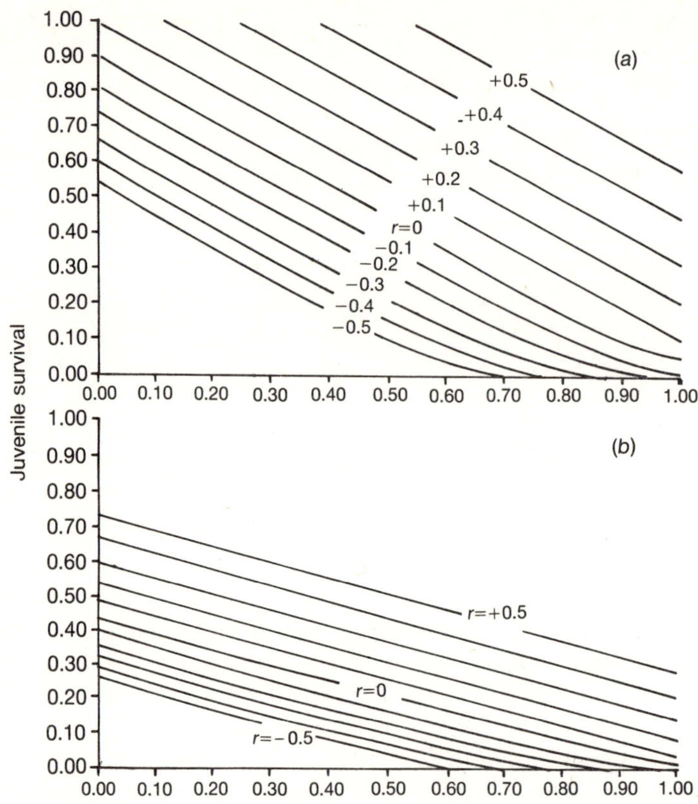

Fig. 15.4. Values for *r*, the exponential rate of population increase (positive values) and decrease (negative values) at different combinations of adult and juvenile survival in an unculled population. In (*a*) there is one breeding attempt per year, producing 1.12 daughters per female; in (*b*) two breeding attempts produce 2.24 daughters per female (calculations by N. J. Harding).

successful, juvenile survival after independence must nearly equal or exceed adult survival rates to achieve population stability. Therefore, either single-broodedness cannot be common, or the survival rates of adults and/or young in such populations are higher than has been assumed in the present model. It is commonly accepted that reduced reproductive effort may directly increase adult survival rates in birds, though field data are limited (Nur 1984). It would be interesting to discover if western African populations of quelea, which have a more restricted breeding season and less chance to breed twice than quelea elsewhere, have significantly greater survival rates than the 50 per cent annual survival recorded in southern Africa. Unfortunately, comparisons are not possible on the present data. As discussed earlier, Manikowski's (1980) estimates for the Lake Chad population should be disregarded, and it must be remembered that the survival figure for South Africa (Jones 1980) was obtained when the breeding attempts of many quelea were greatly disrupted through the efforts of control teams. Nevertheless, quelea population dynamics appear at first sight to be quite sensitive to small changes in these rates of adult survival and fecundity, so that natural populations may be expected to show great changes in size from year to year.

The effects of control

The life table is an idealized statement of the dynamics of a population. The predictions about rates of population increase or decrease that can be made from it depend on the rates of survival and fecundity remaining unchanged for long enough for the population age structure to assume a stable form. This is almost certainly not the case for quelea but it is, nonetheless, useful to experiment mathematically with a model life table to discover the likely effects of perturbations such as large-scale destruction of breeding colonies and roosts.

An immediate problem in such exercises is in knowing whether artificial mortality applied by man to a population is likely to be compensated for by a reduction in mortality occurring naturally at some other stage in the life history. If the natural regulation of quelea populations is brought about in a density-dependent manner by food shortage (Ward 1965*a*), it is likely that the removal of birds from the population by control, prior to the period when density-dependent mortality occurs, will reduce competition for scarce food resources and enhance the survival of birds that escaped control.

We can model two extreme situations: (1) where compensation is total, in which case control will only begin to have an effect on mean adult survival once the proportion of birds killed exceeds those dying naturally, and (2) where natural mortality is not density-dependent, no compensation occurs, and the artificial mortality is additive. Figure 15.5 shows for these two cases the effect of increasingly intense control on the rate of population increase, *r*. The example chosen is a conjectural model of *Quelea quelea lathamii* in southern Africa, where the South African authorities try to eradicate all breeding colonies in South Africa itself. These form the first breeding attempts of the season, after which any survivors are able to breed unchecked elsewhere. It is assumed that 90 per cent of adults in a colony are killed (which is a usual goal of control teams) and no young fledge. The model also assumes that colony size is constant, or that the controlled colonies are a random sample of the total with respect to size. In reality, the smallest colonies are probably overlooked or ignored (it is not known what proportion of a quelea population breeds in small, inconspicuous colonies), and the

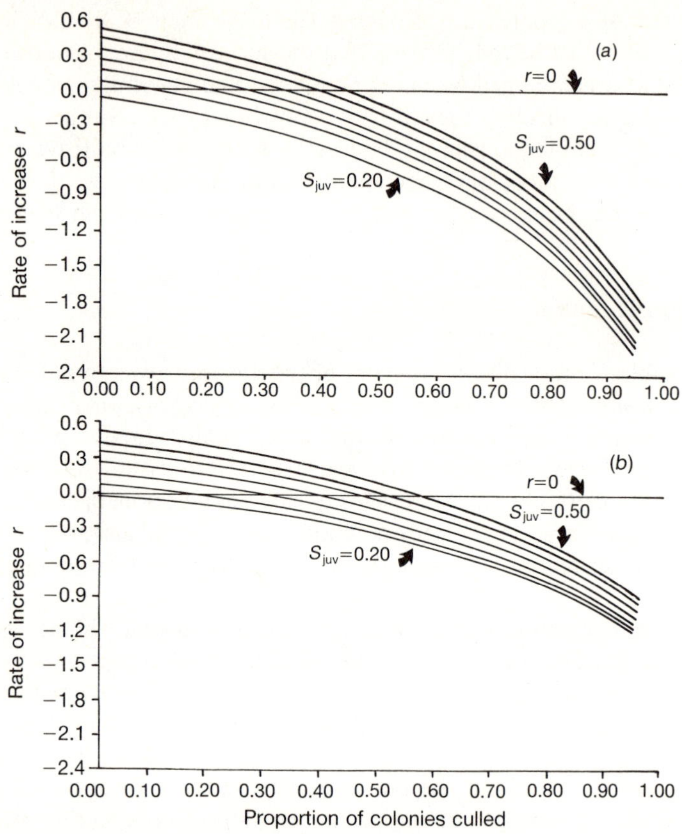

Fig. 15.5. The effect on r, the exponential rate of population increase, of culling breeding colonies at different intensities. Ninety per cent of adults and all young are killed in culled colonies, but surviving adults breed again at a fecundity rate of 1.12 daughters per female. The fecundity of females in unculled colonies is 2.24 daughters per female. Sets of curves are given for values of juvenile survival (S_{juv}) between 0.2 and 0.5 p.a. In (*a*) natural adult mortality at 50% p.a. is additive, i.e. does not compensate for losses through culling; in (*b*) natural adult mortality at 50% in an unculled population fully compensates for losses through culling (calculations by N. J. Harding).

largest may be too large to achieve a 90 per cent kill. If, as before, we take the natural mortality of adult quelea to be 50 per cent per annum under undisturbed conditions, the success of control in achieving a population reduction depends very much on the survival rate of juveniles from undisturbed colonies.

Because the value for juvenile survival is critical for balancing the life table, it is essential that we understand how juvenile survival might vary in a density-dependent manner to compensate for a reduced input of young into the population. The eradication of some colonies may not be of immediate advantage to newly independent young from undisturbed colonies, since such young frequently remain for some time as a cohesive group in the vicinity of the old colony site. This period may be when they are most vulnerable to density-dependent mortality through food shortage, both because of inexperience in finding food and because of the great local depletion of seed stocks during the breeding period (Morel 1968). Their survival during this period may, therefore, remain as low as if no control had taken place elsewhere.

In other situations, where juveniles are more wide-ranging and the feeding areas of young from several colonies overlap extensively, the eradication of some colonies may reduce competition for food among juveniles and enhance their survival. If their survival is doubled compared to that of adults and adult mortality also shows a compensatory decrease, it would be possible to destroy up to 60 per cent of colonies at the first breeding attempt without any decline in the population (Fig. 15.6; curve *a*). If neither juvenile nor adult survival increases to compensate for control, so that they remain at 20–25 per cent and 50 per cent, respectively, as few as 10 per cent of colonies need be controlled for the population to begin to decline (Fig. 15.6; curve *b*). However, the real situation is probably that the survival of both adults and juveniles is greatly enhanced later in the year if some population reduction has occurred during the breeding season, but that the increased survival does not fully compensate for the losses.

A model life table also allows us to calculate what intensity of control would be necessary to achieve particular target rates of population reduction. The rate of decrease in the South African quelea population between 1954 and 1964 of about 20 per cent per annum (Fig. 15.3; $r = -0.20$ to -0.23, at which the population halved approximately every 3 years) was considered a satisfactorily rapid response to parathion spraying. It was immediately seen as a goal that could be achieved elsewhere; but to have achieved this by controlling colonies alone would probably have required the annual eradication of 35–65 per cent of sites, depending on the extent of any compensating increases in survival among the adults and juveniles that escaped (Fig. 15.7). Because of excellent communications and transport facilities, a large proportion of the colonies located in South Africa were in fact destroyed but, as discussed earlier, it is not known what proportion of the total breeding population was actually present in South Africa and vulnerable to control. The subsequent sixfold rise in the number of birds breeding over the next 2 years can only have been due to immigration, not to the reproduction of the small population present in 1964–1965 (Fig. 15.2). It is arguable whether it would have been possible to have consistently carried out a 65 per cent annual cull of the entire first breeding attempt of the southern African

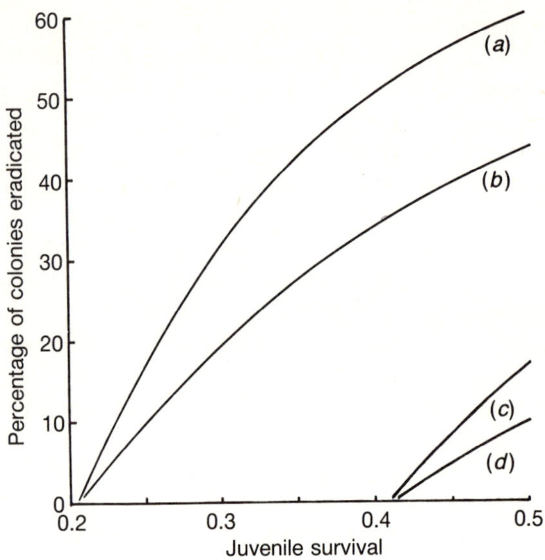

Fig. 15.6. The levels of culling that a quelea population can tolerate to remain stable ($r=0$) in relation to the degree of compensatory response in adult and juvenile survival rates. Density-dependent variation in juvenile survival will occur along the x-axis. Variation in the occurrence of compensatory responses in adult mortality and in the number of broods per year will generate different curvilinear relationships, four of which are shown here:

(*a*) Natural adult mortality of 50% p.a. in an unculled population fully compensating for mortality due to culling; two broods (2.24 daughters per female p.a.) for birds in unculled colonies (see below).

(*b*) Natural adult mortality of 50% additive to culling mortality; two broods p.a. for birds in unculled colonies.

(*c*) Natural adult mortality of 50% p.a. in an unculled population fully compansatory; one brood (1.12 daughters per female p.a.) for unculled colonies.

(*d*) Natural adult mortality of 50% p.a. additive; one brood p.a. for unculled colonies.

In double-brooded populations, only the first breeding attempt is subject to culling. Ninety per cent of adults and all young are killed in culled colonies, but surviving adults breed again at a fecundity rate of 1.12 daughters per female. At values of juvenile survival below 0.21 (two broods p.a.) and 0.42 (one brood p.a.), the population declines even in the absence of culling (calculations by N. J. Harding).

population. In more remote regions of Africa, where other quelea populations regularly breed more than once a season, it is probably impossible to achieve such a consistently high cull.

Where quelea breed only once a year, as may be the case in much of western Africa, significant population reductions may be possible. At the rates of adult survival and fecundity described in the model life table, such

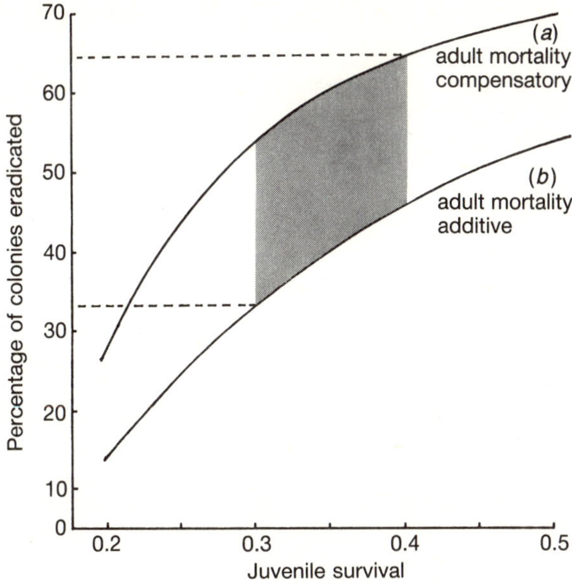

Fig. 15.7. The effect of density-dependent compensation in adult and juvenile mortality rates on the proportion of colonies that must be eradicated at the first breeding attempt for a quelea population that breeds twice a year (e.g. *Q. q. lathamii* in southern Africa—see text), in order to achieve a population reduction of 21% p.a. ($r = -0.23$). Juvenile survival varies along the x-axis and adult survival along the y-axis between the two curves, representing (*a*) a natural adult mortality of 0.5 in an unculled population that completely compensates in a density-dependent manner for losses due to control, and (*b*) natural adult mortality additive to control mortality. The density-dependent responses of adult and juvenile survival are likely to lie within the shaded area, which indicates the probable range of control effort required (calculations by N. J. Harding).

populations begin to decline when, even at juvenile survival rates as high as 45 per cent , only 4–8 per cent of colonies are culled (Fig. 15.6; curves *c* and *d*). Under these circumstances, the permanent reduction of a quelea population may be a feasible control strategy, as, for example, in the Senegal River Valley where the quelea population declined to a much smaller size than it was before control began (Ward 1973*a*). However, the success of control operations is often difficult to distinguish from the direct effects on survivorship of adverse environmental conditions. The quelea population breeding in north-eastern Nigeria was at a very high level in 1958–1960, but by 1970–1975 it had crashed to less than 10 per cent of its former level (COPR 1977). Figure 15.8 details the number of breeding colonies reported, but understates the magnitude of the decline. Breeding in the earlier period took place in vast

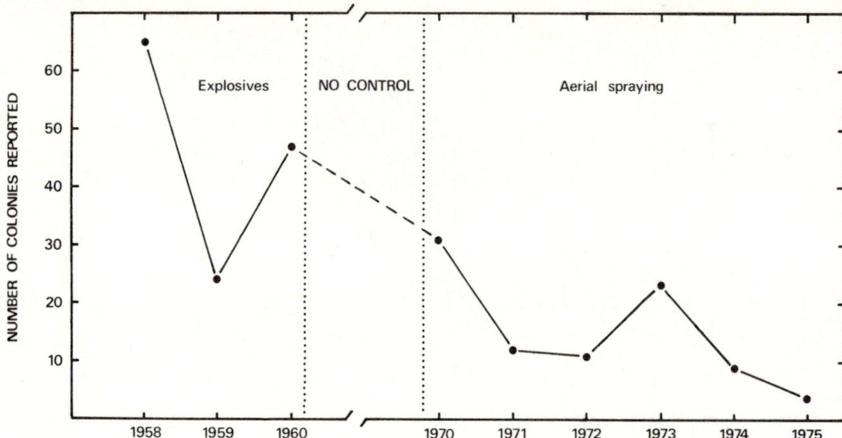

Fig. 15.8. The decrease in the number of reported quelea colonies in north-eastern Nigeria, 1958–1975. The figures for 1958–1960 derive from ground surveys alone, when wet conditions severely limited coverage; they are certainly an underestimate. Later figures for 1970–1975 are much more accurate, deriving from combined ground/helicopter surveys (sources: CORP 1977; Ward 1979; 1958–1960 data supplied by the Federal Bird Control Organization, Maiduguri; 1970–1975 data supplied by E. Dorow).

colonies, whereas in the early 1970s colonies were very small. Although the decline appears to have coincided with the introduction of aerial spraying, it also coincided with the onset of the present severe drought in the Sahel. The dramatic fall seen in quelea numbers in Nigeria also occurred in Cameroon and Chad, at a time when control operations did not occur. A similar decline occurred among other native Sahelian birds that were not controlled, and among European migrants that overwinter in the Sahel (Winstanley *et al.* 1974). While it is impossible to justify the claim that the decline was the result of control activities, control may have caused the low level that was reached to be even lower than if natural factors alone had been operating.

The effectiveness of lethal control depends on the time in the annual cycle it is applied. If it is carried out during breeding, i.e. before the period when density-dependent mortality of adults and young occurs, naturally occurring mortality decreases in compensation, so that control must be much more extensive to reduce the population. Only if control is carried out on roosts late in the dry season, after the 'doomed excess' has already died, is artificial mortality likely to be additive and, therefore, reduce the population at the end of the year to a lower level than it would otherwise have reached. Because the first of the next season's breeding colonies are established soon afterwards, control of roosts in the late dry season is functionally similar to eradication of the breeding colonies themselves, and any density-dependent compensation simply occurs one season later.

It was believed that the combined control of breeding colonies and dry season roosts must be effective and this is what most control teams have tried to achieve. In Senegal, the combined attack on colonies and roosts was probably effective; in South Africa and elsewhere it was not. The difference is attributable to probable differences in fecundity and to differences in the size and accessibility of the areas that the populations exploit. Although an efficient control unit may achieve culling rates of 60–70 per cent of the population annually when the population is small or geographically restricted, such rates are almost certainly not possible when the population is large or when a significant proportion of the breeding occurs in areas remote from effective control.

The dynamics of most quelea populations are resilient enough to absorb most of the perturbations that man can impose. At the same time, they can rapidly respond to the much more pervasive environmental changes that affect survival and fecundity rates. Rare species have generally become rare through habitat destruction rather than by hunting or mass-slaughter (Collar and Stuart 1985), and the habitat destruction consequent upon the Sahel drought has been a major cause of the quelea's decline in western Africa. Elsewhere, habitat changes such as overgrazing and the provision of water impoundments may have helped quelea to increase. Even if man gained short-term control over quelea numbers, the control effort necessary to maintain a permanent population reduction is unlikely to be cost-effective. The existing lethal techniques of quelea control are themselves environmentally destructive. There are more cost-effective and environmentally less damaging methods of immediate crop protection that should now supersede all the old notions of quelea extermination (Dyer and Ward 1977; Ward 1973*a*, 1979).

16

Natural predation on quelea

JEAN-MARC THIOLLAY

Introduction

Predation can be a limiting factor for all natural populations of animals. It may significantly affect their behaviour, distribution, breeding success, or mortality rate at different levels according to the area, season, resources, habitat, and population density. Because quelea have developed prominent adaptations such as flocking, roosting, and colonial breeding, which are now considered by many scientists to be primarily defence mechanisms against predators, one may hypothesize that historically quelea populations may have existed in natural ecosystems under severe, selective predation pressure. The situation may be different in the modified, predator-poor habitats where the species thrives today.

Most information available on quelea predation is related to breeding colonies. The importance of predation outside the nesting season, for example on foraging flocks and roosting birds, is still poorly documented. My data were obtained from three groups of colonies in Mali and northern Cameroon-Chad, hereafter referred to as western and central Africa, respectively (see Thiollay 1975 and 1978*a* for descriptions of study areas and census methods), as well as unpublished observations in Senegal, Niger, and Kenya. Other information comes from Bortoli (1970, 1975), Disney and Marshall (1956), Gramet (1974), Leuthold and Leuthold (1972), Morel *et al.* (1957), Prozesky (1964), Stewart (1959), Vesey-FitzGerald (1958), and Ward (1965*a*, 1965*b*).

Predation on breeding colonies

From many studies on species other than quelea, the main adaptive significance of colonial breeding is now thought to be defence against predators and decrease of predation rate (Hoogland and Sherman 1976; Horn 1968; Lack 1968). There is no direct evidence that colonies are used as information centres or that such information exchange is critical to the evolution of coloniality (Bayer 1982). Colonies may act as early warning systems to increase the likelihood that an approaching predator is detected. Predator density and attack rate do not increase proportionately with colony size. Thus, large colonies swamp local predators and suffer less predation per nest than small colonies (Wilkinson and English-Loeb 1982). They may also increase the reproductive fitness through social foraging that maximizes the birds' feeding efficiency (Emlen and Demong 1975). However, unlike many other species, quelea exhibit little effective alarm or active defence behaviour against predators at the nest and no communal mobbing. Their main defence is to breed in huge, dense, highly synchronized colonies in thorny trees.

Identification of predators

At least 80 species of predatory birds of 15 families have been identified in breeding colonies, mainly from western and central Africa (Table 16.1). The most widespread and numerous species with the greatest impact on quelea nesting success are Cattle Egrets *Ardeola ibis*, Marabou Storks *Leptoptilos crumeniferus*, Abdim Storks *Ciconia abdimii*, White Storks *C. ciconia*, and several raptors such as Black Kites *Milvus migrans*, Tawny Eagles *Aquila rapax*, Wahlberg's Eagles *A. wahlbergi*, Shikras *Accipiter badius*, Chanting Goshawks *Melierax metabates*, Gabar Goshawks *M. gabar*, Lanner Falcons *Falco biarmicus*, and hornbills, especially the migrant Grey Hornbill *Tockus nasutus* (Plate 12*a*).

Numerous mammals also eat young quelea and regularly raid colonies: all savannah species of monkeys (*Papio*, *Erythrocebus*, *Carcopithecus*), galagos (*Galago*), squirrels (*Xerus*, *Paraxerus*), mongooses (*Mungos*, *Herpestes*, *Ichneumia*), foxes (*Vulpes*), jackals (*Canis*), hyaenas (*Crocuta*, *Hyaena*), cats (notably *Felix serval* and *F. libyca*), genets (*Genetta*), and other omnivorous mammals such as the African Civet *Viverra civetta*, the Ratel *Mellivora capensis*, and the Warthog *Phacochoerus aethiopicus*. Even Lions *Panthera leo* and Leopards *P. pardus* have been seen tearing nests and eating young.

Many species of snakes are also regularly found in quelea colonies and prey on both eggs and nestlings. At least eight genera in western-central Africa and seven in South Africa have been identified. Monitor Lizards *Varanus* spp. are among the most frequent nest-robbing reptiles, and the Nile Crocodile *Crocodilus niloticus* may occasionally take a heavy toll on drinking birds (Attwell 1954; Pitman 1957, 1961).

Density of predators

The number of predators varies greatly among colonies and between years. The abundance of most species differs throughout the quelea's range. In

Table 16.1. Avian predators associated with quelea colonies, roosts, or flocks (abundance indicated by + or + +). The recorded predation location is followed by an asterisk (*) when it is present or abundant in more than one-half of the breeding colonies in at least one area. The data are based on published literature (see references), unpublished reports, and personal communications and observations. [a]W, Western Africa (Senegal, Mauritania, Mali, Niger); C, Central Africa (Cameroon, Chad, Sudan); E, Eastern Africa (Kenya, Tanzania); S, South Africa.

Family Scientific name	Breeding colonies					Nocturnal roosts	Foraging flocks
	Eggs	Nestlings	Fledglings	Adults	Area[a]		
Ardeidae							
Ixobrychus sturmii			?		C		
Nycticorax nycticorax		+	+ +		WC		
Ardeola ralloides			+		WC		
A. ibis	?	+ +	+ +		WCES*		
Egretta alba		+	+ +		WC*		
E. intermedia		?	+		WC		
E. garzetta		+	+ +		WC		
Ardea cinerea		+	+ +		WC		
A. melanocephala		+	+ +		WCS*		
A. purpurea			?		WC		
Scopidae							
Scopus umbretta			+		WC		
Ciconiidae							
Ciconia ciconia			+ +		WS*		
C. abdimii		?	+ +		WC*		
C. episcopus		?	+ +		WC*		
Ephippiorhynchus senegalensis		+	+ +		CS		
Leptoptilos crumeniferus		+ +	+ +		WCES*		
Mycteria ibis		?	+		W		
Threskiornithidae							
Threskiornis aethiopicus		+ +	+ +		WC*		

Pandionidae							
Pandion haliaetus			?		W		
Accipitridae							
Torgos tracheliotus		+	+ +		WCES		
Trigonoceps occipitalis		+	+ +		WC		
Gyps rueppellii			+		WCE		
G. africanus			+		WCES		
Neophron percnopterus		+	+		W		
Necrosyrtes monachus	?	+ +	+ +		WC*		
Circus macrourus			?	+	WC	+	+ +
C. pygargus			+ +	+ +	WC	+	+ +
C. aeruginosus			+	+	WC	+ +	+
Polyboroides typus	+ +	+ +	+ +		WCES*		
Terathopius ecaudatus		+	+ +	+	WCES		
Circaetus beaudouini			?		WC		
C. cinereus		?	+ +		WC		
Accipiter badius	?	+	+ +	+ +	WCS*	+ +	+
Melierax metabates		+	+ +	+ +	WCS*	+ +	+
M. gabar		+	+ +	+ +	WCES*	+ +	+ +
Butastur rufipennis	?	+ +	+ +	?	WC*		
Buteo rufinus			+		W		
B. auguralis		+	+ +	?	WC	+	
Lophaetus occipitalis			+ +	+	C		
Polemaetus belicosus			+	?	CS		
Hieraaetus spilogaster		+	+ +	+	ES	+	
Aquila rapax	?	+ +	+ +	+	WCES*	+	
A. nipalensis		+ +	+ +		E		
A. clanga		+	+		CE		
A. pomarina		+	+		C		
A. wahlbergi		+ +	+ +	+ +	WCES*	+	
Haliaeetus vocifer			?		WC		
Milvus migrans	+	+ +	+ +	+	WCES*	+	
Pernis apivorus	+	+ +	+		C		
Elanus caeruleus			+	+	WC	+	

Table 16.1. *continued*

Family Scientific name	Breeding colonies					Nocturnal roosts	Foraging flocks
	Eggs	Nestlings	Fledglings	Adults	Area[a]		
Sagittariidae							
Sagittarius serpentarius			?		C		
Falconidae							
Falco biarmicus			?	+ +	WC*	+ +	+ +
F. peregrinus			?	+ +	WC	+ +	+ +
F. subbuteo				+	C		
F. chicquera			+	+ +	WC*	+ +	+ +
F. ardosiaceus		+	+	+	W	+	
F. tinnunculus			+	+	W		
F. alopex			+	+	C	+	
Gruidae							
Balearica pavonina			?		WC		
Cuculidae							
Clamator jacobinus	?	+			W		
C. levaillanti	+	+			W		
Centropus senegalensis	+	+ +	+		WC*		
Tytonidae							
Tyto alba			+	+	WC*	+	
Strigidae							
Otus scops			+		WC		
O. leucotis			+		WC*	+	
Bubo africanus		+	+ +	+	WCS*	+ +	
B. lacteus			+ +	+	WC*	+ +	
Glaucidium perlatum			+	?	C		

Alcedinidae				
Halcyon senegalensis	+	+	+	WC
H. chelicuti		?		C
H. leucocephala		+	+	WC
Coraciidae				
Coracias abyssinica		+		WC*
C. naevia		+		W
C. caudata	?	?		ES
Bucerotidae				
Tockus nasutus	+	+ +	?	WCS*
T. erythrorhynchus	+	+ +		WCE*
T. flavirostris	?	+		ES*
Bucorvus abyssinicus	?	+	?	W
B. leadbeateri	?	+	?	S
Corvidae				
Corvus albus	+	+ +	+ +	WC*
C. ruficollis	+	+ +	+ +	W

cultivated areas, the number of predators is usually much lower than in natural savannahs. The highest concentrations are found in large national parks such as the Tsavo in Kenya (C. Elliott, pers. comm.) and the Kruger in South Africa where Pienaar (1969) recorded 1200 eagles and 300 Marabou Storks at a large colony. Many predators such as storks and egrets are attracted by water. Therefore, colonies built near pools or rivers have greater numbers of predators than colonies established in dry places.

Up to 80 per cent of predatory bird species and 95 per cent of individuals are African migrants that perform large-scale seasonal movements following the rains (Thiollay 1978*b*). Consequently, their abundance depends chiefly on the coincidence between the breeding period of quelea and their own migration dates. Their dry-season quarters are usually outside the quelea breeding range. They are, therefore, normally only in transit through the breeding areas and if they encounter a colony, may stop temporarily to take advantage of the abundant food. They are likely to concentrate in large numbers in a particular colony if there are no other colonies or major food sources nearby. They are less numerous if, at the same time, many active colonies are scattered over a large area or if locusts, a preferred food for many of them, are swarming in the vicinity.

Predator numbers at a colony are difficult to assess and often underestimated. Some species (hornbills, shikras) remain inconspicuous inside the dense trees. Unless tall, convenient trees are available, large species (storks, egrets, ibis, eagles, kites, buzzards) stay in the colony only when feeding, often 2–3 hours during the morning and then roost elsewhere. Some of them may return later, but the overall number of predators was always greater before 1200 h than afterwards (Thiollay 1978*a*). When fully-grown fledglings were available, the large avian predators became satiated within 30 min and did not feed during the remainder of the day.

The number of predatory birds actually feeding on a single colony ranged between 20 and 800 in western and central Africa. Large colonies attract a greater total number of predators than small ones, but the number of predators per nest may be lower than for small colonies. The number of predators, which is quite low during the incubation stage, quickly increased after hatching and peaked during the fledging period. At this time, it seemed easier to remove young birds from the nest. In Chad, the number of predatory birds counted in a colony of about 200 000 nests increased from 80, after most eggs were laid, to 585, when young began to fledge.

The ratio between the mean density of predators associated with a colony and the density away from it (roadside counts) provide an index of their concentration on the food source. This index is higher for migrants than for territorial residents and higher for regular than occasional quelea predators (see Thiollay 1978*a* for description of index). The number of active nests available per predator gives the potential predation rate which is inversely

Table 16.2. Average number of active nests after hatching per avian predator in various quelea colonies in Chad and Mali.

Country	No. nests/colony	No. nests/predator
Chad	15 000–20 000	500–800
	100 000–200 000	800–2000
	300 000–400 000	5000–6000
	500 000	6000–7000
Mali	< 10 000	25
	10 000–15 000	65
	15 000–20 000	70
	20 000–30 000	130

correlated with colony size; the smaller the colony, the higher the number of predators per nest (Table 16.2).

Diet and behaviour of predators

Ten species of birds and some snakes, monkeys, and squirrels have actually been seen eating quelea eggs. Hornbills and snakes seldom damaged the nest, whereas Harrier Hawks *Polyboroides radiatus* and monkeys often tore it.

Most predators prey upon nestlings (Plate 12*b*). The smallest predators, kingfishers *Halcyon* spp., rollers *Coracias* spp., and Red-billed Hornbills *Tockus erythrorhynchus* are only able to swallow the youngest nestlings, but the larger species readily take the small nestlings and the fully-feathered fledglings. Only a few species (shikras, coucals, kingfishers, hornbills) are able to reach nests inside the dense, thorny trees so that most predation occurs to peripheral nests. Kites, eagles, and storks awkwardly open the nests to remove their contents; many young fall in the process and die or are eaten by ground predators such as jackals, civets, mongooses, hyaenas, ratels, and warthogs. This greatly increases the actual predation rate of these birds. Large birds, including vultures, rarely go under the trees but concentrate on fledglings in or around the accessible nests.

Shikras, goshawks, eagles, and buzzards prey both on nestlings and flying young and occasionally on adults (Plate 13). Harriers and falcons prey mainly on flying adults or juveniles around the colonies. Small raptors, coucals, rollers, and hornbills begin hunting before sunrise. Crows and egrets appear a little later, and large, soaring raptors and storks appear between 0800 and 0900 h. They actively feed during the morning, then retreat for the afternoon into the shadow of large trees. Few of these species hunt in the evening but then snakes and mammals become active.

Daily intake by predators

Analysing crop contents and counting numbers of nestlings removed during feeding probably underestimates the daily food intake because feeding frequency and digestion rates are unknown. In Mali and Chad, it seemed that small species like shikras and hornbills consumed a daily minimum of 4–8 birds (4- to 8-day-old fledglings), whereas medium-sized predators like kites, buzzards, egrets, and ibises consumed 8–16 birds per day; eagles and storks took 30–40 birds daily. These values are halved if fully-grown fledglings (≥10 days old) or adults are taken, and doubled if young pulli (1–3 days old) are consumed. The actual impact of predation is usually much higher because many young, frightened by the disturbance of predators, fall to the ground or get caught and impaled on thorns.

In a quelea colony in Mali that had been sprayed with parathion at 1500 h, the crop and stomach contents of each of 50 dead Black Kites selected randomly from more than 400, contained an average of only 3.8 fledgling quelea (10–12 days old). In 108 stomachs of 21 species examined in western and central African colonies, I never found any prey other than the eggs, young, or adult quelea.

Where spectacular concentrations of predators occur, the number of young quelea predated may be extremely high. In two colonies in Mali, covering 8 and 4 ha and containing 23 000 and 16 000 occupied nests, respectively, an estimated 10 000 and 8000 young were eaten per day by 763 and 678 avian predators, respectively. In Sudan, Gaudchau (1967) estimated that Marabou Storks alone were responsible for the death of about 1 million nestlings in 10 colonies totalling 210 ha and 1.1 million nestlings in 27 colonies totalling 842 ha.

Predation rates

Like the density and diversity of predators, the resulting predation rate varies greatly among colonies. It also increases markedly toward the end of the breeding cycle when nestlings and later on fledglings become available. In 1975 in Chad, studies of the breeding success of seven colonies south of Lake Chad, which contained a moderate number of predators, demonstrated this trend (Fig. 16.1). The overall nesting success was 52 per cent , and predation accounted for 40 and 51 per cent of the losses of eggs and young, respectively. Moreover, predators killed about 0.3 per cent of the adults.

At nine small colonies in Mali, the mean proportion of broods destroyed by predators between laying and the first days following hatching was 32 per cent (range 21–48 per cent). This exceptionally high rate did not include either the numerous unfinished nests on the periphery of the colony or the complete, intact, but empty nests that may have been robbed of their eggs by

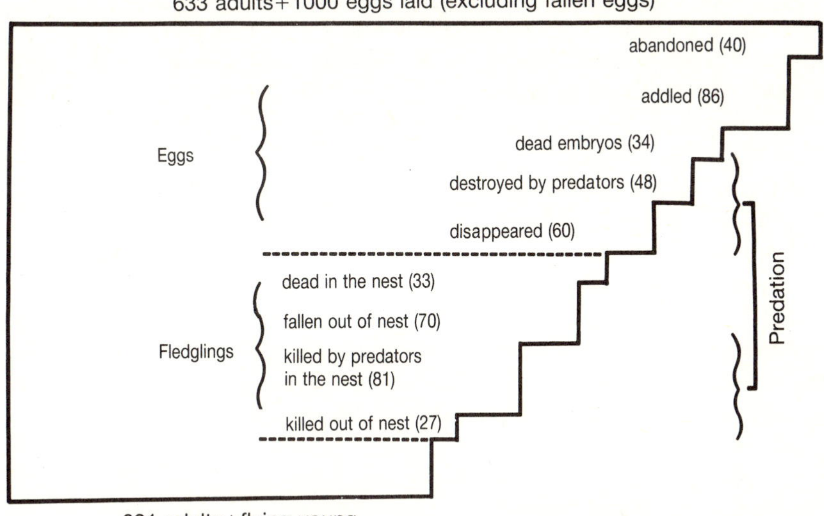

Fig. 16.1. Mean losses and mortality rates in eggs and fledglings during the quelea breeding cycle in colonies in central Chad, 1975.

'careful' predators. After hatching, the daily rate of destruction averaged 5 per cent of the remaining broods per day (range 1–15 per cent) up to the fledging period.

In South Africa, Pienaar (1969) recorded several colonies with 13–35 per cent of the nests destroyed by predators before hatching. Often, 50 per cent of the young were eventually killed, and up to 60 per cent of the broods were destroyed in a large colony. Like Stewart (1959), Van Ee (1973), and Vesey-FitzGerald (1958), Pienaar (1969) mentioned instances where predation was so intense that breeding activity was disrupted and the colony abandoned or even annihilated. Small colonies in high grass seem especially vulnerable to concentrations of egrets, storks, or even baboons. Gaudchau (1967) cited several colonies in Sudan where 20–50 per cent of the fledglings were eaten by Marabou Storks. On the other hand, Disney and Marshall (1956), Morel and Bourlière (1955), Ward (1965*b*), and others never recorded significant predation pressure; however, quelea populations probably had reached their highest peak in the years of their study, and it is reasonable to suggest that the predator populations, limited by other factors, had not paralleled the increase of quelea (G. Morel, pers. comm.).

Colonies are occasionally deserted. Desertion usually occurs early during incubation and probably results from food shortage (Ward 1965*b*). However, for colonies deserted in later stages, predator disturbance is a likely cause. In the only instance I have studied, in northern Cameroon, a colony of more than 100 000 nests was deserted during the hatching period when at least 23 per cent of the nests were destroyed by predators. This predation rate was five times greater than that of the other colonies in the area which were at the same stage and not deserted.

Several features of the biology of breeding quelea may reduce the impact of predation on adults. For instance, killing one parent always causes nest desertion or death of the brood if it occurs during the incubation period, but has no immediate effect if it occurs at the fledgling stage (Barré 1973). During the fledging period, the disappearance of the female is more serious than that of the male. If only one parent survives, the young are underfed (Jackson 1974*a*). Often there is an excess of males and a possibility that fledglings that have lost one or both parents may be fed by other adults or of shortening the fledging period for young whose parents have disappeared (Jackson 1974*a*; Morel and Bourlière 1955; Thiollay 1978*a*; Ward 1965*b*).

Predation on nocturnal roosts

The hypothesis that roosts are information centres where birds experiencing difficulty in finding food may follow successful individuals to profitable feeding areas (Ward and Zahavi 1973; Zahavi 1971) has been only partially (Loman and Tamm 1980) or not at all supported by experiments (Fleming 1981). The statistical significance of reduced heat loss associated with roosting behaviour has often been found to be minimal, and energy saving through thermal benefit is offset by energy costs during flight to and from the roost, especially in tropical countries (Walsberg and King 1980; Yom-Tov *et al.* 1977). Therefore, the main function of communal roosts seems to be protection against predators (Gadgil 1972; Lack 1968; Sengupta 1973).

Twenty-one species of predatory birds have been seen associated with quelea roosts or actually attacking birds in western and central Africa (Table 16.1). No detailed study has ever been undertaken to assess the impact of these predators. The more conspicuous species are diurnal raptors (shikras, goshawks, falcons, and occasionally harriers, buzzards, or eagles). I have regularly counted 2–12 individuals in western African roosts. One or several owls are also regularly present. Arboreal mammals such as wild cats, servals, and genets are commonly recorded as are snakes and terrestrial or scavenging mammals (mongooses, jackals, hyaenas, civets, and ratels). From the number of raptors recorded, the predatory species alone can easily catch about 20–30 birds or more every evening and night, which for the 10 or more months per year that quelea roost would represent a minimum of 6–10 000 individuals taken from large roosts.

As in colonies, the abundance and diversity of predators at roosts varies greatly with area, season, roost size, habitat, vegetation structure, and human disturbance. Predation may seem negligible, but it has probably much deeper results than the direct killing of some birds. Repeated attacks during the hour before dark cause a strong disturbance and, consequently, an energy loss which may be significant during parts of the dry seasons when many birds may be underfed. Moreover, birds can be found impaled in the thorny trees where they roost or can even be killed by collision (Barré 1973; Bruggers *et al.* 1981*c*; J.-M. Thiollay, pers. obs.). Such accidents may be frequent during alarm flights or panics during the night and may be responsible for more deaths than catches by predators.

The fear of predators may influence the choice and then reduce the number of suitable roosting sites which must be safe and concentrate as many birds as possible. Hence, it could increase the distance to and from the foraging grounds, and the energy expended.

Predation on foraging flocks

Numerous studies of the cost:benefit ratio of feeding in flocks have been conducted, often using granivorous birds. Despite the lack of studies in quelea, there is no reason to believe that this ratio in quelea would be different from that of other birds.

Overall group awareness increases and individual vigilance decreases with flock size (Abramson 1979; Caraco 1979); thus, more time can be devoted to feeding (Lazarus 1979). Flocking also confuses attacking predators which become unable to select an appropriate target (Bertram 1978). Hawks experience lower success rates when hunting flocks rather than solitary birds and may be discouraged in pursuing flocks (Morse 1980). Quelea flocks have frequent and unpredictable movements that are thought to be a defence against possible attacks by predators and are called 'protean insurance' (Humphries and Driver 1970). They look like sudden alarm flights and are a source of considerable loss of time and energy. High risk of predation leads to an increase in flock size, group tightness, and scanning rate which influence the birds' time budget (Barnard 1980; Caraco *et al.* 1980).

The frequency of encounters between foraging flocks and predators and the subsequent behavioural changes have not been studied in quelea. I have seen nine species of diurnal raptors attacking quelea outside roosts and breeding colonies: three species of harriers (palaearctic migrants wintering in the Sahel), shikras, two species of goshawks, and three species of falcons. From observations in Senegal and Chad, the Lanner Falcon seems to be the most regular predator of adult quelea by day. It also seems that attacks on quelea flocks are much more frequent during the dry season than during other months, probably resulting from other food sources (locusts, rodents, winged termites) being at their lowest level. At the same time, quelea may begin to experience severe food shortage, and any loss of energy for increased predator defence may then become an additional mortality factor. Quelea often obviously prefer to feed near protective cover rather than in the centre of large, open fields (pers. obs.). This could be interpreted as a behavioural response to predators and would affect both time and energy budgets as well as the pattern of crop damage.

Lethal control risks to predators

From personal observations and numerous reports it seems that any control technique using aerial (Bruggers *et al.*, in press) or ground (Thomsett 1987) spraying of toxic chemicals, even if carefully applied to quelea concentrations, are likely to present a hazard to associated predators. Predators, even when frightened from treated colonies or roosts, will return, sometimes as early as the following morning after an evening spraying. They are mainly intoxicated when eating dead or affected birds that may be available for several days. Likewise, actual mortality rates of predators are difficult to assess because many can die far away from the spray site. In Chad, more carcasses were found within a 2-km radius of a colony treated with fenthion than in the colony itself (pers. obs.).

For these and other reasons when chemical control is the only method considered, I would suggest

(1) not treating small colonies that suffer higher predation rates than larger ones;
(2) not treating large colonies that support large concentrations of predators (this would avoid killing beneficial non-target species that prey on quelea and other pests such as locusts);
(3) treating colonies during the incubation stage, because then the number of predators is less than after hatching;
(4) treating colonies after sunset when many predators are away from the colonies;
(5) not treating colonies established over water where non-target species concentrate;
(6) using less-persistent chemicals to reduce hazards to predatory species (low concentrations may not be a better solution because a greater proportion of quelea are not immediately killed but become affected and are taken in large numbers by predators during a longer period);
(7) cutting large trees on and around colony sites either for nest destruction or for other purposes, which reduces the predation rate because nests in small trees are much less vulnerable to large predators; and
(8) employing lethal control of roosts at the end of the dry season when

predators and other non-target species are less numerous and when natural mortality has already taken most of its annual toll.

Conclusion

Predation pressure is rarely a negligible factor, except perhaps in some agricultural areas where predators have unfortunately been depleted. Its exact influence on quelea population dynamics varies widely among countries, yet remains to be objectively quantified. Even when inconspicuous, predators may significantly affect the success rate of many quelea breeding colonies, accounting in some cases for the loss of 20 to >50 per cent of the eggs laid. However, this figure is lower than most predation rates given for solitary nesting birds in both temperate and tropical zones. The influence of predation pressure on the distribution, behaviour, and energy budgets of roosting and foraging quelea, and the mortality rate they suffer during different seasons, need to be investigated.

The major impact of predation may have been its past influence in natural ecosystems which has shaped the biology of quelea. Today, although strong predation pressure is recorded only in well-preserved natural habitats, most quelea populations probably are faced with significant pressures at some time of their annual cycle or in some part of their annual range. This pressure, even if occasional, may help to maintain some important aspects of the quelea's life history such as foraging behaviour and breeding sites and perhaps limit population increases, range expansion, and crop damage.

The present high level of quelea populations in some parts of its range may be due to both an abundant availability of food supply throughout the year and to a major decrease in the number of natural predators, primarily through intensified agricultural practices. The long-term success of quelea population control programmes must not be lowered by further reduction of predator abundance whose impact is not yet fully appreciated.

17

Feeding ecology of quelea

WILLIAM A. ERICKSON

Introduction

Annual diet and seasonal changes in quelea food habits have been studied in Nigeria's Lake Chad region (Ward 1965*a,b*) and Ethiopia's Awash River Valley (Erickson 1979). In this paper, I review these studies and other data concerning quelea feeding behaviour, diet, damage to cereal crops, and feeding strategy.

Understanding quelea feeding ecology is necessary for assessing their potential adverse impact on cultivated cereals. Quelea are granivorous, feeding principally on grass seeds (Plate 14). Their range extends across approximately 6.5 million square kilometres of semi-arid grassland in sub-Saharan Africa (Crook and Ward 1968; Magor and Ward 1972). Quelea are able to exploit cultivated seeds due to their daily and seasonal mobility, morphological adaptations for seed eating, opportunistic food habits, and ability rapidly to locate ephemeral seed supplies (Dyer and Ward 1977; Wiens and Johnston 1977). Their large numbers and gregarious foraging flocks can result in extensive damage to ripening sorghum, wheat, rice, and millet crops (Allan 1983; Bruggers 1980; FAO 1981*a*; Jaeger and Erickson 1980). The seasonal abundance and availability of seeds, both wild and cultivated, also is critical in regulating populations, seasonal migrations and local movements, and the timing and localities of breeding (Chapters 9 and 13; Ward 1965*a,b*, 1971).

Study areas

Lake Chad region

Ward (1965*a,b*) studied the feeding ecology of *Quelea quelea quelea* over a 22-month period in 1961 and 1962 in Nigeria's Lake Chad region. A vast lake previously covered much of this region so that the present-day plains are covered by a layer of black cotton soil that becomes inundated during the wet season (July–October). Except for local patches of *Acacia seyal*, the region is mostly devoid of trees. Common grasses include *Echinochloa colona*, *Panicum* spp., *Oryza barthii*, and wild *Sorghum* spp. Cultivated grain sorghum is sown from the end of the rains into the early dry season and matures from December to March.

Awash River Valley

Erickson (1979, 1984) studied the diet of *Q. q. aethiopica* over 1 year during 1977 and 1978 in Ethiopia's upper and middle Awash River Valley as part of a broader study of the food habits of five weaverbird species (Ploceidae). The upper Awash River Valley, at 1500–1700 m elevation, is intensively cultivated in dryland cereal crops of Maize *Zea mays*, Grain Sorghum *Sorghum bicolor*, Teff *Eragrostis tef*, Wheat *Triticum durum*, and Barley *Hordeum vulgare*. Heavy rains fall from late June or early July into September with light rains occurring during March and April. Mean annual rainfall is 820 mm.

The middle Awash River Valley, at 750 m elevation, is characterized by semi-arid grassland and Thorn-bush *Acacia mellifera*, *A. nubica*. Cereal cultivation is limited to small fields of maize grown for fodder on a national farm. *Echinochloa* spp. and *Eriochloa meyeriana* grasses abound in dry-season irrigated cotton fields. Heavy rains fall in July and August. Mean annual rainfall is 520 mm.

Feeding behaviour

Morphological adaptations for seed eating

Quelea possess a stout, blunt bill, an adaptation for granivory (Lack 1968; Moreau 1960; Wiens and Johnston 1977). Mean bill length is 14.9 mm for adult male and 14.6 mm for adult female *Q. q. aethiopica* (Table 17.1). Juvenile quelea do not attain adult body size or bill dimensions until about 6–8 months of age. Ward (1965*a*) found seasonal changes in *Q. q. quelea* bill length in Nigeria—increasing slightly during the wet season but decreasing significantly through the dry season due to wear of the outer keratin sheath as seeds were gleaned off the ground. No significant seasonal changes in bill dimensions were found in Ethiopia (Erickson 1984). Body weight (crop empty) of adult male and female *Q. q. aethiopica* in Ethiopia average 19 and 18 g, respectively (Table 17.1), similar to the weights reported by Ward (1965*a*) for *Q. q. quelea* in Nigeria.

Like many granivorous birds, quelea possess a distensible crop for

Table 17.1. Body weights (crop empty) and bill dimensions of *Q. q. aethiopica* collected monthly during a 1-year period in Ethiopia's Awash River Valley (source: Erickson 1984). Juveniles were approximately 2–3 months old when collected.

				Bill dimensions	
Sex	Age	*n*	Body weight (g) $\bar{x} \pm$ SE	Length[a] (mm) $\bar{x} \pm$ SE	Depth[b] (mm) $\bar{x} \pm$ SE
Male	Adult	150	19.0±0.10	14.9±0.04	10.4±0.03
Female	Adult	150	18.0±0.08	14.6±0.04	10.2±0.03
Male	Juvenile	50	17.7±0.12	14.6±0.07	10.1±0.04
Female	Juvenile	50	17.1±0.15	14.3±0.06	10.0±0.04

[a]Bill length was measured from the external opening of the nares to the tip of the culmen.
[b]Bill depth was measured immediately posterior to the nares.

temporary storage of food items (Fig. 17.1). The crop enables them to gather seeds rapidly, thereby minimizing foraging time and reducing exposure to heat stress and predation in open feeding areas (Ward 1978). Seeds pass slowly from crop to gizzard and digest while the birds rest during midday and roost at night.

Daily feeding periods

Quelea feed intensively during early morning and late afternoon. Flocks leave roosts at dawn for feeding sites where they typically feed for 2 or 3 h. When food is scarce, as frequently occurs during the late dry season, or on cool days, feeding sessions often last longer. Intensive bird-scaring by farmers may disturb birds feeding in grain fields and cause them to continue feeding well into late morning or early afternoon (Elliott 1979; Ruelle and Bruggers 1982). Between feeding sessions, quelea congregate in numerous small, scattered day-roosts near their feeding sites. When nesting, adults bring food to their young throughout much of the day and the bimodal feeding pattern is less pronounced or absent.

Quelea must drink water at least once each day (Plate 17). Usually they drink en route to their night-roost (Crook 1960; W. Erickson, personal observation). They also may drink during the hot midday hours when congregating in day-roosts if water is available nearby. Food normally is present in quelea crops by mid-morning, and frequent wetting of the hard, dry seeds may facilitate digestion (Ward 1978).

Fig. 17.1. Quelea can feed quickly by filling their distensible crops, allowing easy diet studies by examination of crop contents (photo: R. Bruggers).

Feeding techniques

Techniques that quelea use to obtain food vary seasonally. During the wet season they perch on grass stems or heads and extract ripening seeds. Small seeds, such as most wild grass seeds, are simply squeezed out from between the bracts. Quelea also supplement their diet with insect larvae and nymphs during the wet season, but they are not adept at catching flying insects. Through much of the dry season, quelea feed on the ground on the dry, mature seeds that are shed in abundance when grasses fruit at the end of the rains. Individual members of a feeding flock search for seeds in a seemingly random and independent manner when seed is plentiful (Ward 1965*a*). If seed becomes scarce, quelea will scratch the soil surface to expose buried seeds; intraspecific competition can be intense and starvation may occur.

Flocks feed in such a manner that they appear to 'roll' across the feeding grounds. Such behaviour, also described for finches (Fringillidae) by Newton (1967), is particularly evident during the late dry season when seed supplies diminish and flock sizes increase. As a flock moves across a feeding site, birds in front deplete the seed supply, and those at the rear must fly to the front to obtain seeds (Jarvis and Vernon, in press-*a*; Ward 1965*a*; W. Erickson, pers.

obs.). Crook (1960) observed similar feeding behaviour by quelea in cultivated rice fields.

Communal food-finding

Quelea are gregarious and feed in flocks of a few hundred individuals to many thousands. Feeding flocks generally are relatively small but numerous when grass seed is abundant and widespread during the early dry season. Seed supplies diminish as the dry season progresses, a result of consumption by many species of granivorous birds, rodents, and insects (Gaston 1973; Payne 1980), and feeding flocks increase in size but diminish in number. Quelea then congregate in areas such as fertile grasslands, marsh fringes, and irrigated or harvested fields where seed density remains high (Erickson 1979; Ward 1965*a*).

Quelea roost and nest in massive aggregations that may benefit all individuals in finding patchily distributed food resources. Night roosts, or 'primary roosts' (Ward and Zahavi 1973), may contain many thousands to several million quelea converging from a wide geographical area at dusk. Roosts form daily throughout the non-breeding (dry) season; and, like flock size, most of the larger roosts are found during the late dry season when seed supplies are scarcest (Jaeger *et al.* 1979; Ward 1965*a*). In Ethiopia, quelea also commonly roost and feed with the Northern Red Bishop *Euplectes franciscanus* during the dry season (Erickson 1984). Roosts can be as far as 50 km from feeding grounds (Ward 1978), but may shift when new feeding or drinking sites are located.

In the breeding (wet) season, quelea congregate in enormous, densely packed nesting colonies (Chapter 14). Large quantities of ripening grass seeds and insects are needed to meet the energetic and nutritional demands of adults and their offspring (Jones and Ward 1976; Morel 1968; Ward 1965*b*). Proper timing of breeding and finding favourable feeding sites are critical for ensuring successful completion of the nesting cycle (Jones and Ward 1979).

Diet

Plant foods

Wild grass seeds Quelea subsist mainly on wild and cultivated grass seeds throughout the year. Wild grass seeds comprised 88 per cent (by dry weight) of their annual diet in Nigeria's Lake Chad region (Ward 1965*a*) and 69 per cent in Ethiopia's Awash River Valley (Fig. 17.2; Erickson 1979). In Somalia, 84 per cent ($n = 3282$) of quelea sampled monthly during a 4-year period ate wild seeds, mostly grasses (Bruggers 1980).

(a)

(b)

Plate 13. (a) The Red-necked Falcon *Falco chicquera* is a frequent predator on quelea in the Sahel, taking them on the wing (photo: M.-T. Elliott). (b) Lanner Falcons *Falco biarmicus* use the same technique but for both species hunting is made difficult by the tight quelea flocks and their manoeuvres (photo: B. Davidson).

Plate 14. Quelea flocks roll across the countryside, feeding on the seeds of abundant wild grasses (photo: J. Jackson).

Plate 15. A trap, simulating water with slit shiny plastic sheeting, caught hundreds of juvenile quelea in a few hours in Chad (photo: J. Jackson).

Plate 16. Varieties of sorghum differ in size, shape, tightness of head, and in their tannin content making some much less susceptible to quelea damage than others (photo: R. Bullard).

Plate 17. Quelea must drink at least twice a day and waterholes may attract a dense concentration causing trees and the water to be fouled for stock. Some birds can be pushed into the water by their neighbours and can be caught by crocodiles or raptors (photo: B. Davidson).

Plate 18. Explosives of gelignite were used in West Africa to control quelea roosts (photo: R. Bruggers).

Plate 19. The fire-bomb consisting of a 200–1 drum of petrol–diesel and gelignite as developed in colonial days produced a sheet of flame from ground level to the tops of 30-m trees to kill roosting birds (photo: R. G. Allan).

Plate 20. A jerrican containing a petrol–diesel mixture, with 3 sticks of gelignite strapped to its bottom, is hauled into a eucalyptus tree—a frequent roost site—in Kenya. The flame and explosion technique can destroy quelea roosts of up to about one hectare (photo: M.-T. Elliott).

Plate 21. Aerial spraying of the organophosphate avicide fenthion is the most frequently used form of lethal control for quelea (photo: B. Johns).

Plate 22. After properly conducted lethal control operations, the ground can be carpeted with dead birds (photo: E. Dorow).

Plate 23. Where quelea roosts are in aquatic habitats, a large variety of non-target water birds can be killed by the spray (photo: M.-T. Elliott).

Plate 24. In Chad, professional trappers use these nets to catch up to 1000 roosting quelea per man per night for subsequent sale as food in local markets (photo: M.-T, Elliott).

Plate 25. After spraying, quelea are sometimes collected by the sackful (5000 birds per sack) to be used as food by the local people (photo: E. Dorow).

Plate 26. In-service training in the field provides a most effective method of learning the routines of quelea work (photo: J. Jackson).

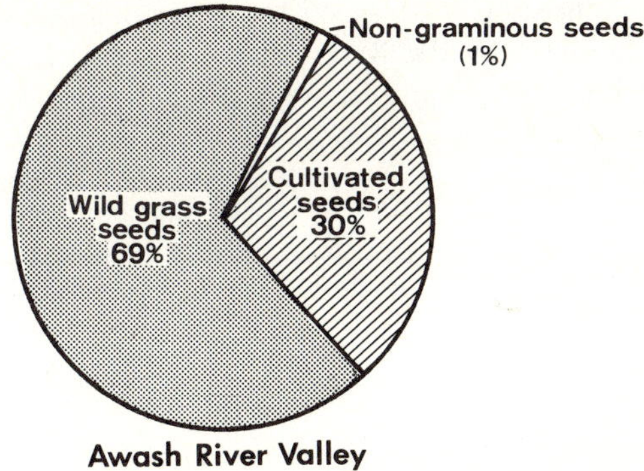

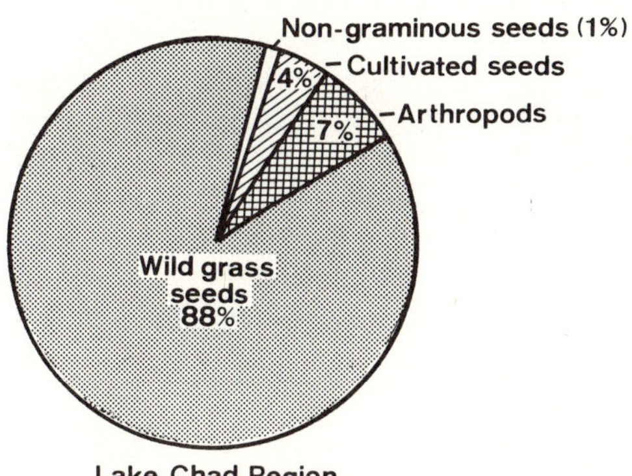

Fig. 17.2. Proportions of major food categories (based on dry weight) comprising annual diets of *Q. quelea* in Ethiopia's Awash River Valley (source: Erickson 1979) and Nigeria's Lake Chad region (source: Ward 1965*a*).

Despite the abundance and diversity of wild grasses in African grasslands, only about a dozen seed types are regularly eaten by quelea in any region. Many are seeds of fast-maturing annuals; these are among the first grasses to ripen after the heavy rains begin (Froman and Persson 1974). Major genera include *Sorghum* (wild sorghums), *Oryza* (wild rice), *Echinochloa*, *Urochloa*, *Setaria*, *Panicum*, *Tetrapogon*, *Eriochloa*, and *Pennisetum* (Erickson 1984; Jarvis and Vernon, in press-*a*; Morel 1968; Ward 1965*a*). Seeds of these genera probably comprise the major portion of the quelea diet throughout uncultivated sub-Saharan grasslands; but the proportion of these seeds in the diet varies from area to area, depending on their abundance and availability. *Echinochloa colona*, because it is widespread and abundant in grasslands and agricultural fields, is important almost everywhere. These seeds are mostly small (0.5–1.5 mg) or medium (5.0–10.0 mg) in size (Table 17.2). According

Table 17.2. Classification by size of the principal seeds eaten by *Q. q. aethiopica* in Ethiopia's Awash River Valley (source: Erickson 1979) and Nigeria's Lake Chad region (source: Ward 1965*a*).

Size class Seed item	Mean dry seed weight (mg)	Mean seed dimensions (mm)
Minute		
Eragrostis papposa	0.1	0.5 × 0.5
Digitaria velutina	0.2	1.0 × 0.5
Eragrostis tef (teff)	0.3	1.5 × 0.5
Schoenefeldia gracilis	0.3	1.5 × 0.5
Setaria verticillata	0.4	1.0 × 1.0
Dactyloctenium aegyptium	0.4	1.0 × 1.0
Small		
Eriochloa meyeriana	0.6	1.5 × 1.0
Setaria pallidifusca	0.7	1.5 × 1.0
Echinochloa spp.	0.8–1.0	1.5 × 1.0
Tetrapogon cenchriformis	0.8	2.5 × 1.5
Urochloa spp.	0.9	2.0 × 1.5
Hyparrhenia anthistirioides	0.9	3.0 × 0.5
Tetrapogon tenellus	1.0	3.0 × 1.0
Setaria acromelaena	1.3	2.0 × 1.5
Panicum sp.	1.3	2.0 × 1.0
Medium		
Sorghum spp.	5.0	4.0 × 2.0
Sorghum purpureo-sericeum	9.8	5.0 × 2.0
Large		
Hordeum vulgare (barley)	15.2	8.0 × 3.0
Oryza barthii	17.5	8.0 × 2.0
Sorghum bicolor (sorghum)	20.8	5.0 × 4.5
Triticum durum (wheat)	24.2	7.0 × 3.5

to Ward (1965a) and Erickson (1979), these size classes comprise 60–95 per cent of the annual quelea diet (Fig. 17.3). Seeds less frequently eaten, but occasionally of seasonal dietary importance, include *Dactyloctenium*, *Schoenefeldia*, *Digitaria*, *Hyparrhenia*, *Lintonia*, *Cenchrus*, *Diplachne*, and *Eragrostis*; most are minute, weighing less than 0.4 mg. Seeds other than grasses, such as *Commelina*, *Indigofera*, *Amaranthus*, and *Portulaca*, are seldom eaten.

Cultivated seeds Cultivated cereals can be a major food source for quelea. Most, except for teff seeds (0.3 mg), are considerably larger than the wild grass seeds that quelea eat. Sorghum, wheat, barley, and rice kernels weigh

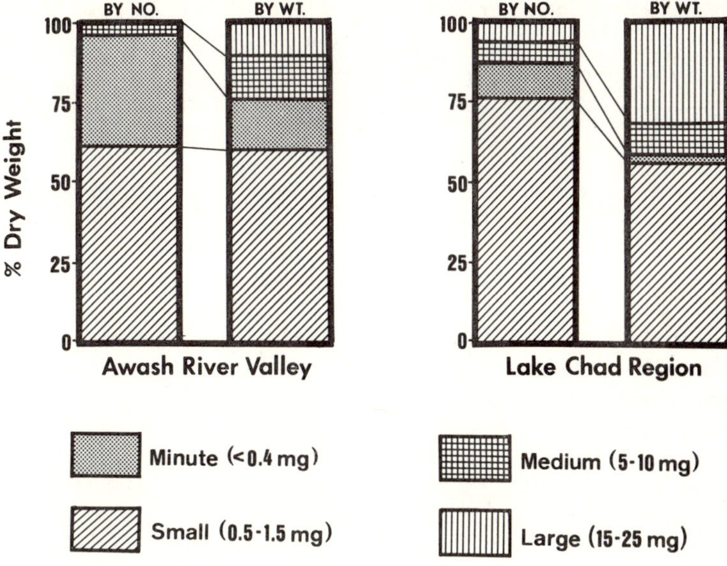

Fig. 17.3. Seed sizes of the annual diets of *Q. quelea* in Ethiopia's Awash River Valley (source: Erickson 1979) and Nigeria's Lake Chad region (source: Ward 1965*a*).

15–25 mg. Maize kernels are much larger and eaten only when soft, after Black-headed Weavers *Ploceus cucullatus* have slit open the husks or when fractured during threshing operations (Bruggers 1980; Erickson 1984; Jarvis and Vernon, in press-*a*). Quelea also ate water-softened kernels available in abandoned maize fields after the Awash River flooded during the early 1977–1978 dry season (Erickson 1984). In the Lake Chad region, ripening sorghum comprised only 4 per cent of the quelea's annual diet (Ward 1965*a*), but teff, sorghum, maize, and barley constituted almost 50 per cent of their diet in the upper Awash River Valley (Table 17.3). However, sorghum was the only cereal crop extensively damaged prior to harvest (Fig. 17.4). Teff is not eaten when ripening during the early dry season because quelea cannot easily perch on the fragile stems and panicles (Erickson 1979). They will eat ripening teff if plants lodge during heavy rains, and they also occasionally glean sown seeds in fields broadcast when the heavy rains begin in July. Most cultivated seeds were gleaned from bare fields, threshing sites, and occasionally around stores after harvest. In Somalia, 15 per cent of all quelea collected throughout the year ate cultivated grain, mostly sorghum and rice. Bruggers (1980) found that they ate most grain during January and September when crops were maturing (Fig. 17.5).

Table 17.3. Diet of male and female *Q. q. aethiopica* in Ethiopia's Awash River Valley in 1977–1978 (source: Erickson 1984). Food items comprising <1.0% of the annual diet are not tabulated.

	% dry weight			
	Upper Awash		Middle Awash	
Food item	Male (19/720)[a]	Female (19/803)	Male (12/514)	Female (12/432)
Eragrostis tef (teff)	18.8	27.1	—	—
Sorghum bicolor (sorghum)	11.7	7.8	—	—
Zea mays (maize)	7.9	4.0	—	—
Triticum durum (wheat)	6.7	5.3	—	—
Hordeum vulgare (barley)	1.6	2.2	—	—
Echinochloa spp.	8.0	5.6	40.9	44.8
Sorghum spp.	22.5	21.3	3.8	2.2
Echinochloa meyeriana	—	—	16.4	16.5
Urochloa spp.	6.6	6.8	4.2	6.2
Tetrapogon cenchriformis	3.8	2.3	7.1	6.4
Eriochloa sp.	2.3	2.5	—	—
Setaria acromelaena	—	—	2.0	2.7
Setaria pallidifusca	2.9	4.4	—	—
Eragrostis papposa	1.9	2.9	—	—
Hyparrhenia anthistirioides	0.8	2.3	—	—
Cenchrus spp.	—	—	2.1	0.9
Unidentified wild grass seed	—	—	1.3	0.7
Commelina sp.	1.0	1.4	—	—
Insects	—	—	2.7	2.2

[a]Nineteen pooled samples of the crop contents of 720 individuals.

Animal foods

Insects constitute most of the animal food that quelea eat (Morel *et al.* 1957; Vesey-FitzGerald 1958; Ward 1965*a,b*). Grasshoppers, mostly nymphs, caterpillars, small beetles, and winged termites are most commonly eaten, but various bugs and flies are also occasionally taken. Jones and Ward (1979) also found several spiders in quelea crops sampled at a nesting colony in Nigeria. Insect availability is highly seasonal. A flush of insect life occurs during the early rainy season but numbers decrease rapidly when the rains stop (Sinclair 1978). Although insects usually account for only a small proportion of the annual diet (Fig. 17.2), they are important at critical times such as when quelea breed and require additional protein (Jones and Ward 1976).

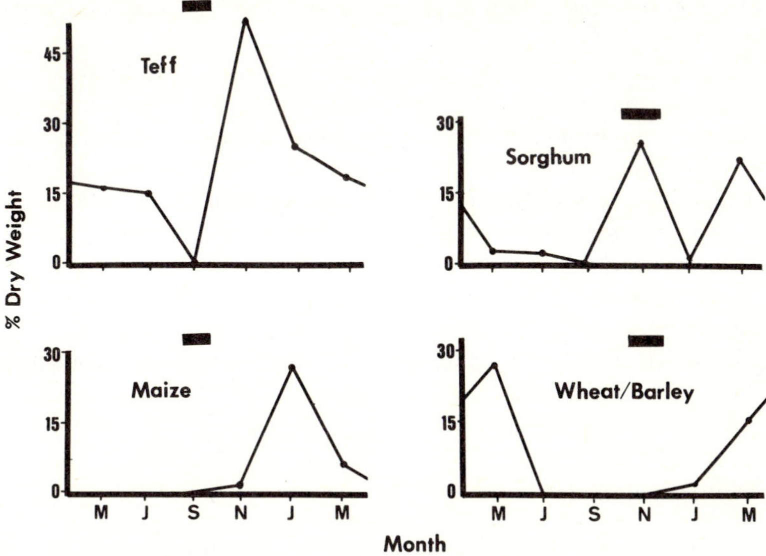

Fig. 17.4. Proportions of cultivated cereals in the diet of *Q. q. aethiopica* in Ethiopia's upper Awash River Valley in 1977–1978 (source: Erickson 1979). The black bars indicate periods when cereal crops were ripening.

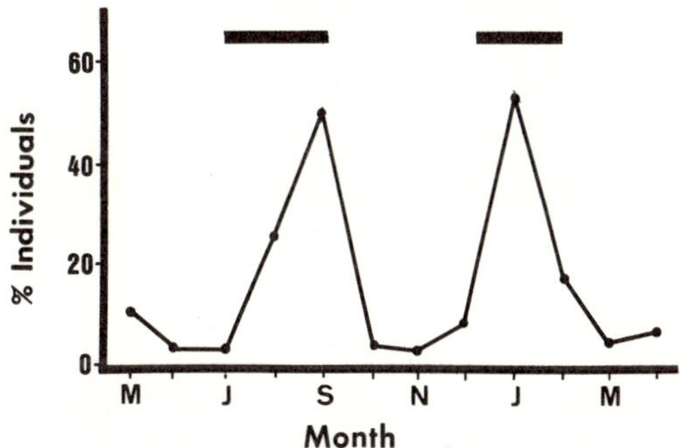

Fig. 17.5. Monthly proportions of *Q. quelea* feeding on cultivated rice and sorghum near Afgoi, Somalia, between November 1975 and July 1979 (source: Bruggers 1980). The black bars indicate periods when cereal crops were ripening.

Quelea opportunistically exploit outbreaks of armyworm *Spodoptera* spp. larvae. Peter Ward observed quelea eating armyworms in a ripening wheat field in Tanzania (Dyer and Ward 1977). Nesting quelea exploited a vast outbreak of armyworms in the southern Ethiopian Rift Valley in 1980; crops of 200 male and female quelea sampled during the incubation period contained armyworms almost exclusively (W. Erickson, unpubl. data). In this situation, ripening grass seed was not available because of a prolonged drought, and it seems unlikely that breeding would have occurred without the armyworm outbreak.

Molluscs are the only other animal food that quelea are known to eat. Jones (1976) found fragments of snail shells and other non-living calcareous material in crops of adults during the egg-laying period. Such calcareous matter probably augments the large amount of calcium needed for eggshell formation (Jones 1976). Small snail shells comprised 1.4 per cent (by dry weight) of the crop contents of 2-week-old fledglings at a nesting colony in Ethiopia (W. Erickson, unpubl. data). The shells were regurgitated, along with green grass seeds and insects, by their parents. Mathew (1976) also found mollusc shells in stomachs of nestling Baya Weavers *Ploceus philippinus* in India. The presence of calcareous shells in their crops is surprising but may be related to requirements for grit (Royama 1970).

Seasonal changes in diet

Wet season Quelea diet varies considerably between the wet and dry seasons, a consequence of major fluctuations in seed and insect availability due to seasonal rainfall. Prior to their early-rains migration, quelea feed on winged termites that swarm for a few days after the rains begin (Ward 1965*a*; Ward and Jones 1977). Termites probably are not essential for premigratory fattening, but they are temporarily abundant in many areas after seeds have germinated.

During breeding, quelea supplement their seed diet with insects, as do other granivorous ploceids (Collias and Collias 1970, 1971; Lack 1968). Grass seeds contain only about 8–13 per cent protein, whereas insects contain from 65 to 70 per cent protein (Doggett 1970; Huffnagel 1961; Jones and Ward 1976). High protein levels are important for egg production and rapid growth and development of nestlings. Adult quelea with insufficient protein reserves may not be able to sustain a nesting attempt and may desert partially completed clutches (Jones and Ward 1979). Nestlings fed only seeds likely will not survive without some insect food (P. Ward, pers. comm.). Grasses importantly utilized during breeding are those ripening earliest in the wet season: *Echinochloa, Setaria, Urochloa, Panicum, Tetrapogon,* and *Pennisetum* (Morel 1968; Ward 1965*b*; W. Erickson, unpubl. data).

Early dry season Geographical dietary differences were most pronounced during the dry season. Cereal crops ripen during the early dry season in the upper Awash River Valley, and they provide an abundant and easily obtainable seed supply. Ripening and waste cereal grains comprised 46–71 per cent (by dry weight) of the quelea's monthly diet through the dry season (November–May in 1977 and 1978); wild sorghums accounted for 19–34 per cent (Erickson 1979). Quelea mainly eat small wild grass seeds throughout the dry season in the semi-arid middle Awash River Valley. In October they disperse to higher elevations, including the upper Valley, and mostly feed on cereals. By December, many begin returning to the middle Valley where *Echinochloa* spp. and *Eriochloa meyeriana* weeds are seeding in irrigated cotton fields. In the Lake Chad region, where cereals are sown during the early dry season, quelea mostly continue eating small wild grass seeds but also feed on later maturing wild rice seeds (Ward 1965*a*).

Late dry season By the late dry season, wild seeds often may be scarce. Although the quantity present varies from year to year, the amount of seed on the ground may be as little as 10 per cent of that produced (Gaston 1973; Payne 1980). Irrigated cereals frequently are severely damaged, but dry-season cultivation is limited in most areas (Bruggers and Ruelle 1981; Jaeger and Erickson 1980).

Minute grass seeds such as *Dactyloctenium*, *Schoenefeldia*, *Digitaria*, and *Eragrostis* feature in the quelea's diet mainly during the late dry season. These minute wild seeds frequently are consumed in large numbers, but because they are so small, they often contribute relatively little to the total biomass of food intake. Considerable time and energy must be expended to gather and ingest these tiny seeds for little energetic reward per item. Evidently, they are mostly eaten when larger seeds are scarce and difficult to locate (Ward 1965*a*). An exception is cultivated teff which is concentrated at threshing sites after harvest; these minute seeds are eaten in large amounts throughout the dry season in the upper Awash River Valley.

Dietary differences between sexes

Dietary differences between male and female quelea are negligible outside the breeding season (Erickson 1984). Both sexes feed on the same wild and cultivated seeds in similar proportions (Table 17.3). Ward (1965*a*) suggested that bill-size dimorphism could result in males taking larger seeds than females during periods of food shortage, especially during the late dry season. Dolbeer (1980) and Linz and Fox (1983) found size-dimorphic differences among Red-winged Blackbirds *Agelaius phoeniceus*, Crase and DeHaven (1978) with Tricolored Blackbirds *A. tricolor*, and Holyoak (1970) with Carrion Crows *Corvus corone*. Male blackbirds possess a bill 11–14 per cent longer and deeper than females (Orians 1961), whereas for quelea the difference is only about 2 per cent (Table 17.1).

During breeding, male-female diets differ by the proportion of insects consumed. Jones and Ward (1976, 1979) and Ward (1965*b*) found that arthropods, principally insects, comprised 10–25 per cent of the females' diets but only 2–10 per cent of the males' diets during laying, incubation, and rearing periods. These differences were most likely due to greater protein requirements of the females when producing eggs (Ankney and Scott 1980; Jones and Ward 1976).

Sexual segregations, reported as common among quelea in the dry season (Elliott 1980*b*; Jaeger *et al.* 1979; Ward 1965*d*), may relate to food competition. In Ethiopia, males dominate flocks and roosts in the middle Awash River Valley and females in the upper Valley throughout the dry season (Jaeger *et al.* 1979). Ward (1965*d*) reported an increasing proportion of males as the Nigerian dry season progressed and suggested that males out-compete females when food is scarce. Laboratory trials have indicated that male quelea can out-compete females when food is limited (Crook and Butterfield 1970; Dunbar and Crook 1975). Wilson (1978) postulated that Red-winged Blackbirds sometimes forage in sex-specific flocks to reduce intersexual competition for food. It is unclear if quelea partially segregate by sex to reduce competition for limited food resources or if other factors are responsible.

Dietary differences between age groups

Diets of adult and immature quelea were compared from roosts in the Awash River Valley during the 1977–1978 dry season. Quelea of both age groups ate the same food items (Table 17.4); however, adult quelea damaged considerably more ripening sorghum whereas young birds gleaned more waste grains of maize and teff (Erickson 1979). Adults were mainly responsible for damaging sorghum in the upper Awash Valley in 1977–1978; but during other years and in other areas of Ethiopia, young birds caused severe damage (Erickson *et al.* 1980; Jaeger and Erickson 1980). In 1976, quelea flocks consisting almost entirely of juveniles (cranial pneumatization indices 0–1) damaged an estimated 51 per cent of 35 000 ha of ripening sorghum in eastern Ethiopia.

Quelea partially segregate by age among roosts and feeding flocks during the dry season (Jaeger *et al.* 1979). Adults abandon their young within days after the young become independent (Crook 1960; W. Erickson, pers. obs.). After young disperse from breeding areas, and even through the dry season, they may remain in roosts and flocks separate from adults (Jaeger *et al.* 1979; Morel *et al.* 1957). Segregation of feeding flocks by age was most pronounced in the Awash River Valley in November, shortly after breeding, when food was plentiful and intraspecific competition was not likely to be intense. Why young quelea remain so segregated is unclear.

Damage to ripening sorghum in the Awash River Basin of Ethiopia

Extent of damage

Ripening grain sorghum commonly is attacked by quelea in the major

Table 17.4. Diet of adult and juvenile *Q. q. aethiopica* from sites where both were collected in November, December, February, and March (1977–1978) in Ethiopia's Awash River Valley (source: Erickson 1984). Food items comprising <1.0% of the diets are not tabulated.

	% dry weight			
	Upper Awash		Middle Awash	
Food item	Adult (6/226)[a]	Juvenile (6/154)	Adult (6/216)	Juvenile (6/221)
Eragrostis tef (teff)	24.9	57.1	—	—
Zea mays (maize)	1.6	1.5	23.6	33.2
Sorghum bicolor (sorghum)	32.6	10.3	—	—
Triticum durum (wheat)	2.6	3.2	—	—
Sorghum spp.	26.8	17.9	8.4	4.3
Echinochloa spp.	0.6	1.0	28.3	24.6
Eriochloa meyeriana	—	—	27.5	33.7
Eragrostis papposa	3.8	2.2	—	—
Hyparrhenia anthistirioides	2.4	1.5	—	—
Cenchrus spp.	—	—	2.8	1.7
Digitaria velutina	—	—	1.7	1.1
Unidentified wild grass seed	—	—	2.3	0.8
Urochloa spp.	—	—	2.3	0.2
Setaria verticillata	0.5	1.9	—	—
Setaria pallidifusca	0.3	1.8	—	—
Commelina sp.	1.2	0	—	—
Portulaca sp.	—	—	1.1	0
Unidentified seed	1.3	1.1	—	—

[a] Six pooled samples of the crop contents of 226 individuals.

sorghum-growing regions (1000–2000 m elevation) associated with Ethiopia's Awash River Basin. Ripening sorghum was the principal food of those quelea collected during early dry seasons between 1976 and 1981 (Jaeger and Erickson 1980; W. Erickson, unpubl. data). Bird damage in an area may vary considerably from year to year, but estimated annual losses of ripening grain, primarily by quelea, averaged a minimum of 10–15 per cent in the 200 000–300 000 ha of sorghum cultivation (Table 17.5), exceeding 40 000 t annually.

Sorghum is not grown extensively in the upper Awash River Valley because of quelea damage. Many farmers are reluctant to grow sorghum, the cereal crop best adapted to the low, erratic rainfall conditions of the area (Brhane Gebrekidan, pers. comm.). Ripening sorghum comprised 25 per cent of the quelea's diet in this area in November 1977 (Fig. 17.4); at one major roost it constituted 80 per cent of the adults' diet (Erickson 1979).

Table 17.5. Bird damage to ripening sorghum in major sorghum-growing regions associated with Ethiopia's Awash River Basin in 1976 and 1977 (source: Jaeger and Erickson 1980).

Location	Estimated no. ha (thousands)	Harvest period	% damage[a] 1976	1977
Chefa	10–15	Nov.–Dec.	18.6	9.2
Robi/Jawa	15–20	Oct.–Nov.	10.3	16.1
Melkassa	0·015[b]	Oct.	41.8	40.3
Afdem/Miesso	20–30	Oct.–Nov.	13.3	NR[c]
Jijiga Plain	1–35	Sept.–Oct.	51.4	NR

[a]Damage levels may be somewhat low because assessments were made 1–2 weeks prior to harvest.
[b]Ethiopian Sorghum Improvement Project research field; treated with methiocarb both years.
[c]Damage was not recorded but reported to be locally severe.

Damage assessments conducted in a 15-ha sorghum research field at Melkassa between 1976 and 1980 indicated annual yield losses ranging from 3 to 22 per cent (Erickson *et al.* 1980).

Sorghum is not currently grown in the middle Awash River Valley. Previous attempts to grow it on national farms and research fields consistently failed due to quelea attacks. In February–March 1977, for example, approximately 80 per cent of the 160 ha of ripening sorghum crop was lost to quelea (M. Jaeger, pers. comm.), despite an abundance of wild grass seeds in adjacent cotton fields (W. Erickson, pers. obs.).

Phenology of damage

Manikowski and Da Camara-Smeets (1979*a*) found damage to sorghum occurring over a 3- to 4-week period, beginning when seeds reached the milky stage of ripeness and continuing until harvest. In Ethiopia, the period of vulnerability to bird damage is longer because in any one area planting dates are highly asynchronous. Ethiopians grow more than 5500 sorghum varieties, and in most agricultural areas many different varieties are sown (Brhane Gebrekidan, pers. comm.). Planting synchrony and sowing fewer, faster-maturing varieties would help shorten the vulnerable period to bird damage.

Because sorghum damage often is most severe when grain is in the milky and soft dough stages, it seems that quelea prefer soft grain and avoid mature seeds, probably because the latter are too large and hard. In the upper Awash

River Valley, however, mature sorghum kernels (21 mg dry weight) were commonly found intact in quelea crops (Erickson 1979). Larger kernels of some sorghum varieties also are broken by quelea as evidenced by partially eaten fragments scattered on the ground in grain fields (Jarvis and Vernon, in press-*a*; W. Erickson, pers. obs.). As pointed out by Dolbeer *et al.* (1984) in relation to blackbird damage to maize in the United States, milky kernels contain up to 75 per cent water and birds need to eat more grain to obtain the equivalent dry-weight biomass of mature seeds that contain only 10–15 per cent water. Thus, the extent of damage may be greater when grain is soft because quelea must eat more to obtain comparable energy and nutrition levels.

Feeding strategy

Optimal foraging theory is based on the premise that natural selection favours those individuals that forage to maximize their net rate of energy intake (Goss-Custard 1977; Krebs *et al.* 1983; Pyke *et al.* 1977). A quelea should be expected to obtain a maximum amount of food energy by expending a minimum amount of time and energy searching for and handling food items. In essence, successful animals must forage efficiently, although certain constraints such as nutrient requirements during breeding or discrimination errors among food items may at times limit their efficiency (Krebs and McCleery 1984; Pulliam 1975). Two important components of quelea's foraging strategy likely are deciding which food items to eat and where to search for them (Pyke *et al.* 1977).

Choice of seed items

Quelea might perceive differences in physical characteristics or nutrient qualities of seed, but few data are available about how such factors affect their diet. Captive quelea discriminated among food pellets differing in colour, size, hardness, and texture (Bullard and Shumake 1979), and they avoided grain varieties high in astringent tannins when other food was available (Doggett 1970; Zeinelabdin *et al.* 1983). Differences in nutrient content among seed species also could influence seed selection, but most wild and cultivated seeds differ little in caloric value per unit of weight (4300–4900 cal/g; Kendeigh and West 1965; Pinowski *et al.* 1972) or in protein level. Ward (1965*a*, 1973*a*) suggested that seed size is important, with quelea preferring small (0.5–1.5 mg) wild grass seeds. However, captive quelea preferred eating millet seeds weighing approximately 8 mg (Manikowski and Da Camara-Smeets 1979*b*), although differences in other qualities among the seeds offered were not examined. Quelea in Ethiopia ate wild and cultivated seed at widely varying size (Fig. 17.3), suggesting that the kind or size of grass seed was less important than seed biomass at feeding sites. Quelea diet probably is influenced more by feeding site selection than by the choice of individual items to eat at a site.

Choice of feeding sites

Foraging studies of birds, including several insectivorous species, indicate that food-patch selection (i.e. where to feed) is extremely important for maximizing their net rate of energy intake (Davies 1977; Royama 1970; Smith and Sweatman 1974; Tinbergen and Drent 1980). Spotted Flycatchers *Muscicapa striata*, for example, switch during the day from feeding on small dipterans and aphids in tree canopies to large dipterans near the ground when the latter prey become more profitable in terms of net energy intake (Davies 1977). Titmice (Paridae) distinguish among food patches with various food densities and spend more time feeding in the most profitable patches (Smith and Sweatman 1974). In laboratory studies, granivorous birds, including quelea and Red-winged Blackbirds, behaved similarly (Alcock 1973; De Groot 1980).

Because seeds are patchily distributed, choosing profitable feeding sites must be a critical aspect of the quelea feeding strategy. The profitability of a feeding site is a measure of total available food biomass, a function both of seed size and seed density (Royama 1970). Quelea may not eat many kinds of seeds simply because most are not available in relatively profitable patches. Seasonal changes in diet probably result from changes in profitabilities of seed patches. Quelea are opportunistic feeders and rapidly respond to temporal and spatial changes in their food supply. Ripening cereal fields provide a high seed biomass per unit area (e.g. yields frequently exceed 1000 kg of seed per ha) and are highly conspicuous. The inclusion of cereals in the quelea's dry-season diet should be expected if they can obtain a higher rate of energy intake by feeding on equally palatable food in cultivated fields.

Locating profitable feeding sites

How do quelea learn to locate profitable feeding sites? Crook (1962, 1964) suggested that the gregarious social systems of open-country ploceids evolved because by grouping they could more easily and efficiently locate and exploit patchily distributed seed resources. Ward and Zahavi (1973) postulated that communal roosts and nesting colonies function primarily as information centres for finding food. They suggested that those individuals foraging poorly in unproductive feeding sites could gain access to more favourable areas by returning to the roost or colony and following more successful foragers to more productive feeding sites. Weatherhead (1983) provided evidence indicating that communal blackbird roosts can function both as information centres for locating food and, as Lack (1968) earlier suggested, for protection from predators.

Laboratory and field studies indicate that birds can rapidly learn the locations of productive feeding areas. Tits, for example, learned to forage for insect prey associated with certain locations or backgrounds (Royama 1970). Starlings quickly learned where to find food and regularly returned to sites where food was most abundant (Tinbergen and Drent 1980). De Groot (1980) experimentally demonstrated that naive quelea can quickly learn where to find profitable feeding sites when in the presence of knowledgeable birds; the means of information transfer were unclear, but a following response seemed to be important for some birds.

Conclusions

Quelea are adapted to exploiting concentrated supplies of grass seeds that are spatially and temporally variable and often unpredictable. Quelea rapidly respond to fluctuating seasonal availabilities of seeds, including cereals ripening during the early dry season. There is little evidence indicating that quelea prefer eating wild seeds and damage ripening cereals only if wild seeds are scarce. Rather, quelea feed opportunistically. If this is indeed their feeding strategy, it would be extremely difficult to predict precisely when and where cereal crop damage by quelea will occur.

18

Traditional African practices for preventing bird damage

EL SADIG A. BASHIR

Introduction

At least 70 per cent of the farmers in Africa are small farmers using essentially traditional farming practices to produce crops. They also use traditional methods to protect those crops from bird attacks from which they have suffered for countless generations. In this chapter, I describe, review, and evaluate these methods and suggest guidelines for their improvement through locally adaptive and biologically oriented research.

Traditional bird control methods are defined as those old, primitive practices that occupy an important part of the daily life of nearly all African farmers, and which have remained almost untouched by modern bird control technologies. In a Pan-African survey (FAO 1980*a*), traditional bird-scaring was the main technique used by 100 per cent of 32 countries (Table 18.1). The selection and prevalance of specific traditional methods vary among countries and ethnic groups. This variation is influenced by such factors as experience, availability of control materials, social and religious attitudes, and prejudices of various ethnic groups toward birds in general and pest birds in particular.

Bird-scaring methods

Various indigenous scaring methods are used to keep pest birds out of crop areas (Fig. 18.1). The most universally practised scaring method employs a combination of visual devices (scarecrows, flags) with auditory ones (shouting, cracking whips, rattling empty cans) reinforced by the hurling of missiles at the birds (Fig. 18.2). The missiles can be delivered by hand or from a sling or as a hard lump of mud flicked off the end of a stick (the latter is common in rice cultivation when there is plenty of mud). The shouting/

Table 18.1. Methods used in different African countries to control bird damage (source: FAO 1980*a*).

			Population reduction techniques			Crop protection	
Region	No. of countries	Crop	Aerial or terrestrial avicide	Explosives	Nest destruction	Bird-scaring	Repellents, nets, agronomic
North Africa	4	Wheat	2	1	4	4	(1)[a]
Sahel	14	Sorghum, millet	9	7	7	14	(8)
West Africa	4	Rice	0	0	0	4	(2)
East Africa	10	Maize	2 and (1)	2	1	10	(3)
Total	32		14	10	12	32	(14)

[a]Parentheses indicate only experimental use.

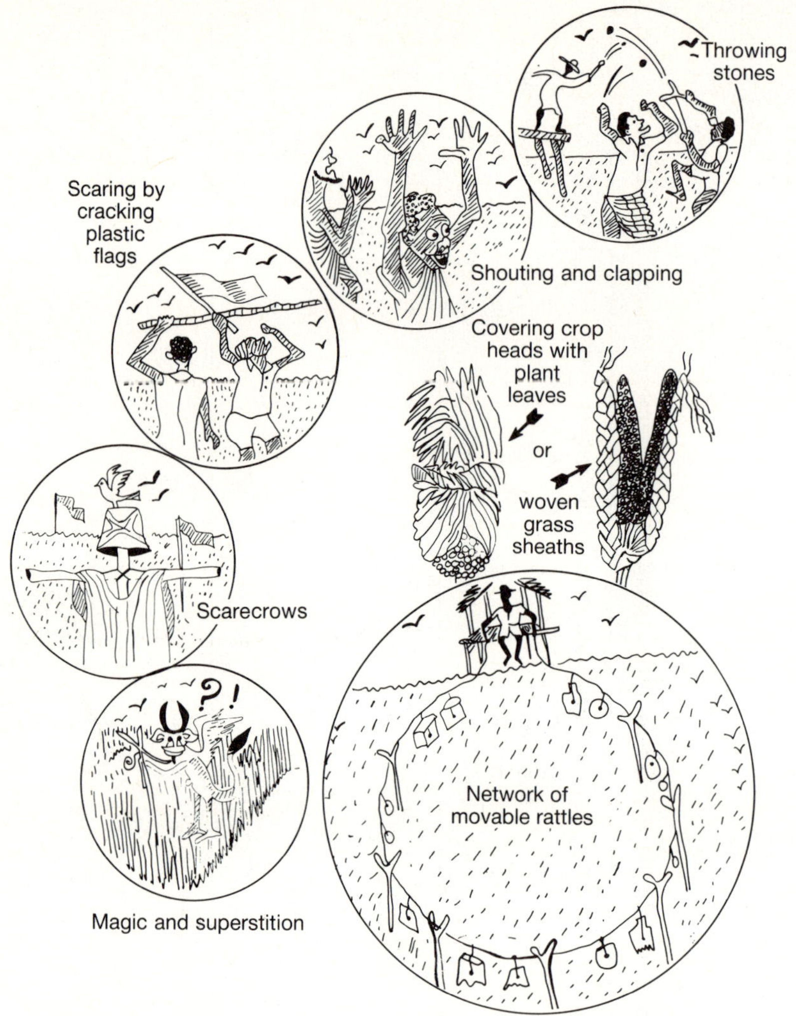

Fig. 18.1. An illustration of various traditional bird control methods used in Africa.

missiles method is sometimes improved by the construction of a platform or by using a natural protuberance such as a termite mound, to bring the scarer up above a tall crop so that he can oversee it properly. The next stage of sophistication encountered all over Africa is to equip the platform with a network of strings stretching to each corner of the field. Empty cans, rattles, and bottles are attached at intervals along the strings and the scarer jerks the strings when he sees a flock of birds approaching. Different sizes and types of rattles are used which can improve the noise produced.

(*a*)

(*b*)

Fig. 18.2. Traditional bird-scaring techniques such as (*a*) string and calabash gourd rattles (photo: R. Bruggers) and (*b*) cracking whips are used throughout Africa by farmers (photo: W. Fitzwater).

In many situations, these traditional techniques can reduce damage (Pepper 1973), but the practice is time-consuming. However, because farming is considered a full-time occupation and a way of life, farmers do not view time as having a monetary value as in industrial sectors. Farmers in Liberia, for example, invariably spend their leisure time involved in some form of economic activity related to farming (McCourtie 1973). Thus, erecting platforms and scaring birds are largely regarded as part of the daily farming routine. What matters to a farmer is the effectiveness of scaring rather than the time involved.

Factors affecting scaring

Farm size and ownership Scaring is more effective in small, privately owned, and persistently guarded farms than in large agricultural schemes guarded by hired labour. Pepper (1973) compared and evaluated five variables (area, surrounding habitat, number of scarers, age of scarers, and farm ownership) in 10 nearly equal-sized sorghum fields in Chad. He concluded that scaring by interested owners was more effective (0–5 per cent damage) than that by disinterested non-owners (80 per cent damage) despite the number of attacking birds. An elderly Chadian farmer and his daughter, for example, suffered 30 per cent loss to a l-ha wheat field compared to less than 5 per cent loss to an adjacent field protected by five active guards (Park 1974). Ruelle and Bruggers (1982) reported that average damage to privately owned sorghum, rice, and maize fields in Somalia ranged from 2 to 9 per cent , compared to between 25 per cent and 30 per cent on non-private farms. It seems that one bird scarer can efficiently protect between 0.5 and 1 ha, but success is greatly influenced by his age, enthusiasm, crop ownership, and crop value. In many African countries where most harvest is sold to the government at a fixed low price compared to the market value, farmers lose interest in protecting their crops from bird damage. In Somalia, Ruelle and Bruggers (1982) observed that farmers stopped protecting sorghum fields when they realized that, by selling the stalks as fodder, they could profit more than by selling grain to the government.

Labour input and economics Cereal crops usually are susceptible to bird damage for 30–35 days. Farm families, including children, must protect their crops from birds every day during the sowing and ripening periods, usually between 0600 and 1800h, for at least 2–3 months in non-homogeneously maturing crops. The economics of bird-scaring relative to crops saved is important. McCourtie (1973) showed that bird-scaring costs were the greatest single labour input of 11 farm operations in Liberia: erecting platforms and scaring birds accounted for 72 man-days/ha out of a total of 173 man-days of adult labour/ha (Table 18.2). In Liberia, the cost for 15–18

Table 18.2. Labour input (man-days/ha of rice) on bird-scaring in Liberia (source: McCourtie 1973).

	Bird-scaring					
	Adults		Children		No bird-scaring	
Activity	No. days	% of total	No. days	% of total	No. days	% of total
Cutting bush	17	9.8	17	16.5	17	18.4
Felling trees	8	4.6	8	7.8	8	8.6
Burning/cleaning land	7	4.0	7	6.8	7	7.5
Broadcasting/scratching rice	15	8.7	15	14.6	15	16.1
Cutting wood and fencing	8	4.6	8	7.8	0	0
Constructing kitchen	3	1.7	3	2.9	3	3.2
Weeding	13	7.5	13	12.6	13	14.0
Making scaffolding (platforms)	2	1.2	2	1.9	0	0
Bird-scaring	70	40.5	0	0	0	0
Reaping rice	19	11.0	19	18.4	19	20.4
Threshing rice	11	6.4	11	10.7	11	11.8
Total	173	100	103	100	93	100

bird scarers to protect 150 ha rice for 30 days in Cape Mount County was US \$1600, yet still resulted in an estimated 35 per cent loss (Bashir 1983). In Ghana in 1984, bird-scaring costs for 35 days for a mechanized-irrigated rice farm were \$120/ha, equivalent to about 8 per cent when compared to a total production cost of \$1533.21/ha (A. Dzakpazu, personal communication; Table 18.3). Labour costs have since tripled and market value has doubled, bringing the cost of scaring to nearly 20 per cent (A. Dzakpazu, pers. comm.). Nonetheless, profits resulted in a cost:benefit ratio of 1:6, indicating that bird-scaring was cost-effective in this situation.

In Sierra Leone (WARDA 1983), bird-scaring for rice paddy production accounted for 7.1 man-days/ha out of about 381 man-days/ha. Ruelle and Bruggers (1982) reported 38 per cent loss in a 2-ha research farm of seed

Table 18.3. Average production cost per hectare for labour inputs for mechanized/irrigated rice in Ghana in Gh. Cedi (source: Agriculture Research Institute, Ghana).

Operation	Approx. cost/ha (US \$)[a,b]
Ploughing	62.85
Harrowing twice	62.85
Harrowing after seeding	31.43
Sown seed (36 kg)	114.29
Broadcasting seeds by hand	5.70
Chemicals for weed control	45.94
Herbicide application	11.42
Fertilizers	85.71
Fertilizer application	11.42
Labour for irrigation and drainage	34.29
Irrigation water	142.86
Bird-scaring for 35 days	120.00[c]
Combine harvesting	171.43
Transportation of paddy rice	22.85
Sacks for paddy	137.14
Sacks for milled rice	114.29
Drying and bagging paddy	28.57
Cost of milling paddy	171.43
Management costs	85.71
Contingencies (5% of above)	73.03
Total production costs	1533.21

[a]Cost of initial preparation, levelling, and bunding is excluded.
[b]Exchange rate 1984: US \$1 = 35¢.
[c]Bird-scaring was later (August 1984) reported as US \$200 or 20% of production costs.

multiplication rice in Senegal, despite 32 man-hours/day of scaring costing \$330/ha. They also reported scaring costs ranging from \$800 to \$1600/ha at research stations and government schemes in West Africa, respectively, and about \$70/ha in Benin. In Cameroon, the cost of hiring scarers for 2 months to protect 0.5–200 ha rice schemes was estimated from \$65 to \$175/month per hectare with losses generally in the range of 0–15 per cent and occasionally as high as 80 per cent (Da Camara-Smeets and Affoyon 1980).

Season Scaring birds from dry-season, irrigated crops is more difficult than from wet-season, rainfed crops because irrigated areas provide a continuous source of seeds and water. In the middle Senegal River Valley, where a complex of several bird species cause serious losses to crops, damage to irrigated rice in each dry season was estimated at more than 5 per cent compared to less than 1 per cent in the main cropping wet season. In this valley, from 1000 to 3000 villagers created a continuous noise throughout the night, using tambourines and drums to prevent birds from roosting, with supplemental noise supplied by firing gunpowder explosives (Busnel and Grosmaire 1958).

The season can also affect the behaviour of the birds. If ripening of the crop coincides with an influx of recently fledged juveniles from a nearby quelea colony, scaring can be extremely difficult. Such juveniles are often apparently so hungry and inexperienced that they scarcely respond to the shouts of scarers. In the Sudan, these unscareable juveniles are called ‘the deaf ones’ and it is sometimes possible for a bird scarer actually to have to shake a sorghum plant to chase off these juveniles.

Surrounding habitat Pepper (1973) found more damage along the edges of sorghum fields surrounded by peripheral vegetation than in fields without peripheral cover, irrespective of intensity or effectiveness of scaring. Birds scared from fields perched on nearby vegetation and reinvaded crop areas. In northern Senegal, damage to 595 and 2956 sorghum head samples from edges and interiors of 18 fields averaged 18 and 11 per cent, respectively (Ruelle and Bruggers 1982). Bird scarers need to concentrate their activities at field edges and, where possible, remove as much peripheral vegetation as they can.

Habituation The key problem in all types of scaring is habituation, whereby birds learn not to respond to repeated scaring stimuli. Thus, the study of the behaviour of pest birds in relation to the components of different scaring stimuli could lead to effective biovisual scarers (Inglis 1980). Biovisual techniques used with a traditional rattle network might improve effectiveness (Bashir 1983). Slater (1980) investigated the various factors affecting habituation and concluded that habituation can be minimized if: (1) stimuli are presented as infrequently as possible during the main feeding periods; (2)

stimuli vary in signal type, location, and temporal pattern, making it difficult for the pest birds to anticipate them; and (3) stimuli are occasionally reinforced by introducing a genuine danger (such as a gunshot, sight of a hawk, or man).

Human-shaped, non-specific bird-scaring devices have become the focus of much research and innovative designs. Crocker (1984) reported that an arm-waving motion may remind birds of a slow-flapping raptor, and that human-shaped scarecrows, whose heads and arms move or who carry artificial guns and periodically pop up from undergrowth, have been found to be effective, particularly if supplemented by a shotgun noise. Martin (1979) reported that a well-designed system combining acoustic, biosonic, and ultrasonic devices effectively scared birds from waste-ponds. However, in the African context it is difficult to imagine that any such devices would have practical merit in the foreseeable future partly because of the probable high cost of importation, and also because the more complex devices require a power-source and maintenance.

Nest destruction

Destroying nests is a traditional co-operative activity carried out collectively by hundreds of villagers directly affected by pest birds. Studies of its effectiveness relative to bird populations and subsequent crop damage are lacking. Nests are destroyed normally using poles equipped with a hooked device for pulling out nests or by actually cutting trees. In western Sudan, where villagers can easily be encouraged to group, manual nest destruction is widely used and is effective under the supervision of the crop protection unit (M. El Mustafa, pers. comm.). During the 1960s, the traditional technique was upgraded by using flame-throwers for nest destruction, particularly in Senegal. Setting fire to nesting colonies in dry river beds prevailed in Chad (Mallamaire 1961).

Nest destruction is a two-edged weapon because in small colonies where birds are threatening cereal crops the operation might be cost-effective and practical in reducing crop losses. But if large-scale tree cutting is involved it might have adverse consequences and intensify desertification.

Factors affecting nest destruction Successful nest removal depends on the number of people involved (and their persistence) relative to colony size, its vegetation type, and accessibility. Ruelle and Bruggers (1982) reported a Golden Sparrow *Passer luteus* colony in which about 125 nests/ha could be removed by one individual at a cost of $0.30/ha. After 1 week, only 6 nestlings/100 nests were found in the destroyed area compared to 68 nestlings/100 nests in the rest of the colony, suggesting that the operation was effective and feasible. In Mauritania, however, 500 quelea nests/ha established in *Acacia senegal* could be destroyed by one individual at a cost of $30/ha; but only 100–200 nests/ha could be removed when the colony was established over water in *Parkinsonia* trees. Although flame-throwers generally are more effective than manual destruction, their use has ceased because they are costly and dangerous, and aerial or ground application of avicides is preferred. Crook (1960) showed that the timing of nest destruction in the breeding cycle influenced effectiveness. For control by flame-throwers, the earlier the burning occurred in the life of a colony the greater was the chance of adults abandoning the colony. Completely burnt colonies were always deserted, but birds from colonies burnt during the first days of brooding nested elsewhere. However, at later stages, birds will continue to fledge from nests in any unburnt trees in a colony. Crook (1960) concluded that the sudden removal of nests containing eggs led to a complete breakdown of defined territories set up during nest building, resulting in a rupture of pair bonds (resulting from increased male aggression), disappearance of a known nest topography, and subsequent wandering of males. He also found that once clutches were complete and brooding had begun, quelea could not tolerate changes that interrupted the breeding sequence. Mutilating nests, beyond cutting small holes, resulted in desertion.

These observations and those I made observing villagers destroying nests in Sudan suggest that quelea nest destruction is most effective at any time between clutch completion up to the time the chicks are about to emerge from the nest, i.e. from about day 6 to day 24 (Fig. 18.3). Nest destruction during nest building (colony age 0–3 days) is useless because adults can repair the nests or shift somewhere else nearby. During incubation (3-14 days), nest destruction will of course destroy the eggs. If nest destruction is supplemented by trapping adults, this will increase effectiveness because it has been shown by Jackson (1974*a*) that, if one adult of a pair is trapped during incubation, the nest will fail. During the chick stage, nest destruction has the advantage of providing a quantity of highly nutritious food for villagers (Chapter 23). By that stage, it is also less likely that the adults will be able to breed again in the same area.

Trapping young birds once they have left the nest may, in theory, greatly reduce subsequent damage to nearby crops. Unfortunately, no method of mass-capture exists. Jackson (1974*b*) tested a trap in one colony and was able to catch hundreds of juveniles in a short time, but applying the technique on a large scale was never tried (Plate 15).

Exclusion and prevention techniques

The placing of some sort of physical barrier between the crop or the grain head and the marauding birds is an obvious way of preventing damage.

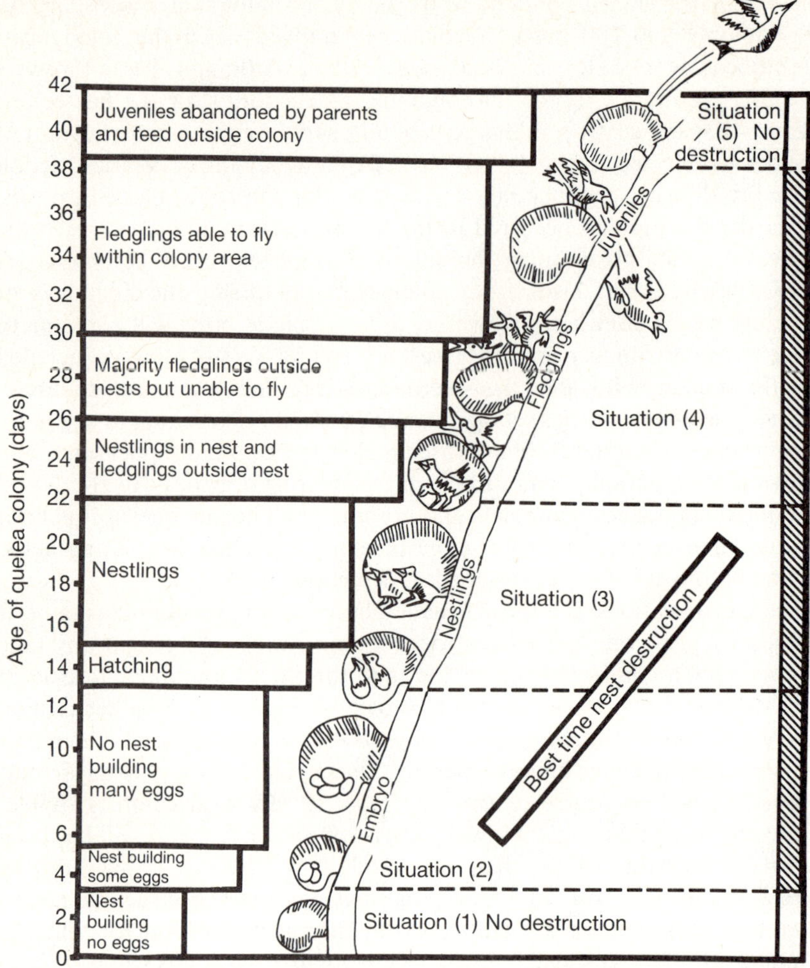

Fig. 18.3. A scheme showing five situations of quelea nest destruction in relation to age and development of colony.

Traditionally, such exclusion techniques are widely used in West Africa and involve covering individual ripening heads with cloth, grass or leaves (Fig. 18.1). A striking example occurs in Senegal where heads of flood-recession sorghum along hundreds of kilometres in the Senegal River Valley are regularly covered with grasses, sorghum leaves or cloth resulting in complete protection (Fig. 18.4*a*). In this valley in 1975, 2 per cent of covered heads were attacked, causing 1–5 per cent damage per head; 33 per cent of the uncovered heads were attacked, causing losses of 5–90 per cent. In 1981, 53 per cent of covered heads were attacked sustaining a 10.8 per cent loss,

(*a*) (*b*)

Fig. 18.4. (*a*) In Chad, individual millet heads are protected from birds by laboriously hand-woven sheaths (photo: J. Jackson). (*b*) In Senegal, sorghum heads can be wrapped in the leaves of the plant (photo: R. Bruggers).

whereas 62 per cent of the uncovered heads were attacked losing 17.6 per cent (Ruelle and Bruggers 1982). In Chad, farmers weave grass sheaths and place them over individual millet heads (Park 1974). Because millet matures asynchronously, these sheaths can be moved and reused on other heads. This technique is mainly confined to western rather than eastern Africa where ethnic groups have, presumably, inherited and mastered stronger, indigenous, and novel traditions of weaving plant materials (Fig. 18.4*b*).

Covering ripening cereal heads is a tedious and time-consuming process that mainly is used in the Sahelian countries of West Africa where severe drought has become an annual phenomenon. Drought compels farmers to protect their subsistence crops at any cost, regardless of crop value or time spent in the operation. Although it is difficult objectively to evaluate the cost-effectiveness of this technique because the time consumed in weaving grass sheaths and covering crop heads cannot be quantified in terms of monetary value, the method usually gives complete protection.

Another preventive technique is that of traditional African 'juju', in which amulets or objects are prepared according to certain religious or magical procedures and placed in the fields to repel birds. The concept varies among ethnic groups depending on religious and superstitious beliefs. In Sudan, for example, some Muslim farmers usually seek the assistance of a religious sheikh for reading particular verses from the Holy Koran or use the verses as amulets to drive birds away. Although 'juju' cannot be scientifically proven, many farmers in many countries claim success for protecting crops at least for short periods: cases have been reported from Liberia, Nigeria, Sudan and Ghana (S. El Bashir, pers. comm.).

Agronomic techniques

Cultivating less susceptible crop varieties, manipulating planting and harvesting dates, and intercropping are techniques used for generations by African farmers to reduce losses to birds. Farmers, through experience with 'bird-resistant' indigenous cereal varieties, are aware of varietal differences in susceptibility to damage, and would tend to plant them in areas where birds are a particular problem. Traditionally, they know that red sorghum varieties (high tannin content) often are less damaged than white varieties (low tannin content), that rice and millet heads with long awns and rice plants with erect leaves are less susceptible than those with short (or without) awns or droopy leaves, and that standing or floating rice with sheathed panicles are more resistant to damage than rice with exposed panicles (Bortoli 1978). However, like farmers all over the world, considerations of taste, yield, and profit may overrule the desire to avoid birds, and in many cases the seeds of resistant indigenous varieties may not be available. In Senegal, Bortoli and Bruggers (1976) reported 5 and 9 per cent damage to 'bird-resistant' red sorghum compared to 18 and 33 per cent to white varieties in two different fields, respectively. In Liberia, damage to awned rice was 2.8 per cent compared to 5.8 per cent of awnless variety (Abifarin 1984).

Likewise, planting crop varieties that ripen simultaneously, overall adherence to similar sowing dates, and continuous weeding contribute to reducing or evenly distributing bird damage. In Senegal and Mali, weed-infested fields with visible gaps encouraged bird damage (Bortoli 1978). Manipulating harvesting times to avoid quelea migration dates in irrigated rice reduced damage in Chad and Cameroon (Elliott 1979). With the exception of few wet, tropical zones, changing crop calendars to avoid bird damage can only be effective in irrigated areas where water supply can be adequately controlled.

Discussion

Considering the hundreds of thousands of people involved in bird-scaring and other forms of traditional crop protection in Africa, it is remarkable how little the methods have been studied. With the exception of those mentioned above for Liberia, Ghana, Senegal, and Chad, few data have been collected on the impact of traditional methods on the birds themselves, nor on the costs of the methods relative to the value of crops saved. Sometimes traditional methods such as nest destruction are applied out of frustration or anxiety, without any real evaluation of the likely benefits. Farmers may mobilize scarers when the threat to crops is still very low or when unspecified flocks of birds, which even may be non-granivorous, are seen in the fields. A national bird control unit can play a significant role in advising and leading traditional bird control activities so that nest destruction is carried out at the most effective time, and bird-scaring energies are applied when they are really needed. Such a unit can also collect bioeconomic data on traditional methods. These data should include information on number of nests, eggs, nestlings, and fledglings destroyed per unit area, number of adult birds deserting a colony in relation to total population, and number of birds involved in crop damage before and after nest destruction, and then be related to management costs and crop values.

Locally adapted research based on farmers' skills and experience in bird control is needed. A human-shaped bird scarer shaking its head and slowly waving arms might be effective, but probably is too sophisticated for traditional farmers. Simply importing complicated, expensive foreign bird control technologies, without regard to culture and cost, should be discouraged. Experience shows that techniques suggested by traditional farmers and field-tested by biologically oriented investigators can give promising results.

Traditional bird-scaring is an area where information on the behaviour of birds toward different stimuli is lacking. The danger of the scaring technique is that birds are repelled only from protected into unprotected fields. Therefore, bird-scaring needs to be organized co-operatively so that birds are scared out of a farming area into alternative wild food supplies elsewhere. Otherwise the only result will be a re-distribution of damage and not a reduction of damage. Alternatively, trapping pest birds, as opposed to scaring them should be seriously investigated. Trapping during times of high bird pressure can reduce crop losses, provide farmers with a good supply of animal protein, and reduce the number of birds likely to attack unprotected fields. Therefore, realistic crop protection approaches that fit the socio-economic situation and the psychological fabric of a poor farming society are needed. This is a challenging problem facing African ornithologists and crop protection specialists. Without adopting locally adapted and biologically oriented approaches based on the wealth of biological and ecological information available on bird pests, no significant improvement in traditional bird control methods can be achieved.

19

Assessment of bird-repellent chemicals in Africa

RICHARD L. BRUGGERS

Introduction

Chemical repellents have been used to protect crops since the onset of agriculture. Wright (1981) reviewed chemical repellents and gave a brief history of their use, derivation from natural plant substances, and an account of the sensory mechanisms by which birds perceive chemical stimuli. Repellents began receiving attention in Africa in the mid-1970s. This chapter reviews and summarizes the results of research evaluating chemical repellents for protecting crops from pest birds in Africa and suggests possible directions for future research.

Phenomenon of repellency

Rogers (1978*a*) defined a repellent as 'a compound or combination of compounds that, when added to a food source, acts through the taste system to produce a marked decrease in the utilization of that food by the target species'. Schafer (1981) characterized repellents as 'tactile (chemical modification of surfaces), taste (chemical modification of flavours), physiological (chemical-induced illness), and other (primarily chemically caused behavioural changes over large areas not directly treated)'. Rogers (1978*a*) stated that, historically, no consistently effective chemical repellents have been developed for use against vertebrate pests, primarily because, until recently, efforts were directed towards finding 'bad-tasting' chemicals. This search was predicated on an anthropomorphic-based assumption that the human taste experience was directly transferable to birds. Although birds possess the same basic sensory mechanisms as mammals, it is naive to assume that their perception of stimuli corresponds to that of humans (Wright 1981). This approach led to laboratory screening efforts that resulted in many candidate chemicals for field testing.

Most researchers believe that a co-ordinated effort among chemists, biologists, behaviourists, physiologists, and other disciplines is needed if chemical repellents are to function effectively (Rogers 1978*a*). Understanding a chemical's mode of action, the sensory physiology of the bird, and its subsequent behaviour has become important to demonstrating efficacy data to obtain Environmental Protection Agency (EPA) registration. Registering chemicals is such a lengthy and costly process—estimated at US $15 million (Weidner 1983)—that it is impractical to develop new chemicals for minor uses on specific avian species (Schafer 1981).

Using repellents to protect crops from highly gregarious, grain-eating birds in Africa is based on the premise that not all individuals in feeding flocks must be affected to inhibit feeding. Because birds will learn to avoid food that makes them ill (Wright 1981), in some cases for 4 weeks or longer (Rogers 1978*b*), chemicals that cause a conditioned taste aversion would seem to show more promise as repellents than chemicals that simply taste bad (Rogers 1980). Conspecifics in feeding flocks also can learn to avoid treated food by observing affected birds (Mason and Reidinger 1981).

Repellents tested in Africa

Chemical repellents have been evaluated in more than 40 demonstrations or replicated trials in 15 African countries between 1975 and 1984. The principal repellents tested in Africa included one that apparently affects the taste sense, Curb[R] (synergized aluminium ammonium sulphate, SAAS), and two that apparently cause an illness-induced, adverse physiological reaction, methiocarb [3,5-dimethyl-4-(methylthio)phenyl methylcarbamate, Mesurol[R]] and trimethacarb (80 per cent 3,4,5-trimethylphenyl methylcarbamate and 20 per cent 2,3,5-trimethylphenyl methylcarbamate). These studies determined the susceptibility of different pest species to the repellents and the effective application rates for seed dressings and ripening panicles. Application procedures were also improved and made more economical and residue degradation patterns were determined.

General species sensitivity

Considerable variation exists in the sensitivity of African bird species to the repellent and toxic effects of the three chemicals (Table 19.1). All species were more sensitive to methiocarb than the other chemicals in R_{50} tests—the concentration at which 50 per cent of the birds are repelled in a single-choice test. Sensitivity data paralleled field experience, for higher application rates of Curb and trimethacarb were needed to obtain field efficacy than for

Table 19.1. Comparison of chemical repellent LD_{50} (mg/kg) and R_{50} (%) for several pest bird species. Sensitivities for all species not denoted with superscripts were determined in Bruggers *et al.* (1984*b*).

Chemical	*Quelea quelea*	*Passer domesticus*	*Psittacula krameri*	*Ploceus cucullatus*	*Agelaius phoeniceus*	*Passer luteus*	*Euplectes orix*
Methiocarb (Mesurol®)							
LD_{50}	4.2[a]	14.1	7.1	7.5[b]	4.7[c]	5.62[b]	5.62[b]
R_{50}	0.02[d]	0.11	0.18	0.06[e]	0.08[c]	0.18[b]	0.13[b]
Trimethacarb							
LD_{50}	50.9	33.6	11.9	11.3	13.3[f]	ND[g]	ND
R_{50}	0.24	0.22	0.34	0.14	0.12[f]	ND	ND
Aluminium ammonium sulphate (Curb®)							
LD_{50}	ND	ND	ND	ND	> 100[f]	ND	ND
R_{50}	ND	ND	ND	ND	> 1.0[f]	ND	ND

[a] Schafer *et al.* (1973).
[b] Shefte *et al.* (1982).
[c] Schafer and Brunton (1971).
[d] Shumake *et al.* (1976).
[e] Garrison and Libay (1982).
[f] E. Schafer, unpubl. data.
[g] ND, not determined.

methiocarb. Quelea are 4.2–11.8 times more sensitive to methiocarb (R_{50} = 0.015 per cent) than Black-headed Weavers *Ploceus cucullatus*, Golden Sparrows *Passer luteus*, or Red Bishops *Euplectes orix*. Garrison and Libay (1982) determined R_{50} values for methiocarb of between 0.036 and 0.057 per cent for three species of mannikins (*Lonchura malacca*, *L. punctulata*, *L. leucogaster*) that are pests to rice in the Philippines. The R_{50} for *P. cucullatus*, an occasional pest to rice in African countries, is most likely in the same general range. Sultana *et al.* (1986) also determined an R_{50} = 0.18 per cent for methiocarb for Rose-ringed Parakeets *Psittacula krameri* in Bangladesh where the species is an important pest of maize, as it is in some African countries.

Trimethacarb

Trimethacarb is a carbamate mixture registered with the EPA as an insecticide. DWRC originally screened trimethacarb in 1967 on Red-winged Blackbirds *Agelaius phoeniceus* and European Starlings *Sturnus vulgaris* and determined LD_{50} levels of about 10 mg/kg and > 100 mg/kg, respectively (D. Cunningham and E. Schafer, unpubl. data). In R_{50} tests, the compound was rated as marginal but designated for possible reconsideration (R. Brunton, D. Cunningham and E. Schafer, unpubl. data). Since then, promising results from studies showing conditioned aversion by common crows to treated eggs (Nicolaus *et al.* 1983) reduced damage to newly sown corn seed by Ring-necked Pheasants *Phasianus colchicus* and reduced damage following topical applications to cherries (P. Kleyla, unpubl. data). Additional laboratory studies with technical grade material of better quality have generated renewed interest in the compound.

Trimethacarb has been evaluated only twice in Africa: on ripening rice in Mali in June 1983 and on ripening wheat in Zimbabwe in July 1984. These initial demonstrations and other studies conducted in Haiti, Bangladesh, the Philippines, and India suggest that it has potential both as a seed treatment at levels between 0.125 and 0.25 per cent (by seed weight) and as a head spray at application rates > 4.0 kg/ha (Bruggers *et al.* 1984*b*). However, germination of wheat, millet, rice, and sorghum seeds treated at levels > 1 per cent was significantly retarded compared to untreated seeds.

Degradation No studies have yet been conducted on trimethacarb degradation in Africa. However, in Haiti, following 4- and 8-kg (a.i.)/ha application rates in rice, average residues of 9.4 p.p.m. and 465 p.p.m., respectively, declined to 0.1 p.p.m. on the seed (both application levels) at harvest (Bruggers *et al.* 1984*b*). Half-life was 5.2 days for the 4-kg a.i./ha application and 4.2 days for the 8-kg a.i./ha application. Residues exceeding the R_{50} of most bird pests remained on the seeds-glume sample and the seeds-only sample for 21 days after spraying. Chemical on seeds at harvest was less than the amount shown to inhibit germination, which suggests that sprayed seeds could be used as seed stock.

In the Philippines, trimethacarb residues on grain sorghum seed were 0.18 p.p.m. about 4 weeks after a 3-kg a.i./ha application and 0.68 p.p.m. after a 4-kg a.i./ha application. Comparable residues at harvest were found on seed sprayed either once (0.68 p.p.m.) or twice (0.42 p.p.m.) at 4 kg a.i./ha (Bruggers *et al.* 1984*b*). If more detailed and comprehensive repellency studies demonstrate efficacy, trimethacarb could presumably become more widely available and be marketed competitively (P. Kleyla, pers. comm.).

Curb Curb is a relatively non-toxic formulation with reported repellent properties to rodents, deer, and many species of birds (Stone 1976). In Europe, Curb occasionally has been used successfully against birds when applied as a seed dressing to corn (Leinati 1968) or as a spray to fruit buds (Stone 1976) and vegetables (Dar 1974) at application rates of 20–40 kg/ha. Alum-based formulations are bitter, highly astringent, and taste bad (Schafer 1981). In the United States, it has not been registered for use, nor do those North American bird species tested show much sensitivity to Curb.

Curb has been evaluated during the wet and dry seasons in Sahelian and wet coastal Guinean ecosystems, on several types of ripening cereals, and under varying levels of attack from at least 10 species of birds other than quelea that were resident during the trials (Bruggers 1979*a*). In studies at Bambey, Senegal, and Mitro, Benin, Curb provided some protection for between 7–14 days when applied at 8–10 kg/ha in one or two applications. At Sinthiou Maleme, Senegal, weights of treated sorghum heads at harvest were greater than weights of untreated sorghum heads ($P < 0.05$). All trials in Chad and Cameroon in which Curb was applied against quelea at rates of 4, 8, and 16 kg/ha have resulted in very little, if any, protection (Park and Adam 1976; Park *et al.* 1975; Park and Assegninou 1973). In laboratory and limited short-term tests in the United States, a solution of 0.5 kg Curb in 2 l water provided some protection to grapes from House Finches *Carpodacus mexicanus* (Ewing *et al.* 1976).

Methiocarb

Seed treatment Methiocarb has proved to be a consistently effective broad-spectrum repellent when applied as a seed dressing and head spray in many countries worldwide (Besser 1973; Calvi *et al.* 1976; Crase and DeHaven 1976; Dhindsa and Toor 1980; Duncan 1980; Guarino 1972; Holler *et al.* 1982; Mott *et al.* 1976; Poché *et al.* 1980; Rogers 1974, 1978*a*,*b*). In Africa, most seed treatment trials were conducted on rice in Senegal in the Senegal

River Valley, where Ruffs *Philomachus pugnax* were the main pests. Their numbers annually swell from about 10 000 in September to more than 100 000 during December and January (Treca 1976). White-faced Tree-ducks *Dendrocygna viduata* and Black-tailed Godwits *Limosa limosa* also eat newly sown rice. Seed treatment applications of methiocarb at 0.25–0.83 per cent (by seed weight) have resulted in more rice plants and fewer birds in treated than in untreated plots in all trials (Bruggers 1979*b*). The number of plants and the yield at harvest in plots sown with treated rice seed also were greater ($P<0.05$) than those in plots sown with untreated seed (Ruelle and Bruggers 1979). At rates of 0.25–0.50 per cent, no sick or dying birds were found; dead birds were found when methiocarb was applied at 0.83 per cent (Bruggers and Ruelle 1977). In another rice trial in Sudan, seed losses to crested larks and sparrow larks were 3.8 times greater in untreated plots than in treated plots (Hamza *et al.* 1982). Methiocarb residues from the dried samples of seeds and seedlings were all <1 p.p.m. (the level of detectability of the analytical method used) after 15 days, a 99.9 per cent decrease from pre-sowing level of 1100 p.p.m. Holler *et al.* (1982) also found rapid degradation with only 16 per cent remaining after 3 days and 6 per cent after 21 days.

Ripening cereals Initial studies in West Africa of methiocarb as a head spray to ripening crops at research stations were erratic yet encouraging (Table 19.2; Fig. 19.1). The results demonstrated how the success of a repellent is affected by the application rate, the type of crop, its location and maturation process, the bird species and their numbers, and the season of the year.

Protecting crops during the dry season at agricultural research stations is difficult. These stations are in effect oases that provide a predictable supply of water and food, and therefore attract large numbers of birds from the surrounding countryside. Although one application of 6.5 kg a.i./ha to millet in Senegal early in the 1975 dry season resulted in treated heads (4.0 g) weighing nearly twice as much as untreated (2.2 g) heads, some damage occurred to almost all heads in the field. Two applications of 1, 2, and 3 kg a.i./ha applied at the end of the 1976 dry season to ripening millet at the same location were completely ineffective. Three days post-application, 50–100 per cent of the heads in all plots were completely denuded and most of the crop was devoured before it reached the early milk stage. Such acute damage is characteristic of the end of the dry season, a period in the Sahel when natural food supplies are scarce (Ward 1965*a*).

During the normal growing season in Senegal, 2.5 kg/ha of methiocarb applied over the entire field protected ripening sorghum at Darou and ripening rice in the Senegal River Valley in 1975 (Table 19.2). However, applying methiocarb to rice in alternate bands was unsuccessful in the Senegal River Valley in 1977, owing to abundant weed seed in the rice fields,

Table 19.2. Summary of methiocarb studies to protect ripening crops from bird pests in Senegal from 1975 to 1977 (source: Bruggers 1979*b*).

Crop, location, (date), trial[a]	Applications		Yields or % damage	Main pest species
	No.	kg a.i./ha		
Millet				
Bambey (Feb. 1975)				
Treated	1	6.5	4.61 g/head	*Passer luteus*, *Passer griseus*, *Euplectes orix*; 200–300 total in small resident feeding flocks
Untreated	–	–	2.19 g/head	
Bambey (Apr. 1976)				
Treated	2	1.0, 2.0, 3.0	93–100%	*Quelea quelea* (30%), *Passer luteus* (33%), *Euplectes orix* (23%); roost population of 1000–3000 birds
Untreated	–	–	100%	
Sorghum				
Darou (Nov. 1975)				
Treated	2	2.5	4%; 49.4 g/head	*Ploceus cucullatus*, *Ploceus melanocephalus*, *Lamprotornis chalybaeta*, *Vinago waalia*, *Bubalornis albirostris*, *Ploceus* spp.; from breeding colony of 500–1000 birds
Untreated	–	–	67%; 15.8 g/head	

Rice				
Richard Toll (Sept. 1975)				
Treated	2	2.0, 2.5	2.05 g/head 206 ± 8 g/0.5 m^2	*Euplectes orix*, *Quelea quelea*, *Ploceus* spp.
Untreated	–	–	1.91 g/head 149 ± 50 g/0.5 m^2	
Richard Toll (Sept. 1977)				
Treated	3	2.0	38%, 63%, 75%	*Quelea quelea* (80%), *Ploceus* spp. (5%), *Euplectes orix* (15%); 600–2000 birds each period all from pre-nesting or nesting colonies
Untreated	–	–	93%, 100%, 100%	

[a] Trials were conducted on 1–3 treated and untreated plots of between 0.04 and 0.80 ha.

Fig. 19.1. Mesurol can be sprayed onto ripening sorghum heads to repel quelea (photo: R. Bruggers).

extended maturation period, and drought that simulated dry-season conditions. Both rice studies demonstrated the influence of differences in feeding behaviour of pest species and crop maturation on repellent effectiveness (Bruggers 1979*b*).

The initial field experiments with methiocarb in West Africa (Bruggers 1979*b*) and in eastern Africa (De Grazio and Shumake 1982) involved application rates of 3–6 kg/ha applied over entire fields and again gave variable, yet encouraging results. From these early field experiments, it was evident that in many situations bird damage was localized in fields along the borders or sheltered bush areas and often never reached the interior. Subsequent studies in eastern Africa employed economical application techniques such as edge, band, or spot-spraying (Bruggers *et al.* 1981*b*). Results from these studies also were encouraging (Table 19.3). Protection of rice at the Libya–Somalia (LIBSOMA) scheme in Somalia, sorghum in Ethiopia, and wheat at West Kilimanjaro, Tanzania, was particularly good. In these locations, the crop was either the only one in the area or was in the vulnerable stage at the onset of the trial. Inconsistent, although less satisfactory results occurred in a rice trial at the Agricultural Research Centre in Afgoi, Somalia, and a wheat trial at the Arusha Foundation Seed Farm (AFSF) in Arusha, Tanzania, primarly because untreated fields were immediately adjacent to treated fields and held birds that might otherwise be repelled.

Cost:benefit analysis In the eastern African trials, birds usually began attacking crops along field borders or in spots and then moved into the fields, making edge or spot applications appropriate and economical (see Bruggers *et al.* 1981*b* for detailed cost explanations for each trial). The cost of applying methiocarb to bands at the 3-ha rice field at the LIBSOMA scheme was approximately US $45/ha. Using bird scarers raised the cost an additional $50/ha. With a harvest value of approximately $375/ha and a 5 per cent loss to birds, LIBSOMA still showed a benefit of nearly $200/ha, this is particularly good when compared with $7.50/ha benefit from a 36-ha field that was 90 per cent damaged just before the trial, when bird scarer costs totalled $143/ha. Crop protection using bird scarers accounted for 15–43 per cent of total rice production costs at LIBSOMA (Bruggers 1980).

At the AFSF in Tanzania in 1979, the yield for the treated field of variety Trophy averaged $53/ha more than the untreated field, and the number of bird scarers in all trial fields was reduced by 50 per cent. In 1979, bird scaring costs for 60 ha of trial fields were between $650 and $820, compared to approximately $1250 in 1978 when the AFSF lost 60 per cent of its crop (M. Mmari, pers. comm.).

At West Kilimanjaro in 1979, spot-spraying areas where birds were seen feeding cost only $176 for the entire 1012-ha farm. The value of the yield was approximately $70 000. The value of yields in 1976 and 1977 when methiocarb was also used and 5 per cent was lost to birds was $267 000 and $317 000, respectively, greatly exceeding the $47 000 value of the 1978 crop when methiocarb was not used.

Degradation Using repellents on food crops requires an understanding of their degradation patterns to ensure permissible residue levels at harvest. Methiocarb degradation patterns have been determined on sorghum in Senegal and on wheat in Tanzania. In Senegal, a 2-kg a.i./ha application rate (with 60 ml of Triton AE adhesive per 100 l water) resulted in residues of <3.5 p.p.m. on seed after 20 days and a half-life of 6–7 days (Gras *et al.* 1981). Initial degradation was very rapid, and metabolic effects began to appear on the second day. The methiocarb $R_{50} = 0.015$ per cent or 150 p.p.m. for quelea (Shumake *et al.* 1976), an amount equivalent to that remaining on grain 3 days after spraying. The time to treatment inhibition was 20–23 days (Gras *et al.* 1981). This time period corresponded to actual field application methods. Second applications usually are applied within 7–10 days of the first, leaving approximately 3 weeks until harvest in a normal 4- to 5-week maturation period.

Table 19.3 Harvest results from methiocarb studies in eastern Africa between 1974 NS 1980 (sources: Bruggers *et al.* 1981*b*; De Grazio and Shumake 1982; Hamaza *et al.* 1982)

Location, crop, (date), trial	Fields		Application technique			Yield or % damage	Pest species
	No.	Area (ha)	kg a.i./ha	No. applications	Method		
Somalia							
ARC—rice (July–Aug. 1979)							
Treated	1	0.15	1	2	Applied to $\frac{1}{3}$ area in edge and centre bands	1600 kg/ha	70% *Quelea quelea*; 25% *Ploceus cucullatus*; 5% *Euplectes* spp.
Untreated	1	0.15	–	–		1700 kg/ha	
LIBSOMA—rice (Aug.–Sept. 1979)							
Treated	1	3	1	2	Applied to $\frac{1}{3}$ area in edge and centre bands	5%; 1500 kg/ha	90% *Quelea quelea*; 10% *Euplectes* spp., and *Ploceus cucullatus*
Untreated	1	36	–	–		90%; 160 kg/ha	
Tanzania							
AFSF—wheat (June–Sept. 1979)							
Treated	2	12, 15	1	2	Applied to $\frac{1}{3}$ area in edge and centre bands	291 and 316 g/m^2 1072 and 1125 kg/ha	95% *Quelea quelea*; 5% *Ploceus rubiginosus*
Untreated	2	18, 12	–	–		280 and 200 g/m^2 1293 and 639 kg/ha	

Simba/Poverty Gulch—wheat							
Treated (July–Sept. 1979)	25	1125	0.005	As required	Spot application	< 5%; 556 kg/ha	95% *Quelea quelea*; 5%
				As required	only to areas being		*Passer domesticus*, and
(1977)	25	1125	0.005	As required	attacked	< 5%; 1938 kg/ha	*Ploceus rubiginosus*
(1976)	25	1125	0.005			< 5%; 993 kg/ha	
Untreated (1978)	25	1125	–	–		86%; 311 kg/ha	
Rujewa—wheat (Feb. 1974)							
Treated	1	0.05	3	1	Complete	5%	*Quelea quelea*
Untreated	1	0.05	–	–	coverage	51%	
Ethiopia							
IAR, Melkassa—sorghum							
Treated (Sept.–Oct. 1980)	1	12	5.3 and 1.2	2	Alternate row	14.2%	99% *Quelea quelea* all
(1979)	1	12	5.3 and 1.2	2	(5 m) application	22.1%	years; 1% *Ploceus*
(1978)	1	12	5.3 and 1.2	2	to heads of $\frac{1}{2}$ area	5.7%	*melanocephalus*,
(1977)	1	12	5.3 and 1.2	2		< 2–3% after treatment; 23% before treatment	*Streptopelia* spp., and *Euplectes orix*
Untreated (1976)	1	12	–	–		42%	
Kenya							
Nanyuki—wheat (Jan. 1974)							
Treated	1	0.012	3	1	Complete	6%	*Quelea quelea*, *Ploceus*
Untreated	1	0.012	–	–	coverage	50%	*rubiginosus*, *Euplectes progne*
Sudan							
ARC, Shambat—wheat (1980)							
Treated	2	0.25	1	1	Edge; ($\frac{1}{3}$ of the	8.4%	*Passer domesticus*, *Ploceus*
Untreated	2	0.25	–	–	area)	16.2%	spp., *Euplectes orix*

Residue degradation following a 1.8-kg a.i./ha application to ripening wheat at elevations of 1500–2000 m in Tanzania (Ndege 1982) followed a similar pattern to that observed in Senegal. Much of the chemical and its metabolites (sulphoxides and sulphones) was deposited on glumes, and very little was found on the seed itself. Methiocarb residues in glumes decreased rapidly to about one-seventh of initial levels within 2 days. Chemical levels sufficient to induce repellency to quelea ($R_{50} = 150$ p.p.m.; Shumake *et al.* 1976) were estimated to remain on the glumes for about 3 days after the first application but then fell to less than 15 per cent of the R_{50} level 5 days post-treatment. The half-life was calculated as 4 days. Methiocarb is registered federally with the EPA as a seed treatment for corn (maize) (Schafer 1979). Registration with EPA of a 0.25 per cent a.i. Mesurol 75 per cent Seed Treater on aerially sown rice seed is pending.

Additional investigations

Foliar and lure crop sprays, impregnated string, and baits

Several innovative uses have been suggested for employing chemical repellents (methiocarb in particular) to protect crops from birds in Africa. These include: spraying roost vegetation or nests, spraying lure crops that ripen before the main crop, impregnating string with methiocarb and aerially applying it in nesting areas, distributing methiocarb-impregnated baits in crop fields, and enhancing repellent sprays with sensory cues.

Some of these ideas have been tried on quelea and other avian pest species. Because quelea will avoid white ($CaCO_3$)-coated seeds in the laboratory (Elmahdi 1982), consideration has been given to moving birds from day roosts near crops by spraying roost vegetation with this material. Ruelle (1983) has shown that damage to rice in Senegal and The Gambia can be reduced by spraying carbofuran on early maturing varieties in bordering plots; repellents could also be tried. S. Shumake (pers. comm.) has demonstrated that caged quelea will weave nests with string and he has established application rates for several chemicals that will impregnate the string and kill quelea constructing nests with it. A dispenser that attaches to aircraft and cuts string into desired lengths is used in insect control. Jaeger *et al.* (1983) found that methiocarb-impregnated sunflower seed baits would not stop Red-winged Blackbirds from damaging maturing sunflower; it seems unlikely that the technique would be any more effective on quelea.

Sensory-cue enhancement

Of those speculative uses for repellents in Africa, enhancing the effectiveness of head sprays (particularly methiocarb) with sensory cues, has received the most attention. No primary repellent, including wattle tannin (Zeinelabdin 1980; Zeinelabdin *et al.* 1983), has provided consistent, economical protection when topically applied to ripening cereal crops. Apparently the toxic, emetic, or nauseating effects associated with a secondary repellent, like methiocarb, are necessary stimuli consistently to reduce feeding by birds (Bullard *et al.* 1983*a*). However, the conditioned aversion formed in response to the adverse effects of secondary repellents can be enhanced with the addition of inexpensive, tactile, visual, or olfactory sensory cues that are, or can become, primary repellents because of subsequent association with methiocarb.

Most research to enhance repellent effectiveness with sensory cues has been laboratory work. Quelea will avoid seed treated with methiocarb and taste cues of citric-acid (Shumake *et al.* 1976), wattle tannin (Bullard *et al.* 1983*b,c*) and colour cues of calcium carbonate (Bullard *et al.* 1983a). In fact, visual cues may be more important than taste cues in forming conditioned aversion (Capretta 1961; Gillette *et al.* 1980; Mason and Reidinger 1982). Starlings and many other species readily learn to associate visual cues with illness (Czaplicki *et al.* 1976; Rooke 1983; Schuler 1980). Red-winged Blackbirds display food aversion that lasts about 2 weeks (Mason and Reidinger 1982) as a consequence of observational learning (Mason and Reidinger 1981). Because opportunistic feeders such as Red-winged Blackbirds (Dolbeer 1980) readily learn to avoid foods associated with aversive consequences (Klopfer 1958), it is possible that quelea, as highly gregarious, opportunistic foragers, would exhibit similar response patterns.

Initial attempts to investigate sensory-cue enhancement of methiocarb to protect ripening cereals have been encouraging (Table 19.4; Bullard *et al.* 1983*c*). In field enclosure studies conducted in the Sudan in 1979, the median percentage damage to ripening heads of sorghum and millet treated with 1.0 per cent methiocarb or 0.5 per cent methiocarb/1.0 per cent wattle-tannin combinations was minimal and significantly less ($P<0.001$) for both grains than damage to untreated heads. For both sorghum and millet, there was no significant difference in damage between the methiocarb and the methiocarb/wattle-tannin treatments. Methiocarb and methiocarb/wattle-tannin also provided comparable protection to wheat from bird damage in the Sudan. The methiocarb treatment reduced damage by 80.3 per cent and 82.9 per cent at the milk and dough stages of the wheat crop, respectively; the methiocarb/wattle-tannin formulation reduced damage by 85.1 per cent and 97.8 per cent at the milk stage and dough stage, respectively. At Babougon Seed Farm in the Office du Niger, Mali, methiocarb/wattle-tannin suspension provided some protection to milk-stage sorghum from quelea. Bird damage at harvest was 75 per cent in a 250-m^2 untreated plot and <25 per cent in a 100-m^2 treated plot.

Table 19.4. Summary of field trials[a] comparing 0.5% methiocarb/1.0% wattle tannin and 1.0% methiocarb topical applications (both applied at the rate of 3 kg a.i. methiocarb/ha to ripening cereal in developing countries. Both repellent formulations contained 0.05% Rhoplex AC-33 adhesive (source: Bullard *et al.* 1983*c*).

Crop, location, (date)	Area (ha or m^3)		Application technique	No. applications	% damage			Main bird pests
	Treated	Untreated			Methiocarb/ tannin	Methiocarb	Untreated	
Sorghum								
Mali (Mar. 1981)	0.001 ha	0.025 ha	Applied over entire field	2	<25	NA[b]	75	*Quelea quelea*
Philippines								
IRRI (Mar. 1982)	0.012 ha	0.012 ha	Applied to damaged heads	2	<2	NA	28	*Passer montanus*
UPBL (Apr. 1982)								
Site 1	0.03 ha	0.03 ha	Applied over entire field	2	6	10	32	*Passer montanus*
Site 2	0.03 ha	0.03 ha	Applied over entire field	2	4	34	10	*Passer montanus*
Sudan								
ARC (Nov. 1979)	8 m^3	8 m^3	Enclosure test	1	1	4	100	*Quelea quelea*
Wheat								
Sudan								
ARC (Mar. 1980)	0.008 ha	0.008 ha	Applied to heads on one side of plot	2	0.7	4.5	26.4	*Euplectes orix*, *Passer luteus*, *Passer domesticus*, *Ploceus* spp., *Quelea quelea*
Millet								
Sudan								
ARC (Nov. 1979)	8 m^3	8 m^3	Enclosure test	2	20	10	100	*Quelea quelea*
ARC (Nov. 1979)	0.1 ha	0.1 ha	Applied to heads on one side of field	2	2	NA	15	*Euplectes orix*, *Passer luteus*, *Passer domesticus*, *Ploceus* spp., *Quelea quelea*

[a] Trials conducted on one treated and one untreated plot at each site.

[b] NA, not applicable.

Although additional field studies and analyses of chemical breakdown are needed before recommending a methiocarb/wattle-tannin combination, initial work indicates that it shows promise for economically protecting ripening cereals from birds. The addition of cues that are detectable by the olfactory, visual, or tactile senses of a bird apparently enhances the effect of methiocarb. Dimethyl anthranilate (DMA), an inexpensive human food flavouring agent, is apparently unpalatable to birds (Mason *et al.* 1985) and may be another such cue that could be used with a chemical repellent or perhaps alone. By finding ways to lower the costs of chemical repellents without losing efficacy (for example, reducing the amount of methiocarb in the spray formulation and adding inexpensive sensory cues) and employing economical field application techniques (for example, edge, alternate row, or spot spray), traditional farmers in developing countries may be able to consider using repellents in some situations to protect their crops from birds.

Recommendations for use

Using repellents successfully necessitates a thorough knowledge of the specific bird problem and actual treatment timing and procedures. Adhering to certain guidelines should improve the effectiveness of chemical repellents.

(1) Some damage must be expected during the conditioning period of the pest population; 100 per cent protection is not a reasonable expectation (Rogers 1980).
(2) The likelihood of successfully protecting a crop decreases as the dry season progresses and natural food becomes scarce.
(3) Effective repellents are not necessarily bad tasting; a pest learns to associate a particular taste with an adverse physiological reaction (Rogers 1980).
(4) Because of differences among crops, repellents will not be uniformly effective. Better protection can be expected when treating millet or grain sorghum than when treating rice, because it is easier to deliver the chemical to the exposed grains of these crops. Because rice panicles often are surrounded by leaves, much of the repellent does not land on the panicle. Likewise, the seeds are protected by glumes that are not ingested by feeding birds, further diminishing the amount of repellent to which a bird is exposed.
(5) With most cereals, at least two applications (one during the early milk stage and one during the soft-dough stage) will usually be necessary.
(6) For a successful spray application, one must consider the physical characteristics of the plant (height, pubescence, panicle type), agronomic practices (row spacing, flooded or dry field), the climatic conditions

of the area (temperature, evaporation rate, rainfall, wind), and then choose and prepare the spray system and apply the chemical accordingly. Martin and Jackson (1977) have discussed these procedures and related practical application problems.

(7) Considerable attention should be given to the timing of the application relative to the intensity of bird attack. Delays of only 1 week may be critical.

(8) A particular repellent is unlikely to be effective against all species. Better success seems to be achieved with bishops, Golden Sparrows, and Black-headed Weavers, which despite being less sensitive to repellents than quelea, are still gregarious, yet feed in smaller flocks and inflict less damage before being repelled.

(9) The repellent principle of learned aversion will work better in the presence of resident rather than transient species.

(10) The possible effect of adhesives in masking repellency (Hermann and Kolbe 1971) has not yet been satisfactorily evaluated. Evidence suggests that this may, indeed, occur when a high ratio of adhesive to repellent, for example 1:3, is used at low application rates to protect ripening cereal grains (J. Besser and D. Elias, unpubl. data). Conversely, adhesive to repellent ratios of 1:1 have provided protection to seeded rice in flooded paddies at repellent application rates (Besser 1973). The results of the trials in Senegal confirm that repellency can occur when adhesives are used, and that they probably are necessary under conditions of heavy rainfall or aerial irrigation.

(11) Trial designs in fields having untreated plots adjacent to treated plots seemed to retain birds that might otherwise be repelled, with additional resulting damage. Small, closely associated treated and untreated plots are unsatisfactory for the most effective repellent use. Additional considerations of test design and efficacy evaluations in developing countries are discussed by Martin and Jackson (1977) and Bruggers and Jackson (1981).

(12) Birds often begin attacking crops along field borders or in patches along borders and later damage areas further into the field. Economical application techniques employing edge, alternate band, or spot sprays applied at the onset of damage often can protect the field.

(13) Repellents can be more effective in fields free of insects and weeds. Insecticidal properties of methiocarb apparently play an important role in reducing crop damage by birds, at least in maize (Woronecki and Dolbeer 1980; Woronecki *et al.* 1981). An insect-free field would presumably be less attractive to pest species that sometimes eat insects or rely heavily on them to feed their nestlings. A shift by the adults from insects to grain and the subsequent use of the field by fledglings would seem to be quite easy and not unexpected. Likewise, damage is less

(Luder 1985*a*) and methiocarb protection is better (Bruggers 1979*b*) in properly managed weed-free fields than in weedy fields.

Implications for use in Africa

Despite the encouraging results, research on and use of repellents in Africa have only been localized. Since 1979, a time when advances were being made to use methiocarb economically (Bruggers *et al.* 1981*b*), little additional work has been undertaken. There seem to be several reasons for this lack of interest.

First, the priority of the two UNDP/FAO grain-eating bird projects in eastern and western Africa has been to assist and strengthen regional lethal control operations. Although both projects stressed the need to look into 'alternative methods of control' (FAO 1980*c*), neither project has recently advocated using repellents. In West Africa, repellents have been dismissed as 'too costly', and efforts have been directed towards poisoning crops, employing ground sprays, destroying nests, baiting with poisons, and trapping birds (FAO 1982*b*).

Second, the inconsistent results that characterized the initial field screening trials in West Africa have impeded the acceptance of repellents, despite evidence from later trials in eastern Africa demonstrating more consistent protection.

Third, the feeling persists that repellents, when they work, only redistribute damage so that overall losses in an area remain constant. However, in many of the studies conducted in eastern Africa (Bruggers *et al.* 1981*b*), the protected crop was the only one in the area or in a vulnerable stage at the onset of the trials. Under these conditions, it can be assumed that 'repelled' birds shifted to wild seeds and perhaps insects, the only other available foods. In perhaps the only damage situation in which behaviour of 'repelled' birds has been systematically monitored, flocks of radio-equipped Red-winged Blackbirds frightened from vulnerable sunflower fields in North Dakota fed next, on 43 per cent of 56 occasions, in stubble fields, weed patches, non-vulnerable sunflower fields, corn fields, and swathed wheat; 27 per cent of the flocks visited cornfields but inflicted only negligible damage. The remaining 30 per cent of the flocks visited vulnerable sunflower fields usually within 19.3 km of the roost (Besser 1978; Besser *et al.* 1979). Similar studies need to be conducted in Africa to understand better the movements of 'repelled' birds and the relationships of alternate food sources and adjacent cropping areas to crop protection efforts.

Redistributing damage may have merit. Farmers are more likely to produce annually or even expand a particular crop if damage levels are tolerable. Similarly, it has been shown that many crops such as grain sorghum (Beesley 1978), maize (Woronecki *et al.* 1980), and sunflower (J. Sedgwick *et al.*, unpubl. data) may compensate through increased weight of remaining seeds for minimal amounts of damage at early periods in maturation. Finally, a reduction in sprouting does not necessarily infer similar yield reductions because many small-grain crops can compensate for stand reduction by increased tillering. Reduction in yield of both barley and oats is not proportional to the number of grains removed (Feare 1974). Given that growth compensation is apparently common, it probably occurs in other cereals that quelea damage.

Irrespective of arguments supporting continued evaluation of repellents, individuals and organizations best able to promote their use seem to feel that because repellents will not protect crops in all situations, crop protection efforts should be directed towards improving lethal control methods. Because of this attitude, most farmers are unaware that repellents exist. This is unfortunate because repellents can be cost-effective. Those who have been instrumental in evaluating them recognize that they are not a panacea and that they will probably not supplant lethal control to reduce crop damage by quelea in Africa. Many situations exist in Africa where repellents can be used alone or with lethal control to protect crops (Bruggers and Jaeger 1982; Erickson *et al.* 1980). Repellents could also be especially useful to individual farmers, agronomists at research stations, and individuals at seed multiplication or production schemes when facing damage by small populations of quelea or other pest species for which lethal control may be impractical, uneconomical, or unwarranted. The importance of having both lethal and non-lethal methods available to protect crops from bird pests in Africa cannot be overstated.

20

Agronomic techniques to reduce quelea damage to cereals

ROGER W. BULLARD and
BRHANE GEBREKIDAN

Introduction

Careful attention to planting schedules, cropping schemes, husbandry practices, and selecting less susceptible cereal varieties can reduce quelea damage to cereal crops in Africa. Linked to the ingenuity and skills of African farmers, agronomic methods may reduce the dependence of traditional farmers on national or regional lethal control units to protect their crops.

Cereals are important members of the grass family Gramineae, which together with Leguminosae comprise most of the world's food sources (Harrel and Dirks 1955). Wheat *Triticum aestivum*, Barley *Hordeum vulgare*, Rice *Oryza sativa*, Maize *Zea mays*, Sorghum *Sorghum bicolor*, and Millet *Panicum* spp. are all found in Africa and suffer bird damage.

Wheat and barley were domesticated about 7000 BC in the Middle East, before coming to Ethiopia and the rest of Africa (Doggett *et al.* 1970). These crops are now grown in the more temperate zones of Africa. Sorghum cultivation began later and probably originated in north-east Africa (Doggett 1970). The tolerance of wild sorghums to hot lowland conditions may have favoured their development; sorghum cultivation supposedly spread west and then south.

Many sorghum and millet types are found on the continent because early agriculturists in Africa were able to select the characteristics most suitable for the various altitudes, rainfall conditions, tolerance to insects and diseases, and soil types. Because sorghum and millet have a greater yield stability over such a wide range of environmental conditions compared to other cereals, they have had a major place in African agriculture. In addition to the widely varying sorghum varieties, culture and uses differ appreciably in eastern, western, and southern Africa. From Ethiopia and Sudan southward, sorghum is considered an important cereal for human food consumption and brewing. Plant height, panicle and grain types, and maturation period vary greatly among regions and sorghum cultivars (Doggett *et al.* 1970).

Great variability also exists in plant, head, and seed characteristics of African millets. Pearl millet *Pennisetum typhoides* is the most widely grown African millet (S. Clarke, pers. comm.) and one of the most drought-tolerant of all cereals. Early-season millet matures within 60–95 days, whereas late-season millet matures within 130–150 days (Hulse *et al.* 1980). Finger millet *Eleusine coracana* is found primarily in eastern and central Africa, and probably is the next most important variety. In addition, a number of minor, localized, small-seeded cereals exist that usually are listed in the millet category (S. Clarke, pers. comm.).

The exposed seeds, the strong stalk, and leaf characteristics make sorghum and millet highly vulnerable to quelea damage. The sorghum varieties usually grown for food have white, corneous grains that can be ground into white flour. Quelea also prefer these varieties, and in some areas of Africa damage has been so severe that farmers have turned to maize instead. For example, sorghum is not widely grown in the hot subtropical zone of the Awash River Basin of Ethiopia, even though it is the cereal most suited to the low and erratic rainfall (Erickson 1979). The same is true for much of the rest of the Great Rift Valley and other lowlands of eastern and southern Africa.

Plant breeding concepts of less susceptible varieties

Breeding cultivars that are less susceptible to or less preferred by granivorous birds is a popular means to minimize losses. Harris (1969) defined bird resistance as 'that mechanism or characteristic of a variety that when given a choice of feeding material birds will not normally depredate.' By this definition, bird resistance is affected by alternative feeding sources, bird population levels, and several other factors. Therefore, even the least susceptible varieties may be destroyed when bird pressure is high.

One strategy of plant geneticists has been to impart morphological or chemical characteristics to a plant or seed that will make it less attractive to birds as a food source (Plate 16). Unless birds become 'sick' (Rogers 1978*a*), it is improbable that cereal varieties can be developed that will totally resist attack when birds are hungry and have no alternate food; but losses can be substantially reduced in many situations. Doggett (1957) observed that 'varieties may be bred which are unattractive to birds, and which are attacked only as a last resort.' Preferences in feeding behaviour can be a powerful tool if applied properly. An effective way to make a variety less susceptible to bird damage is to combine several characteristics to make it as difficult as possible for birds to eat the grain.

Chemical factors

Astringent tannins are the best known chemicals associated with bird resistance in sorghums and millets. The astringency, producing that mouth-puckering sensation, results from binding of the proteins of the saliva and mucous epithelium by combination with the tannins (Joslyn and Goldstein 1964). Plant breeders working on bird resistance in sorghums have emphasized tannins and African farmers have traditionally selected brown sorghums containing tannins as a means of reducing bird damage. If a choice is available, quelea will seldom prefer the astringent varieties. A direct relationship has been shown between condensed tannin content of sorghum grains and repellency to Red-winged Blackbirds *Agelaius phoeniceus* and quelea (Bullard *et al.* 1980), but in the absence of alternative food, even those sorghums may be ravaged.

The condensed tannins, composed of procyanidin oligomers are located in the grain testa and the pericarp of some varieties. Their biochemical properties such as enzyme inhibition, grain weathering, tanning of hides, deleterious nutritional effects, and astringency are all related to protein-binding properties. Tannin composition and properties are inherited traits. The capacity of tannins to form strong cross-links with proteins is broadly related to size, structure, and shape of the tannin (Goldstein and Swain 1963; Quesnel 1968) and of protein molecules (Hagerman and Butler 1980). More specifically, binding depends on the number of separate sites on the tannin molecule that can bind with sites on the particular protein.

Astringent tannin oligomers are present in the early milk stage of grain development (Bullard *et al.* 1981). The synthesis process apparently begins as chlorophyll develops in the pericarp (Gupta and Haslam 1980). Usually, tannin concentration increases during the milk and early dough stage and then decreases during grain hardening and ripening. Even within the same variety there are variations depending on temperature, sunlight, date of flowering and soil conditions (Hoshino and Duncan 1981; Mabbayad and Tipton 1975). Too much protein-binding activity often remains in the harvested brown sorghum grain and affects its palatability, digestibility, and nutritional quality (Bullard and Elias 1980). These negative factors have given brown sorghums a bad reputation and generally resulted in lower market values.

Sorghums are classified into three groups (I, II, and III) on the basis of their polyphenolic properties. The non-tannin types without a testa are classified as Group I. The testa-containing sorghums are classified as either Group II or III, based on differences in response to vanillin and modified vanillin assays (Price *et al.* 1978). Group III sorghums have similar values for both assays, whereas the modified vanillin values are much higher than those for the vanillin for Group II varieties. Apparently, tannins in Group II sorghums express their protein-binding activity only in the immature stages, not in the ripened, physiologically mature grain. Group II sorghums are nutritionally equivalent to non-tannin Group I varieties (Hartigan 1979; Oswalt 1975). Bullard (1979) and Bullard *et al.* (1981) have found tannin activity to be consistently lower for Group II sorghums at all maturation stages. Recently, purple testa sorghums were recognized as belonging to Group II (York *et al.* 1981, 1983).

The expression of polyphenolic properties in ripening Group II and III varieties is quite different. Eight ripening hybrids were compared by three chemical, three biochemical, and quelea preference assays (Bullard *et al.* 1981). All hybrids showed an increase in polyphenol activity that peaked in the dough stages and then dropped sharply in mature grain. Group II tannin protein-binding activity tended to peak earlier during grain development and then drop much lower in the ripened grain. Recently, purified tannins from a Group II sorghum (IS 8768) and Group III sorghum (DeKalb BR-64) were compared, and no significant differences were found in their properties (Asquith *et al.* 1983). Apparently, other grain components had an influence on the expression of protein-binding properties in IS 8768.

The major obstacle yet to be overcome is the development of Group II varieties that have enough tannin activity in their immature stages to deter attack from moderate to high bird feeding pressure. At least three Group II hybrids have been found with tannin activity in the milk and early dough stages which is comparable to activity of the most repellent Group III hybrids, but then exhibit the loss in tannin activity charcteristic for the mature stage (Bullard and York 1985).

Processing techniques for Group III sorghums

Numerous efforts, other than plant breeding, have been directed toward improving the palatability and nutritional quality of harvested Group III sorghum grain. Some of these efforts may result in methods that can be integrated into economic food processing systems characteristic of those used in other cereals. Mechanically dehulling sorghums has been practised in Africa for years. The pericarp (bran) is removed by vigorous pounding in a mortar and pestle. Sorghum usually decorticates into quite large flakes (Shepherd 1981). Hand-pounding is tedious, taking about an hour to process 2 kg of sorghum (Munck *et al.* 1982). Thus, there is a ready market for small, diesel-driven village mills serving individual farmers.

The dry milling process for sorghum grain varies from cracking to bran and germ removal. In a pilot village-scale sorghum milling operation in Nigeria, abrasive-type mills have proven superior to attrition-type mills (Reichert and Youngs 1977*a*). Subsequently, a small batch mill was developed for local milling of sorghum (Munck *et al.* 1982). Further modifications to permit continuous milling have been used successfully in Botswana (Eastman 1980). Sorghum dehullers are being tested in several East African countries including Ethiopia, Sudan, Kenya, and Tanzania.

The current milling practices may not provide a satisfactory solution to eliminate tannins. First, the techniques are expensive. Second, the milling properties of brown sorghum are generally inferior to corneous endosperm sorghum without testa. Third, important grain constituents are lost in the dehulling process. Reichert and Youngs (1977*b*) found that 31–51 per cent of 9–18 per cent protein was lost in mechanical dehulling, while 7–21 per cent oil and 5–9 per cent protein was lost in the traditional mortar and pestle technique in Nigeria. Chibber *et al.* (1978) also found that approximately 12 per cent of the grain was lost in each of the three dehulling replications. Up to 37 per cent of the grain and 45 per cent protein was lost in this process that removed up to 98 per cent of the tannins. Furthermore, dehulling caused a decline in the content of lysine, the most limited amino acid in sorghum.

Some African farmers soak Group III sorghum grain in wood-ash or lime during food preparation, to reduce tannin activity. Chemical dehulling can be achieved with a 20 per cent solution of NaOH at 75°C, which removes the pericarp in 4–8 min and leaves the endosperm and germ intact (Blessin *et al.* 1971). Mixing dilute NH_4OH into whole Group III sorghum seeds or dilute K_2O_3 with ground grain (Price *et al.* 1979), extracting with aqueous alkali followed by washing in hot water (Armstrong *et al.* 1974), and soaking seeds in dilute formaldehyde solution (McGrath *et al.* 1982), H_2O, HCl, or NaOH followed by storage under CO_2 atmosphere (Reichert *et al.* 1980) have all led to reduction in tannin activity (Mitaru *et al.* 1983). Likewise, certain methods of processing grain sorghum, such as steam-flaking, reconstitution before grinding, or micronizing (rolling and dry heat treatment) may enhance the food value of the grain (Farris 1975).

With these techniques, it is unlikely that any tannin is lost. It probably either becomes too highly polymerized to remain active, or it becomes tightly bound to other grain constituents (Bullard *et al.* 1981). Unless milled, the ground product will still be red to brown in colour, as would be the case with Group II sorghums. Presently, most consumers tend to treat these products as being inferior because of their colour (Rooney and Murty 1982). Hopefully, with appropriate extension programmes these attitudes may be changed.

Some Finger Millet varieties have testae that contain tannins. Ramachandra *et al.* (1977) found that tannins, phenols, and *in vitro* protein digestibility (IVPD) of several varieties with white seeds had low total phenol and tannin values; dark brown seeds had high values. Two African varieties (IE 927 and IE 929) had high tannin values and had extremely low IVPD values compared to other cultivars.

Morphological factors

For all cereals, plant morphological characteristics can influence bird damage by grain-shielding or perching effects which makes feeding difficult, by 'social' effects of plant foliage creating a feeding environment that is 'stressful' to the bird, and by grain characteristics such as size, shape, hardness, and colour that influence acceptance. These morphological characteristics can be bred into a variety to make it less attractive than alternative food sources.

Stalk characteristics are not seriously considered as a deterrent for quelea. As feeding pressure increases, perching becomes a negligible factor for quelea flocks. Ward (1965*a*) has described large flocks as advancing across a field like a 'gigantic roller', flattening the crop as it is being stripped. The height of the cereal ears or heads also can be an important factor in choice situations with some species; short varieties can be advantageous. Manikowski and Da Camara-Smeets (1979*a*) observed that some birds prefer to feed on the tallest stalks. Dawson (1970) observed that the highest ears of wheat and barley suffered the greatest loss to House Sparrows *Passer domesticus*, and no grain was taken from ears lower than 30 cm.

Panicle types

Panicle structure influences susceptibility, especially for sorghums and millets. There is a wide range of variation in the size and shape of panicles, particularly rachis length, number of nodes per rachis, length of seed branches, angle of insertion of these branches, and number and distribution of branches and spikelets in sorghum. Sorghum breeders have introduced the open-headed or lax varieties as one means of reducing bird damage, primarily from blackbirds, but lax panicles are thought to be much more effective deterrents to feeding by large (>50 g) birds. Doggett (1957) observed that quelea seem to be able to perch on the most slender panicle branches of sorghum. Lax panicles would be expected to have some effect on sorghum damage, but probably only in a choice situation where alternative food is readily available. Similarly, Pearl and perhaps Foxtail Millet *Setaria italica* panicles would be easier for quelea feeding than Finger Millet, Japanese Barnyard Millet *Echinochloa frumentacea*, or Common Millet *Panicum milaceum*.

Panicles of some sorghums become recurved after emergence from the sheath. These goose-necked types appear to be less susceptible to birds (Doggett 1957), again because of the inconvenience to feeding. Damage is often on the top side of the curved panicle where it is easier for quelea to perch and feed. Also, in the Horn of Africa and other areas where sorghum grain matures under low humidity, farmers often plant very compact, large-seeded, goose-necked sorghums of the durra race; Muyra, Abdelot, and Degalit are widely grown Ethiopian varieties of this type.

Awned (bearded) types

Awns, the slender bristles attached to the lemma that covers cereal grains, also can be a deterrent to birds. Adesiyun (1973) observed that 'awns of some varieties of wheat act as natural barriers to attack by birds'. Studies have indicated that awnless varieties are more vulnerable to bird attack than strong-awned types (Jowett 1967; Perumal and Subramanian 1973). In Zimbabwe, quelea preferred the awnless to awned types of wheat growing in large fields containing numerous small plots of different wheat varieties (Plowes 1950). In Liberia, where Black-headed Weavers *Ploceus cucullatus*, Chestnut Weavers *P. rubiginosus*, and Vieillot's Black Weavers *P. nigerrimus* are the most abundant bird pests (Bashir 1984), an awnless rice variety (TOX 502–13-SLR) was damaged significantly more than a variety (ROK 16) having awns with a mean length of 62 mm (Abifarin 1984).

Doggett (1957, 1970) reported that the awnless sorghums are eaten first by quelea, but afterwards the awned types are taken quite readily. Jackson (1971) observed in a test with six caged quelea on awnless, weak-awned (1 cm), and strong-awned (2 cm) varieties of Pearl Millet, that the latter appeared to be comparatively highly resistant to feeding. A field study of awned and awnless varieties of bajra (Indian Pearl Millet) gave similar results (Beri *et al.* 1969).

In addition to lessening bird damage, awns can also be an important factor in increasing grain yield, especially when moisture is scarce (Chen and Li 1980; Shannon and Reid 1976). Studies in both wheat and barley have indicated that awns are an important photosynthetic centre that can contribute significantly to grain growth in the head (Lawlor *et al.* 1979). In wheat, awns seemed to increase the incorporation of N-compounds from vegetative organs to the caryopsis through increased transpiration (Pavlov and Kolesnik 1979). Because awns are a nuisance in mechanical harvesting, plant breeders have selected against this characteristic, possibly at the expense of yield and resistance to bird damage. It may be advantageous for farmers in some areas to return to awned varieties.

Large glumes

Sorghums with large glumes that envelop the grain are widespread in Africa. Perumal and Subramaniam (1973) observed a highly significant difference for glume length in studies where two cultivars with short glumes were more susceptible to bird damage than one with long glumes. Under low bird pressure, large glumes seem to provide bird protection. Large glumes also are often combined with other traits in attempts to breed less susceptible varieties. One sorghum cultivar, Bishinga Worabeisa, that is found in the Chercher Highlands of eastern Ethiopia, combines large glumes that completely cover the seeds and a very compact, goose-necked panicle. It does not seem to be bothered by birds at all, despite the fact that the seeds are pearly white and of excellent quality. In the United States, lax panicle, awned lemma, and large-glume traits are often combined in brown sorghum hybrids.

Ripened grain size and hardness

If availability and nutritional factors are not considered, quelea probably prefer small grass seeds; therefore size and hardness of cereal grains can influence food choice. Some studies have indicated that quelea prefer grass seeds each of about 1 mg, but eat increasing amounts of smaller (0.3–0.5 mg) or larger (14–30 mg) seeds as 1-mg seeds become scarce. Erickson (1979) observed 'that seed selection is also determined to a great extent by other factors, including the quantities and possibly the qualities of items available. If other seeds are available and the alternate food is ripened cereal grain, the size and hardness can have a definite effect on foraging behaviour.'

Quelea damage can be deterred by choosing varieties that produce very large or hard grains. Doggett (1957) observed that caged quelea would spit out hard grains and suggested that some grains are too large for the beak gape of small birds so that birds have difficulty consuming them. In Chad and Cameroon, Elliott (1979) observed that rice was eaten by quelea at all maturation stages. In laboratory studies, quelea preference for various whole cereal grains was less for wheat than either rice or a medium-sized red sorghum (Elmahdi *et al.* 1985). Size and hardness characteristics apply only to ripened grain. In studies of two Northrup King sorghum hybrids in four stages of maturity, kernel size had little effect on quelea feeding preference during the milk or dough stages, but the smaller grain (NK-1467) was highly preferred over the larger (C-21219) at the ripened stage.

Familiarity and nutritional factors

Food preference behaviour in birds is often influenced by recognition. In some laboratory studies, quelea often refuse unfamiliar food in the presence of a familiar one (Bullard and Shumake 1979). Caged quelea rejected red- or green-coloured millet grain when offered the choice of normal millet, but in the absence of sufficient light to discriminate colour, feeding became random (Shumake *et al.* 1983). This result corroborates that of Da Camara-Smeets and Manikowski (1979), who found a preference in quelea for naturally coloured grain.

Cereal breeding programmes specifically designed for bird resistance are rare. In view of the seriousness of the problem in Africa, selection for bird resistance should be given high priority, at least as part of overall crop improvement. Breeding programmes for bird resistance must consider several factors among which are useful sources of resistance, effective screening techniques, the inheritance and mechanisms of resistance, and methods of breeding and selection. Sources of resistance, involving the various morphological and chemical factors, could be obtained from various national programmes and the world sorghum collection at the International Crops Research Institute for the Semi-Arid Tropics, but more sources of resistance need to be identified. Major efforts are needed to combine the various sources of resistance into cultivars that have acceptable agronomic, sensory, and nutritional characteristics. Developing standardized screening methods for bird resistance breeding are essential. The most widely used method currently is screening under field conditions where damage usually occurs. The inheritance and mechanisms of bird resistance have not been sufficiently investigated, and more concerted efforts are needed if progress is to be expected.

The best bird-resistant cultivar is perhaps one that carries as many resistance-conferring traits as possible. In addition, the cultivar must be agronomically acceptable in its target area of production. A breeding programme designed to develop such a cultivar must by necessity work with several traits simultaneously. When this is the case, especially when each trait may be polygenic in its inheritance, it is a slow and challenging process. In such a case, perhaps one of the most promising approaches is the use of a recurrent selection scheme. The base population in such an approach must contain as many different types of bird resistance traits as possible and have the desirable range of plant height, maturity, plant type, grain type and size, and other resistance traits. The target variety would have reduced susceptibility to bird damage and produce a high yield of a palatable and nutritious grain.

Crop phenology

Changing crop phenology, for example, planting and maturity dates, is the most discussed agronomic technique for reducing quelea damage to cereal grain crops in Africa (Jackson and Jackson 1977). The bird migration pattern in many areas, where different varieties and planting times can be selected, is such that crops may mature when birds are absent or when an abundant supply of natural food is present (Crook and Ward 1968; Doggett

1957; Plowes 1950; Ward 1973*a*). Elliott (1979) referred to this as the 'harvest-time' method of bird damage control. Doggett (1982) observed that in northern Tanzania a short-cycle crop planted in February avoided damage at maturity, but in Uganda, crops planted in March–April were later decimated. In the Teso District of Uganda, August plantings of 100- to 120-day sorghum varieties escaped bird damage. Rice grown in the north-eastern corner of Nigeria also escaped damage because it ripens at the end of the rainy season, when the natural food supply is abundant despite the presence of vast numbers of breeding birds (Magor 1974; Ward 1965*c*).

Manipulating planting dates can be a very practical technique to reduce losses to birds in the semi-arid zone of Africa, if irrigation facilities are available. For example, in the lower Awash River Valley of Ethiopia, irrigated sorghum is planted in September and harvested in December without any quelea damage because the birds are not present during grain development and maturity. In the same area, sorghum planted in December and maturing in March can be completely destroyed by quelea. Elliott (1979) has presented quantitative data indicating that the damage to irrigated rice at Bongor, Chad, and at Yagoua, Cameroon, can be solved in some years by timing the vulnerable stages of the crop to fall between the period of mid-May to mid-June when quelea are absent. Conversely, cereals that mature in the dry season are likely to suffer damage if grown in regions that naturally support large bird numbers (Magor 1974). Areas under dry-season irrigated cultivation often provide the only seed and water in a region and therefore attract birds (Da Camara-Smeets and Manikowski 1979; Ruelle and Bruggers 1982).

Short-cycle varieties most often are used to manipulate harvest times to precede the arrival of birds. Early-maturing varieties have also been important in protecting crops from young quelea. In Bongor and Yagoua, millet, rice, and sorghums usually are attacked between November and January (Jackson 1974*c*), often by young birds. Quick-maturing varieties occasionally escape this damage because they mature when wild grass seeds are still available. The greatest advantage of short-cycle crops probably is that they are exposed to bird attack for a briefer period during the immature stages. Wild quelea eat about 2.5 g/day (dry weight) of seed, but in a cultivated crop they can destroy more, especially in the immature stages of grain development (Chapter 3; Jaeger and Erickson 1980).

Farmers must plan carefully, co-operate, and know the characteristics of the varieties they use or face severe problems with short-cycle varieties. First, growers whose crop matures much earlier or later than the average usually suffer the most damage from granivorous birds in an area (Meanley 1971; Raju and Shivanarayan 1980). Second, in many areas of Africa, short-cycle varieties must be photoperiod insensitive and resistant to grain moulds (Doggett 1982). Curtis (1968) found that yields of local sorghum varieties are associated with heading dates. Varieties indigenous to an area flower each year at a time that is associated with the mean date of the end of the growing season. New varieties that mature sooner often face fertility problems and greater exposure to sucking insects and moulds because the rainy season has not ended (Curtis 1968; Doggett 1982).

Several techniques can be used to reduce the period of exposure to birds. Artificial dryers or drying techniques can save valuable grain without reducing seed viability. In Botswana, it was found that sorghum could be harvested as much as 2 weeks early to prevent dove *Streptopelia* spp. losses without impairing germination (COPR 1975). Dimethipin, a plant maturity regulator chemical, when applied during the late season to rice foliage, and 7–14 days before harvest, will reduce seed moisture content at harvest time without affecting yield or milling quality (Blem *et al.* 1983). Finally, C. Elliott (pers. comm.) suggested swathing as a practical early harvesting technique in wheat. In Tanzania, 'swathed wheat appears to be unattractive to quelea which prefer the standing crop or shattered grain.' However, on the 20 000-ha farm in northern Tanzania, where some of these observations were made, he observed that the swathed grain attracts lines of Egyptian Geese *Alopochen aegyptica* and Knob-billed Geese *Sarkidiornis melanotos* but not in sufficient numbers to make a significant impact on the crop.

Other cropping practices

Crop substitution or crop diversification can also be used to decrease bird damage. Maize can be substituted for more susceptible cereals in areas with adequate rainfall requirements. In some situations, damage can be diluted by farmers synchronizing their crop schedules so that damage is spread over a larger area (Feare 1974). Conversely, farmers in some areas of Africa use wide planting dates and stagger their crops so that large numbers of fields in the same maturation stage do not attract massive numbers of birds by the 'local enhancement' phenomenon (Manikowski and Da Camara-Smeets 1979*a*). Staggered planting can also result from logistical constraints or expected rainfall patterns. Quelea are often more numerous where food is more abundant (Ward 1965*a*), and local enhancement among the highly gregarious quelea has an important role in food searching (Manikowski and Da Camara-Smeets 1979*a*). In some cases, crop diversity may result in less bird damage. For example, in the United States, Red-winged Blackbirds gathered in larger flocks in monocultures, flew shorter distances to feed, and caused significantly greater damage than in nearby areas where crops were more diversified (Dyer and Ward 1977).

Planting buffer or diversionary (sometimes called lure or decoy) crops to attract birds away from important cultivars (Farris 1975) may sometimes be economically justifiable. Another diversionary technique has been to plant high-tannin brown sorghums around the more preferred varieties (Ruelle and Bruggers 1982). Delaying plowing of early-harvested fields until all fields in an area have been harvested will provide alternative feeding sites (such as grain stubble), and possibly thereby make it easier to protect the unharvested fields. Planting seed varieties that mature uniformly can minimize the period of susceptiblity and will generally result in a more valuable product.

Clean farming is another way of reducing bird damage (Meanley 1971). Brushy field borders should be cleared to eliminate potential day roosts that may be used during the heat of the day. Brushy areas, especially near rice fields, also harbour insects that attract birds (Woronecki and Dolbeer 1980). Removing brush usually improves the effectiveness of bird-scaring techniques. In Tanzania, a close correlation was found between the presence of weeds in wheat and high levels of quelea damage (Luder 1985*a*). Clean fields were less attractive. In the USA, blackbirds have been observed to prefer weedy rice fields (Meanley 1971).

Conclusions

The plant genetic and agronomic techniques that we have discussed are potential 'tools' for farmers to use in increasing their food production. Each locality has its own set of conditions that affect farmers. Specialists, trained in most of the above techniques, are needed in each country to help farmers make the proper choices and develop effective programmes. Bruggers and Jaeger (1982) discussed current efforts to transfer the appropriate scientific and technological advances through assistance programmes to government control organizations and agricultural institutions. These efforts need to continue.

21

Lethal control of quelea

WOLFGANG W. MEINZINGEN,
EL SADIG A. BASHIR,
JOHN D. PARKER,
JAN-UWE HECKEL, and
CLIVE C.H. ELLIOTT

Introduction

This chapter will review the evolution of lethal methods used to control quelea after more than 60 years of development. Most quelea control operations presently are carried out by aerial spraying. However, prior to the arrival of Europeans in Africa, traditional methods were used to alleviate quelea damage to subsistence farming. Of the many traditional methods used (Chapter 18), only nest destruction killed sufficient numbers of quelea (but only nestlings) to be considered a lethal control technique. With the commencement of larger scale cereal production, around the beginning of this century, farmers began to develop a number of ways of killing the birds *en masse* to prevent crop losses. Poison baits were used in South Africa in 1925 for quelea control (Lourens 1957). From then until 1954, various forms of ground-based control techniques were employed with a coincidental emphasis on particular techniques in the three regions of quelea activity: poisons in South Africa, fire-bombs in eastern Africa (and Nigeria), and explosives in western Africa. Flame-throwers were widely used for the specific purpose of destroying quelea breeding colonies. In all regions, experiments were carried out from time to time with other techniques such as poisoning water holes with cyanide and mass-capturing birds with nets in Sudan (Yahia 1957). During the period 1920–1950, mass lethal control was used only on a large scale beginning after the Second World War. During the early 1950s, control reached what seems, in retrospect, extraordinarily intense levels. For example, in Senegal in 1956, 5050 kg of explosives were used to destroy 27 quelea roosts and 162 580 l of diesel were used in flame-throwers to burn 3231 ha of breeding colonies (Sonnier 1957).

In South Africa, where poisons, particularly parathion, were commonly used, the search for new methods led to a natural progression towards aerial spraying. Lourens (1957) described experiments for spraying the poison directly on to standing wheat, using a tractor-mounted sprayer. When the results were unsatisfactory, attempts were made to apply the chemical directly on to the birds. First, a quelea roost in Blue Gum *Eucalyptus* spp. trees was sprayed by ground machinery, then a breeding colony in *Acacia karroo* was treated, in the latter case using a dosage of 2250 l/ha of a 1.5 per cent emulsion of parathion. By present spray standards (usually < 10 l/ha of fenthion) such applications seem to be a veritable drowning of the birds, with enormous potential environmental damage and serious possible consequences to operators. Experiments on aerial spraying were begun in 1955 on quelea roosts in South Africa, using parathion and boom and nozzle equipment. The first trials gave dramatically successful results that gradually led to the adoption of aerial spraying for quelea control all over Africa.

During the 1960s and 1970s, aerial spraying underwent great changes as more became known about spray droplet dynamics and bird behaviour. Fenthion, which is less toxic to mammals, replaced parathion and application rates diminished remarkably (GTZ 1987).

History and development of control techniques

Toxicants

The use of poisons other than in ground or aerial spraying was common practice in many countries in the 1960s. It was very effective during the dry season when huge flocks of roosting birds had only limited drinking sites. Natural waterholes were fenced off or artificial ones created by laying plastic sheets over depressions near vulnerable crops. In Sudan, once quelea began regularly drinking at the site, it was poisoned usually with sodium cyanide but sometimes with phosdrin, alpha-chloralose, fenthion, or parathion. After treatment, the dead quelea were collected and buried and the ponds filled with dirt. Yahia (1957) reported that 3 657 198 quelea were killed in this manner in the winter of 1956/57 in Sudan. The practice has ended because of its danger to humans and livestock, if the site is not properly guarded, and its indiscriminate killing of many non-target animals that penetrated the fencing.

Other poisoning techniques, such as grain mixed with strychnine or parathion, were commonly used in South Africa in the 1930s for killing birds during the dry season and sometimes for treating breeding colonies (Naude 1955*a*). The practice has now almost completely ceased. Dry-season control efforts came to an end with the eventual realization that total population control could not be achieved even using such techniques. The strategy was replaced by one that concentrated on controlling birds actually threatening crops or breeding close to cropping areas as described in Chapter 22.

However, farmers still sometimes revert to the old poisoning methods. For example in Zimbabwe, farmers apply a poison designed for aphid control (azodrin = monocrotophos) to wheat field edges to kill quelea. The practice is strongly discouraged by the authorities (P. Mundy, pers. comm.). None the less, in recent years field trials have been undertaken to test again the efficacy of applying poisons directly to the crop. Ruelle (1983) applied Furadan to small parts of rice paddies and placed bushes near them as perches to increase their attractiveness. His application rate was 0.1–0.3 g active ingredient (a.i.)/m^2 and resulted in between 2.5–22 dead target birds/g a.i. Mayo and Lesur (1985) applied 60 per cent fenthion to wheat at a rate of about 4 ml/m^2. Some birds fed and died but the others avoided the treated areas and fed elsewhere. It was concluded that the whole crop would have had to be sprayed, which would have been prohibitively expensive.

Shooting, flame-throwers, nets

Shooting quelea when they are feeding on the ground using a shotgun with small shot (No. 7–9) can achieve high kills. P. Ward (pers. comm.) claimed a record of 150 with one shot. In Sudan, tribal farmers in remote rural areas still use the method, often employing locally made muzzle-loaders (E. Bashir, pers. obs.). However, as a lethal control method, shooting is both impractical and expensive.

The use of flame-throwers for controlling quelea breeding colonies is another technique that has long since been discarded (Fig. 21.1). Army surplus equipment or modified stirrup pumps were used by trained personnel who walked during the daytime through the breeding colony burning nests. The work was dangerous for the operators and created bushfires. Moreover, only the eggs and nestlings unable to fly could be destroyed; adult quelea escaped, continuing to cause damage to crops and sometimes renested elsewhere.

Yahia (1957) reported that mass-capture was used as a technique for eliminating a colony in Sudan. Clapnets, 20 m long by 4 m high, were set up in a clearing near a colony. After dark the quelea were driven into the nets, although on moonlit nights the birds avoided the nets. Yahia said that a team of 60 men cleared a nesting site of 60 km^2 in 10 days, but presumably he meant 60 ha. The method was also used in the Fung area, southern Blue Nile Province, Sudan at that time (E. Bashir, pers. obs.) with results evaluated in terms of the number of 5-gallon tins filled with killed birds. As many as 100 tins full of birds were reported collected on one occasion, to the extent that the head of the unit, the late Mr. El Hewaris, suggested that the birds should

Fig. 21.1. Flame-throwers were often used in the early days of mass quelea control to burn the nests (photo: FAO).

be processed as canned food (Chapter 23). Jackson (1974*b*) developed a quelea trap in Chad which was very effective at catching fledgling quelea. The Australian crow trap (Boudreau 1975), which was modified (Bashir 1979) for capturing Rose-ringed Parakeets in Pakistan, might also, with further modifications be useful for quelea. Neither trap has ever been tested as a control technique. In western Africa, some attempt has been made to use mist-nets in rice cultivation to control pest birds (Ndiaye 1979). The method has not been encouraged because of its indiscriminate capture of birds and the unlikelihood that non-target species would be released unharmed.

Explosives and fire-bombs

Explosives and fire-bombs were widely used in the 1950s (Plate 18). The original method using only explosives was probably first developed by E. C. Wilson in Sudan sometime in the late 1940s (Naude 1955*a*). Yahia (1957) mentioned that about 30 tonnes of gelignite were used to blast swarms of quelea on the rivers Rahad, Dinder, and Blue Nile in 1949. The South Africans tried to improve the technique by adding a 1:1 mixture of diesel and petrol in containers varying in size from bottles to 200-l drums. The drum method then became routinely used in South Africa and was also adopted in Tanzania and Kenya. It was used exclusively in Nigeria between 1958 and 1959 and reached a peak in 1963 when 60 roosts were treated (GTZ 1987). From then until 1967 there was a big drop in the number of sites needing treatment, which GTZ (1987) implied meant the Nigerian quelea population had been reduced overall. The control effort was then relaxed but after only two seasons the quelea population greatly increased and control was resumed in 1970.

According to Naude (1955*a*), the fire-bomb method required only 4 per cent of the quantity of gelignite that had to be used if gelignite was used solely as the explosive. Nevertheless, the method using gelignite alone continued in francophone Africa well into the 1970s. C. Elliott observed it being used between 1975–78 in the Lake Chad Basin by OCLALAV staff. Night roosts in acacia and mimosa and day roosts in isolated trees around the periphery of rice cultivation were treated. Although many birds were killed, results were not convincing because only a small proportion of each roost was affected. The activity was very popular with local people who collected the poison-free dead birds for food.

The fire-bomb method has been largely displaced by aerial spraying but is still used regularly in some situations in Kenya (Plate 19). In parts of the Rift Valley where farming is intensive, virtually all the natural bush has been removed and some farms have planted windbreaks of exotic eucalyptus trees. Quelea roost in these trees often at a height of 15–25 m. Since colonial days, roost control in these situations has been achieved by placing 200-l drums spaced out by 20 paces throughout the roost area (F. Kitonyo, pers. comm.). Each drum contains 70 l of petrol and 70 l of diesel and is rolled over a hole dug in the ground containing 36 sticks of gelignite. The network is connected using Cordtex instantaneous detonating fuse and detonated, using safety fuse and detonators, about 30 min after sunset when all the roosting birds have settled in. The explosion produces a sheet of flame through the trees which carries above the topmost branches to at least 40 m. The flame extinguishes itself within a few seconds but often the grass understory and any dead branches are ignited so that the flames have to be controlled before they spread. The force of the explosion and flame is such that most birds within about a 20-m radius of each drum are killed. Local people usually collect the dead and injured birds for food, sometimes in dense roosts by the sackful. Despite the force of the blast, the trees seldom suffer any permanent damage and recover within about 2 months. The advantages of the method are that few birds escape, there is no chemical contamination of the environment, and local people obtain uncontaminated proteinaceous food (Chapter 23). Its disadvantages are that it can only be used in small (<2 ha), densely populated roosts where there are no major fire hazards and that it is also potentially dangerous to the operator if the vegetation is dense. Dense vegetation can allow someone to re-enter the site unseen by the officer-in-charge before detonation. In Kenya, the Blue Gum tree roosts are open with good visibility and no one has ever been put at risk by the technique. Even so, careful checks have to be made to make sure that the area is clear before detonation. The technique should not be used in roosts where non-target birds have concentrated.

Using the 200-l drum method costs about US $500 for each 0.5 ha covered (1987 prices). In recent months, F. Kitonyo and J. Ngondi (pers. comm.) have been experimenting with lifting of petrol bombs into the trees to the actual level of roosting birds (Plate 20). This has reduced costs to about $150/0.5 ha and resulted in much less fire at ground level.

We feel that the fire-bomb, particularly the refined method being developed by F. Kitonyo and J. Ngondi, has a useful role in the specific context of Blue Gum tree roosts in Kenya. These sites are difficult to spray with ground sprayers because of the height and are awkward to spray by aircraft on account of the small area and surrounding farmland. The fire-bomb technique might also be useful if it could be adapted to aquatic habitats, since chemical spraying in such places is undesirable. Although presently used only in Kenya, the method should be useful in similar situations in other countries that are politically stable enough not to be overly concerned with the use of explosives for civilian purposes.

Aerial spraying

The objective of aerial spraying is to kill quelea using a toxic chemical (Plate 21). Effective aerial spraying is a complex technology. It involves such diverse aspects as the aircraft, spray gear, the physics of the spray cloud and the component droplets, the birds' behaviour, the physiological process of poisoning, the mortality of the birds, the spray 'success', and the environmental consequences. A summary of many of the documented spray applications of the most used chemicals is provided by Manikowski (1988).

The avicide In the 1950s, parathion was the most commonly used avicide. In South Africa it was relatively inexpensive (US $3.60/l; Lourens 1957) and was selected from a variety of chemicals including arsenious oxide, Phosdrin, paraoxon, chloropicrin, and phosgene. The first ever aerial spray of quelea was carried out in South Africa using a 23.3 per cent parathion emulsion at a rate of 6.7 l/ha (Lourens 1957). However, because the emulsifiable concentrate (EC) formulation did not penetrate the birds' feathers, the formulation was diluted with diesel oil and much better results were achieved.

Parathion is highly toxic to birds (Hudson *et al.* 1984). The LD_{50} for quelea is 4.2 mg/kg (FAO 1979*b*). Its main drawback however is its toxicity to mammals including humans. The oral LD_{50} for female and male rats is 3.6 and 13.0 mg/kg, respectively (Worthing 1979). FAO (1979*b*) suggested that 0.21–0.35 g would be enough to kill a 70-kg man. A number of fatal accidents to spray operators eventually resulted in a search for alternative chemicals. Nevertheless, parathion was used in western Africa into the late 1970s mainly because stocks existed. In South Africa in 1986 it was still the main chemical used, primarily because it was a quarter of the price of fenthion (Tarboton 1987).

The search for less dangerous alternatives to parathion began seriously in 1958. Professor Schmutterer from the University of Giessen, West Germany, was the first to experiment with the organophosphate insecticide fenthion on quelea roosts. When quelea control in Sudan began in 1962, fenthion was the principal avicide used (GTZ 1987). From that time, fenthion has gradually become the accepted avicide for quelea control and in 1988 is used across Africa (Manikowski 1988).

The oral LD_{50} to quelea of fenthion is reported as 1.8 mg/kg (Schafer *et al.* 1973), 2.0 mg/kg (MacCuaig 1984), and 2.6 mg/kg (Cisse 1981; FAO 1979*b*) making it slightly more toxic than parathion. It is considerably less toxic to rats than parathion (FAO 1979*b*) and, therefore, safer to spray operators. We have never seen a mammal killed as a result of a fenthion spray operation for quelea control. Fenthion inhibits cholinesterase activity at nerve endings. On absorption into the skin, it is metabolized to the more toxic *o*-sulphone derivative which becomes the active anti-cholinesterase ingredient (FAO 1979*b*). Fenthion is marketed for quelea control as Queletox[R] in a 60 per cent a.i. miscible-in-oil formulation. Although an improvement over parathion, fenthion poses secondary non-target toxicity hazards and must be handled with care by spray operators. Because of its indiscriminate effect on all birds, efforts have continued to try to find an avicide that is specific to quelea. Of the 68 compounds mentioned in the Avicide Index (MacCuaig 1986), 48 have been tested against quelea in the laboratory. Of these chemicals, only cyanophos (Jaeger and Erickson 1981) and phoxim (Pope and King 1973) have shown promise. While both are less toxic to mammals than fenthion (FAO 1979*b*), both are more expensive and their use has never caught on. A major constraint to the development of a selective avicide is that the potential sales market in Africa is too limited.

Quelea normally are controlled in the evening between sunset and dusk, and a successful operation results in most of the birds being dead by morning. Cage experiments (GTZ 1987) have shown that birds first stop eating and/or drinking, become paralysed, and then die slowly. Pope and Ward (1972) found that death from fenthion poisoning could occur between 12 h and 72 h, depending on the dose received. Studies of the protein and fat content of dead birds suggested to Pope and Ward (1972) that those dying up to 12 h died of the effects of direct paralysis of the poison while those that died after 12 h died from the indirect effects of starvation. Pope and Ward (*loc. cit.*) also concluded that quelea could not survive for more than 33 h without food but with water. Dorow (1973) made similar observations on the speed with which quelea die from fenthion spraying when he tested birds in Nigeria. Cage studies (GTZ 1987) also showed that quelea can recover from sublethal fenthion poisoning after 4 days even after their body weight reached as little as 13 g; normal body weight is between 18 and 21 g.

In the field, the time over which quelea die depends on the spray technique, the application rate, and possibly the body condition of the birds. Those with a higher fat content could survive longer. This extended time to mortality following low application rates also has been described by Mierzejewski (1981) and Bruggers *et al.* (in press). The latter found young quelea dying up to 7 days after spraying 2.5 l Queletox/ha, which they found surprising in view of Pope and Ward's (1972) starvation experiments. Possibly, contaminated quelea take longer to die than starved quelea because the poison renders them immobile and may lower their basal metabolic rate. The fact that fledglings took longest to die may also reflect the tendency of fledglings to have a high fat content (up to 9.9 per cent of live weight; Ward 1965*b*). Thomsett (1987) also observed quelea dying for possibly as long as 9 days after a low-volume spray operation. This information has an important bearing on the environmental consequences of spraying quelea with fenthion.

The target The quelea is a good target for aerial spraying because it is so gregarious and at night concentrates in great numbers in roosts, breeding colonies, or combinations of the two. A frequent misconception among farmers is that the birds will be sprayed in the crop fields while attacking them, leading to queries about crop contamination. Birds are never sprayed in the crop because of contamination effects and because they are insufficiently concentrated, moving, and only present during the day. In fact, aerial spraying is seldom now carried out during the day except at colonies in the fledgling stage when the workload for dusk spraying has become too great. Roost targets are usually a minimum of several hundred metres from the crop and can be up to about 30 km away. Colonies close to crops are very unusual (the exception being colonies in sugar-cane adjacent to irrigated rice). Mostly they are at least several kilometres from human activity.

Quelea behaviour during spraying Aerial spraying is presently carried out in the 20–30 min between sunset and darkness, at the time when the quelea have returned to their roost or colony for the night, and when there is still enough light for the pilot to see. Breeding birds in the early stages of the cycle return a bit earlier and will be settled in within about 5 min of sunset (GTZ 1987). Roosting birds may fly in until it is almost dark. For roosts it is essential that birds are sprayed during the second half of the dusk period to avoid their escape.

During aerial spraying, the aircraft passes several times over the target about 5 m above the vegetation. The reaction of the quelea to the spray aircraft and noise differs if the target is a roost rather than a colony. Quelea roost in all kinds of vegetation, including acacia bush, reedbeds, sugar-cane, Blue Gum trees, tall grass, and maize. Roosting quelea frightened by the aircraft often have the possibility of moving to an alternative site nearby. Quelea colonies are by contrast fixed in one place by the birds' attraction to their nests for the duration of the breeding cycle. Roosts are therefore much more unstable. Formerly, quelea targets were sprayed in the pitch dark to reduce the tendency for birds to escape (Lourens 1957). In Zimbabwe, the practice continued until 1977 (LaGrange, in press-*b*). The necessity for night-time spraying was partly a function of the large droplet sprays applied, as will be explained later.

The first pass of the aircraft usually produces only a visible agitation in the birds, but by the second pass, the birds fly up in a dense swarm. Collisions with the aircraft are expected but pilots usually try to minimize them by climbing above the bigger flocks. If the timing of the spray is correct, many birds will be contaminated immediately. Those that escape will fly back to the roost within 30 s to about 5 min. Breeding birds seem to return more quickly to the colony after being disturbed, sometimes within 15–30 s after each pass. Meinzingen (1980) indicated that adults are most attracted towards their nests from the time the clutch is completed until the chicks are 4 days old. Nest attraction decreases slowly thereafter until adults begin to behave like roosting birds and may even start to form roosts separate from the juveniles. However, when spraying is finished, all birds usually will be back in the colony within 3–5 min (GTZ 1987).

Spray applications have generally been more successful when a fast, quiet aircraft rather than a noisy helicopter is used. Because birds may start to leave the target area when a helicopter is still a kilometre away, helicopters are seldom used for roost spraying.

Spray droplet behaviour The behaviour of the pesticide droplets is important for effective quelea control. Droplet behaviour is closely related to the type of avicide formulation used and, particularly, to the size of droplet produced. Important components of droplet behaviour include their descent rate (i.e. the time that the droplets remain airborne in the target area), their drift, the extent to which they are filtered out on vegetation, the efficiency with which they impact on flying and perching birds and the method by which they are absorbed into the bird's body.

The rate at which droplets settle out through air—the sedimentation rate—is proportional to the square of the droplet's diameter which varies with droplet size (Table 21.1). Small droplets remain airborne for longer than large droplets. Under certain meteorological conditions, droplets smaller than 50 μm, with a sedimentation rate of less than 7.5 cm/s in still air, may have a negligible fall rate and may remain airborne indefinitely. Such behaviour normally occurs under conditions of very low wind speeds and thus, paradoxically, the downward drift of small droplets is usually greatest under low wind conditions.

The efficiency with which a droplet impacts on the bird is also directly proportional to the square of the diameter. Large droplets impact more efficiently than small ones. Ward and Pope (1972) concluded that because large droplets will contain more pesticide than small ones, large droplets would be more desirable in bird control operations. Ward and Pope (1972) flew quelea through a curtain of droplets of various sizes in a homemade 6-m flight tunnel. They found that droplets of 180–250 μm volume mean diameter (vmd) impacted better than those of 30–75 μm vmd. However, this conclusion overlooked the fact that many more droplets are produced at a lower vmd and that their sedimentation rate is slower (Table 21.1). By remaining airborne longer, small droplets are likely to impact on birds returning to the target site after the aircraft has left.

Very little work has been done on where droplets impact a quelea and how the bird subsequently dies. It is not known whether quelea receive a lethal dose by inhaling tiny droplets, by preening the chemical off their feathers, or by dermal contact. Because the feathers are dead structures, the chemical cannot kill the bird by landing on them. Ward and Pope (1972) investigated the points of impact by fitting paper targets to various parts of a bird and

Table 21.1. Spray droplet behaviour (source: Zaske 1973).

Droplet diameter (μm)	Number /ml	No. required for 1 LD_{95} dose[a]	Release from 10 m in 1 m/s windspeed	
			Time to fall[b]	Drift downward[c] (m)
50	12 279 000	2546	2.2 min	133
100	1 909 900	318	35 s	36
150	565 880	94	23 s	23
250	122 230	20	8.5 s	8.5

[a]Assuming an LD_{95} of 0.1 mg a.i. fenthion, and the spraying of the 60% formulation.
[b]In still air conditions and assuming droplet does not impact on vegetation.
[c]In conditions of laminar air flow.

then estimating the per cent coverage based on the presence of a dye in the spray formulation. Many droplets impact on the beak, face, upper breast, and front edge of the wings of the bird. However, the quelea could reach no more than 30 per cent of normal flight speed in the tunnel, so the bird's angle to the spray may have been abnormal. Results from studies, when fluorescent particle markers (Chapter 5) were applied by aircraft in exactly the same manner as a control operation, showed the greatest concentration of marks on the underside of the primaries (C. Elliott, pers. obs.). Birds marked using ground sprayer equipment showed 75 per cent of the marks on the wings and none on the breast (Meinzingen and Latigo 1986). The chemical probably gets from the wings into the body by collecting at the base of the feather shaft and penetrating the skin. GTZ (1987) tested the sensitivity of various parts of the quelea's body to 0.1 mm^3 of 60 per cent fenthion and found the ocular and axillary areas were the most sensitive. Perhaps the droplets collected by the primary feathers reach the axillary area when the bird is resting with folded wings.

Weather also is an important influence on the success of aerial spraying. Fortunately, the period from sunset until dawn is a time when the wind often drops to $\leqslant 1$ m/s. Low wind speeds can be accompanied by an atmospheric temperature inversion, in which the temperature close to the ground is lower than it is above ground level. This inversion layer usually forms 2–4 m above the ground and is 3–5 m deep (GTZ 1987). The effect of this inversion layer is to help suspend the droplets and increase the time during which they can contaminate the birds. As wind speed increases, progressively greater turbulence is produced, increasing the vertical movement of small droplets both upward and downward. This increases the possibility of droplets impacting on vegetation or the ground and reduces downwind droplet drift. If the wind speed is greater than about 5 m/s, there will be so much turbulence that small droplets will remain airborne in the target area for less than a minute. In this situation, either the operation has to be postponed or adjustments must be made to the spray equipment to produce coarser droplets.

Initial application techniques Quelea concentrations were first aerially sprayed in South Africa and then in Sudan, using boom and nozzle equipment. Although dosages varied, both parathion and fenthion usually were applied as the 60 per cent formulation diluted to about 25 per cent with diesel. In Zimbabwe, usually 45–60 l/ha, but sometimes as much as 200 l/ha, were applied at night using droplets in excess of 100 μm (LaGrange, in press-*b*). The droplets probably had a vmd of around 150 μm, resulting in droplet sizes of 60–350 μm. The pilot, flying in a Piper Supercub, used hurricane lamps, flares, or fires and was guided by a ground controller using two-way radio to assure the spray was in the target area. In Sudan, application rates of 30 l/ha of 25 per cent diesel-diluted fenthion applied by Piper Supercub and

Cessna 180 or 185 aircraft became common (GTZ 1987). The number (usually 44–48) and sizes (D8, D10, and D12 with size 45 cores) of hollow cone nozzles depended on the aircraft and application rate. Because of droplet dynamics, it was obligatory to spray roosts in the dark so that the birds did not have a chance to escape. Because spray equipment produced droplets of 150 μm vmd, most would fall to the ground within 23 s (Table 21.l). Given the range of sizes produced, 97.7 per cent of the volume of chemical sprayed remained in the air for less than 60 s after leaving the aircraft (Meinzingen 1983). The pilot sprayed at right angles to the wind direction, placing the first swath on the upwind edge of the target, until the whole target was covered. Because of the large droplets, spraying under wind conditions of ⩾5 m/s was still possible. In light winds, the first swath could be placed over the centre of the target. The same aerial spraying techniques were used for spraying fledglings in breeding colonies in daylight. Colonies were sprayed beginning on about day 28 of the breeding cycle until the time the young fledged. When a colony was sprayed in systematic swaths, control of juveniles often was excellent, but adults usually abandoned the colony.

Bell 47 helicopters were occasionally used in early quelea control. They had the advantage of high manoeuvrability but required that spraying be restricted to moonlit nights. These first helicopters had no artificial horizon and roosts had to be sprayed at night because of the noise. Presently, the two most commonly used helicopters for quelea work are the piston-engined Bell 47 used in Sudan, Nigeria, and Mozambique, and the Bell Jetranger used in Tanzania. Their versatility, slow flying, and vertical take-off and landing make these helicopters ideal for control of quelea breeding colonies, but their effectiveness is offset by the flying cost which is at least three times greater than for a fixed-wing aircraft. Helicopters are also highly effective for surveying and marking colonies (Chapter 4; Elliott 1980*c*). All large-scale quelea control operations (e.g. Tanzania, Sudan, Lake Chad Basin) should have the services of a helicopter, if not on a full-time basis, then at least during the early part of the quelea breeding season when accurate survey work and rapid spraying is necessary.

Ultra low volume (ULV) technique A breakthrough in aerial spraying techniques came in 1971 when a rotary atomizer was used to produce much finer droplets than those produced by boom and nozzle equipment. Six two-disc rotary atomizers were fitted on the booms of a Bell 47 helicopter. The atomizers were electrically driven from the helicopter's power supply, at a rotational speed of about 16 000 r.p.m., producing a spray of 45 μm vmd (Meinzingen 1980). By the mid-1970s, this equipment was replaced by wind-driven micronairs consisting of a rotating cage of wire mesh (Fig. 21.2, 21.3). The chemical is released at the centre of the cage by a deflector and is atomized by the centrifugal forces driving the chemical through the wire

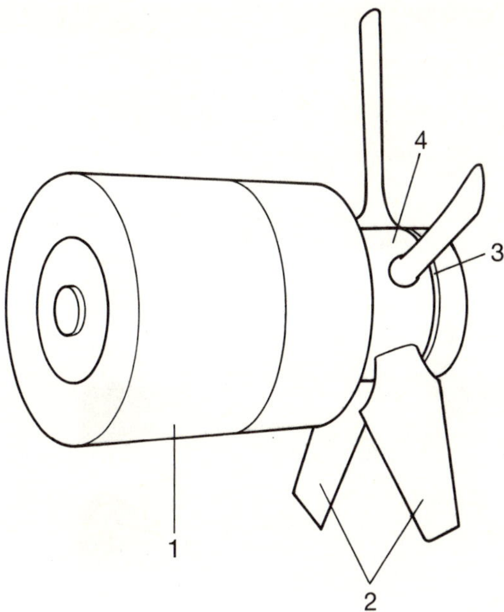

Fig. 21.2. Micronair Atomizer AU3000 as used on many aircraft for spraying quelea. Constituent parts include: the rotating gauze cage that achieves the final stage of droplet atomization (1); the propeller blades that rotate the gauze cage at a speed related to blade angle and aircraft speed (2); the brake disc that allows the pilot to stop the micronair when spraying is completed (3); the anodized aluminium alloy hub on which the propeller blades are clamped (4).

mesh. The cage is rotated by the effect of the aircraft's slipstream on plastic propeller blades placed at the front end of the cage. Rotational speed is a function of the speed of the aircraft and the angle at which the propeller blades are set. For quelea spraying, normally one micronair is placed on each side of the aircraft, but better droplet distribution is achieved with two on each side.

A number of different models of micronair have been produced including the AU3000, AU4000, and AU5000 that differ in their size and the length of the cage. For slow-flying helicopters, longer propeller blades are used so that rotational speed is maintained. The main advantage of micronair over boom and nozzle equipment is that droplet size produced is both small and more uniform in size (Fig. 21.4) resulting in less chemical wastage. The most current spray system is the pod (Fig. 21.3). A self-contained unit of spray tank and micronair is hung under each wing. For quelea control, the pod system is used mainly on the De Havilland Beaver aircraft in eastern Africa. The system does not influence flight performance, removes the chemical from the cabin, and permits the aircraft to be used for survey and transport

Fig. 21.3. The development of the Micronair pod system took the chemical out of the aircrcaft, making the cabin safer for the pilot and improving the flight characteristics of the aircraft (photo: M-T. Elliott).

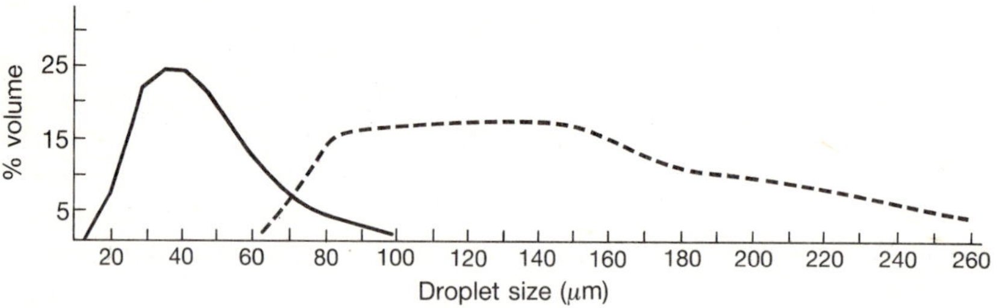

Fig. 21.4. The distribution of droplet sizes by percentage volume for a spray of 45 μm vmd as in ULV quelea control (solid line) compared to that of a spray of 150 μm vmd (dash line), as used in the early days of quelea aerial spraying (source: GTZ 1987).

activities. Other aircraft commonly used for quelea spraying include the Piper PA18 (Lake Chad Basin) and the Thrush Commander (Zimbabwe).

With the introduction of the rotary atomizer, it became possible to switch to ULV spraying, which we define as an application rate of $\leqslant 5$ l/ha. Instead of attempting directly to hit birds with large, fast-falling droplets and being obliged to spray at night, a cloud of small droplets could be laid over the

(a)

(b)

Fig. 21.5. The GTZ ground sprayer produces a fine cloud of droplets which drifts across the path of quelea coming in to roost. The sprayer consists of a sprayhead, a hydraulic mast, spray tank, and generator (photos: J.-U. Heckel).

target site at dusk. For ULV spraying, the formulation of the chemical is important. The solvent must not be highly volatile or the droplets will decrease in diameter when they fall through the air as the solvent evaporates.

Since the early 1980s, it has been found that effective kills can be achieved by ULV spraying of an undiluted standard 60 per cent formulation of Queletox (Plate 22). This spraying technique consists of laying swaths of 50–60 μm vmd droplets at right angles to the wind direction. For colonies, spraying can start at early dusk, giving nearly 30 min of spraying time. For roosts, spraying must start at late dusk, giving an effective spraying time of only about 15 min. Since the objective of ULV spraying is to lay down a cloud of slowly sinking droplets into which the birds will return once the aircraft has gone, better results are achieved if the pilot sprays the first swath down the centre of the target and then flies a broad circuit away from the target. The pilot sprays the next swath between 2 and 4 min later. This allows time for birds to return to the site between swaths and decreases the chances of a mass departure to an alternative roost. Because of the birds' attraction to their nests, it is important to spray a colony systematically. Roosting birds present more of a moving target so that the pilot has to place his swaths where he sees the greatest swarms of birds.

Methods have been developed for the objective estimation of mortality (Elliott 1981*b*). Excellent results have been achieved with ULV spraying in many parts of Africa (Manikowski 1988). On occasion results are very poor. This can be attributed to a number of reasons such as pilot skill and experience, inadequate or incomplete ground support, or excessive wind. ULV applications of 50–60 μm droplets at wind speeds of more than 3 m/s often fail.

The development of ULV spraying permitted colonies as large as 100 ha to be controlled in a single evening, and enabled quelea colonies to be sprayed at almost any stage of the breeding cycle. Adult and juvenile birds can be sprayed when both are together in the colony. However, spray results tend to be poor at the stage when chicks have left the nest but are not yet able to fly and, thus, do not collect a lethal dose of chemical; it seems that small droplets of ULV spraying do not impact well on a stationary bird. At this age, chicks also may be able to find their own grass seed, with the result that while those being fed by adults may die because their parents were killed, others may survive.

Meinzingen (1980) has suggested that during nesting, spraying is likely to be more effective than during roosting due to the increased physiological stress associated with breeding described by Ward (1965*b*) and Jones and Ward (1976). Likewise, controlling nesting colonies stops crop damage being caused by adults and prevents damage often caused by dispersing juveniles. The development of the ULV technique opened up possibilities for the controversial idea of attempting to spray all breeding sites in order to eliminate the quelea once and for all as an agricultural pest (Chapter 22).

The application rate necessary to achieve successful ULV quelea control is debatable. Meinzingen (1984) and GTZ (1987) have suggested that 1–3 l/ha is sufficient and that greater quantities are a waste of chemical. Nevertheless, control units frequently spray more. In Zimbabwe, application rates are 4–20 l/ha (LaGrange, in press-*b*). In 1984 in Tanzania, 119 sites of between 1 and 250 ha were aerially sprayed at rates of between 0.8 and 10.1 l/ha (Ndege and Elliott 1984). Rates of 2–6 l/ha are not used for small, dense roosts that may cover only 2 ha, because the entire amount of chemical will be released too quickly to permit the pilot to place swaths or space them out during the spraying period. Ndege and Elliott (*loc. cit.*) concluded that with such a variation in site size, a sliding dosage was appropriate; 1–5 ha (8 l/ha), 6–40 ha (4 l/ha), >40 ha (2 l/ha) for colonies; and 1–4 ha (8 l/ha), >4 ha (4 l/ha) for roosts. There is no doubt that the ULV technique has allowed a drastic reduction in dosage and that the trend is for control units eventually to bring the dosages close to the ULV rate of 1–3 l/ha.

The conventional method for deciding the quantity of chemical needed for an area is to multiply the recommended application rate by the size in hectares of the target (FAO 1979*b*). For roosts the area can sometimes be misleading because some roosts of only 1 or 2 ha can hold hundreds of thousands of birds, while others may be up to 10 ha in size with a dispersed, less dense population. An alternative method based on the number of birds in the area was proposed by Latigo and Meinzingen (1986). Their Guided Application Dose (GAD) depends on close contact between a ground controller and the pilot. In brief, the controller advises the pilot on the behaviour and density of bird swarms rising from the roost as the aircraft passes over so that the pilot can appropriately apply swaths to achieve maximum contact. The effectiveness of GAD has yet to be thoroughly field tested. The concept is, nevertheless, useful to demonstrate how small a quantity of chemical actually is needed to kill a large number of quelea, and thereby encourage control units to reduce application dosages.

Ground spraying

The two greatest operational drawbacks to aerial spraying are the expense and the dependence of the bird control officer on a technology that he himself cannot use. Effective ground spraying equipment could control quelea at a fraction of aerial spraying costs and would be easy to handle by the control officer. A first attempt to develop ground spraying equipment for bird control was to use Exhaust Nozzle Sprayers which were being used for controlling Desert Locust *Schistocerca gregaria* (Kaske 1970). These sprayers were fitted to landrovers and the spray was produced from the top of a pipe about 2 m above the ground. The landrover was driven along the upwind edge of a quelea roost to allow the spray to drift across the target. Results were poor because the large spray (120 μm vmd) droplets dropped quickly to the ground, and the spray failed to get above the roost vegetation.

In 1968, a farmer in Zimbabwe invented the 'static line' method in which pipes were placed over the upwind edge and down the centre of a quelea roost (LaGrange, in press-*b*). Spray nozzles were fitted along the pipes and Queletox, diluted to 30 per cent a.i. was pumped through the system after dark producing a blanket of droplets of 160 μm vmd that settled over the roosting quelea. The system sometimes worked perfectly but was erratic, being influenced both by wind direction and bird behaviour; quelea often moved away from the static lines.

In 1971, W. Meinzingen (pers. obs.) tested a swing-fog machine for control of quelea roosts and colonies in Nigeria with poor results. In 1983, Parker and Casci (1983) tried out a TIFA 100E fogger on quelea sites in Tanzania. Foggers use heat to vaporize the avicide which then condenses to form a dense cloud of very small droplets. Although some of the avicide is probably lost through thermal degradation, field tests indicated that this was not significant (Parker and Casci, *loc. cit.*). In the TIFA fogger trial, a highly visible white fog cloud was produced from the machine. Returning flocks of quelea avoided the fog resulting in only 2 per cent mortality; nestlings suffered 90 per cent mortality. It was concluded that the fogger could control only sedentary quelea. No attempt was made to try it on a colony in the dark. A disadvantage of this fogger was its high noise level.

In 1972, a high-volume ground sprayer was developed in Nigeria (GTZ 1987). Queletox was diluted 9:1 with water to produce a 6 per cent a.i. emulsion. The mixture was contained in a large tank on the back of a Unimog truck and was pumped out under high pressure through 220-m-long hoses fitted with spray guns. Application rates were between 200–300 l/ha, and in small, dense roosting sites (<5 ha) more than 80 per cent mortality was usually achieved (GTZ 1987). The disadvantages of the method were that extensive site preparation was necessary to get hoses into the roost and that the application method tended to splash chemical on the operators.

In 1977, the Zimbabweans modified a Kenkelder mistblower normally used for spraying orchards into a quelea ground sprayer (LaGrange, in press-*b*). The avicide is pumped under pressure into six venturies where it meets a fan-produced wind, blasting at 160 km/h, which shears it into about 80 μm vmd droplets and deposits it for a distance of about 100 m downwind. The sprayer is carried on a tractor and uses a power take-off as the energy source. The tractor is driven slowly along the upwind edge of the roost soon after dark. The application rate is about 20 l per 100 m forward travel, giving a rate of 20 l/ha for Queletox diluted with diesel to 30 per cent a.i. The mistblower is now used as part of the standard weaponry for quelea control in Zimbabwe, being employed wherever possible in preference to aerial spraying because it is relatively inexpensive (LaGrange, in press-*b*). In principle, only sites that are inaccessible to a tractor are aerially sprayed. Quelea mortality induced by the mistblower is often as impressive as that of aerial sprays. Good kills of quelea and another bird pest, the Chestnut Weaver *Ploceus rubiginosus*, have been obtained in Tanzania using the mistblower (C. Elliott, pers. obs.).

The next advancement in ground sprayers was the development in Nigeria of the GTZ ground sprayer (GTZ 1987; Fig. 21.5). This ground sprayer consists of a generator, a compressor, a spray tank, a 10-m telescopic mast, and an atomizer head. Droplets of 35 μm vmd are produced within a narrow range provided that the 60 per cent ULV formulation of Queletox is used (GTZ 1986). The horizontal distribution of the Queletox is achieved solely through the natural movement of the air and is aided by the presence of thermic or inversion layers in the target area. The flow rate required to build up a good spray cloud is about 0.2–0.3 l/min, and the amount of avicide applied can be kept at about 4–6 l per sprayer.

The sprayer is transported in a Unimog truck to the target where it is set up with the generator placed up to 100 m away to minimize noise. The mast is adjusted to a height whereby a spray could well form at the height of incoming quelea flocks. Spraying is normally carried out during the last 20 min before dark when flock movement is at a maximum (GTZ 1986).

Two factors greatly influence the success or failure of this sprayer. It is essential to survey the target area properly over at least two evenings to determine incoming flight patterns. It is also mandatory that wind strength be sufficiently light to drift the chemical across the path of incoming flocks. In Nigeria and Niger where the sprayer was first tested, the wind in the flat, even terrain prevails almost always from one direction and is light. In eastern Africa, the influence of mountains and the sea is such that the wind can change both its direction and its strength from one evening to another. Wind speeds over 3 m/s destroy the shape of the spray cloud and require a postponement of the operation. Experience in Tanzania and Somalia has shown that quelea roosts in reedbed habitats are difficult to control with the GTZ sprayer because the vegetation retains heat at dusk and no inversion layer exists to support the build-up of spray cloud. By contrast, many quelea roosts in acacia over a sandy substratum in Niger and Somalia have been treated successfully. The sprayer has mainly been used for controlling quelea roosts of <5 ha, but colonies up to this size have also been successfully controlled (J.-U. Heckel, pers. obs.). Although it is in regular use in Niger and Somalia, it is expensive (US $12 000). In the early 1980s, FAO began the search for a more economical and appropriate sprayer. This led to the development of the FAO Migratory Pest Sprayer which can be used for controlling quelea, armyworm caterpillars, and locusts. It follows the same basic design as the GTZ machine, but uses an aerial-spraying-type micronair AU7000 as the sprayhead, backed by a fan for blowing the spray, a pressurized avicide tank, and the alternatives of electric or petrol motors to rotate the micronair. The sprayer can be mounted on the back of a landrover giving it a maximum operational height of 5 m. In light winds of less than 3 m/s, the FAO sprayer produces droplets of 50 μm vmd at a micronair speed of 12 000 r.p.m. In stronger winds, the rotational speed can be cut to give droplets of 55–65 μm vmd. Its mobility, flexibility, and its 50 per cent less cost are its chief advantages over the GTZ model. Its main disadvantages are that the effective spray height is 5 m, which often is not enough to get the spray cloud clear above the site vegetation, that it produces a high-pitched whine, and that the machine has yet to be fully field tested.

Because of its mobility, the machine can be driven to different spots in the colony or roost beginning along the upwind side of the site. Spraying is started when the first sizeable flocks enter. In roosts this can be 20 min before dark but for colonies spraying can start as much as 45 min before sunset. Spraying is best carried out from a fixed position, but after spraying for 1–3 min the sprayer can be moved to other locations of high bird concentrations. Colonies of 3 ha and 40 ha have been controlled successfully using this sprayer in Magadi, Kenya. However, spraying attempts using a 10-m mast in an unstable roost, have ended in failure (W. Meinzingen, pers. comm.). These variable results suggest that the FAO sprayer still requires extensive field testing before use recommendations can be given.

In 1985, FAO commissioned the Cranfield Institute of Technology in England to design and develop a portable, inexpensive ground sprayer for quelea control. A prototype mast sprayer was developed consisting of a battery and a positive displacement electrically driven back-pack pump and a 9-m mast made up of 1.5-m sections of pipes on top of which was carried a spray boom fitted with five small 'mini-ulva' electrically driven atomizers (Parker 1986). The mast can be positioned in a colony or roost, the spray can reach 10 m in height, and the mast can be quickly repositioned without disassembling if the wind direction changes. The Cranfield Mast Sprayer is nearly silent and is now being evaluated in Senegal.

We feel that ground sprayers in their various forms are cost-effective devices for quelea control. Once purchased, the running costs are only that of the avicide, the transport, and the personnel. Using ground sprayers eliminates the costs of aircraft, pilots, airstrip preparation and lighting, fuel, and extra transport. However, ground spraying requires much more physical work, than aerial spraying, especially if the target is large (> 3 ha). Therefore, if quelea control units have the financial resources to support aerial spraying, it is likely that it will be preferred.

Environmental consequences of quelea control

Many early control techniques were abandoned or discouraged for the very reason that they involved unacceptable environmental contamination. On the other hand, the fire-bomb, a technique known for 30 years, may be the least environmentally hazardous method presently available for quelea control. In 10 fire-bomb operations in 1986/87 in Kenya, only once was a non-target bird of prey found—a Black-shouldered Kite *Elanus caeruleus* (C. Elliott, pers. obs.). However, the apparently insignificant effects of fire-bombs on non-target birds may be related to their current use in Kenya on roosts in tall eucalyptus stands. When used in acacia bush in Nigeria in the old days, sometimes many birds of prey were killed in the blast (W. Meinzingen, pers. obs.).

The potential for adverse environmental effects exists with any of the avicide application methods. The most obvious concern has been the occasional slaughter of conspicuous numbers of non-target birds (Plate 23). This is related to the high toxicity of the chemicals used to all species of birds. Grue *et al.* (1983) found reports of 31 confirmed incidents of wildlife mortality (26 unintentional, 5 intentional) due to organophosphate poisoning in North America, and 747 incidents (387 unintentional, 360 intentional) in other parts of the world. In the unintentional poisonings in the United States, birds were most frequently reported as being affected. More specifially in Africa in the early 1970s, J. Thiollay (Chapter 16) found over 400 Black Kites *Milvus migrans* dead in a colony in Mali after it had been sprayed with parathion. R. Bruggers (pers. obs.) counted more than 100 dead raptors after a quelea spray near Gedarif, Sudan in 1980, and Nikolaus (1981) also reported more than 200 White Storks *Ciconia ciconia* being poisoned by eating quelea after a spray in Sudan. In 1978–79 in Tanzania (C. Elliott, pers. obs.) and Somalia (R. Bruggers, pers. obs.) each found over 20 Marabou Storks *Leptoptilos crumeniferus* killed in a similar manner at quelea colonies. Thomsett (1987) reported big reductions in raptor populations, particularly Tawny Eagles *Aquila rapax*, Augur Buzzards *Buteo augur*, Black-shouldered Kites, and Black Kites, following ground and aerial spraying in Kenya. Following aerial sprays conducted at two colonies in Kenya specifically to study environmental effects of what we have defined as ULV spraying techniques, 16 of 23 raptors collected post-spray were found, based on reduced cholinesterase, to have been exposed to the chemical (Bruggers *et al.*, in press). Most recently, an incident has occurred in Botswana in which over 60 raptors were found dead after a high-volume (105 l Queletox, 105 l diesel oil) ground spray operation (E. Bashir and R. Bruggers, pers. obs.).

Although these numbers appear spectacular, we feel that such incidents probably are infrequent. They usually occur when colonies are sprayed that have attracted an unusually high concentration of predators or scavengers during the short period (about 10 days) when the chicks are just coming out of the nest. Avoiding spraying during this stage could minimize these disasters. The incident in Mali was also partly due to daytime spraying, a time which is now no longer contemplated. The incidents in Kenya and Botswana were caused by the irresponsible use of ground sprayers. As a result of the outcry at the effects of spraying on raptors, it is unlikely that the same mistakes will be repeated.

Some data have shown that the avifauna occupying the acacia roosts or colonies can also be seriously affected. Quelea colonies in acacia can be rich in avifauna. Morel and Morel (1978) recorded 112 species (not including Accipitridae and Falconidae) in only 25 ha in Senegal over 8 years. Bruggers *et al.* (in press) identified 94 species of birds in and around the two colonies they studied in Kenya. However, only a careful and systematic search after spraying shows the variety of species killed. For example, after a successful quelea spray operation near Shinyanga, Tanzania more than 200 villagers, in collecting the poisoned birds for food (Chapter 23), found 16 birds of 12 species, including weavers, waxbills, kingfishers, shrikes, a woodpecker, a quail, and a goshawk (C. Elliott, pers. obs.). In Zimbabwe, when a quelea roost next to a forest was sprayed, 162 birds of 43 species were killed, including bee-eaters, hoopoes, sunbirds, and flycatchers (Jarvis and LaGrange, in press). In the two colonies studied in Kenya (Bruggers *et al.*, in press), 44 birds of 17 species were found after 71 man-hours of searching. There may be a correlation between the openness of the target site and the number of non-target passerines killed. In open sites with spaces between the bushes, non-target birds may be more likely to be killed than in dense vegetation, where birds can be protected from the spray.

Another factor which exacerbates the effects of spraying, especially ULV spraying, is that contaminated quelea can disperse over wide areas. Bruggers *et al.* (in press) found dead quelea in a 35-km^2 area around target colonies, and J. Ash (pers. comm.) found dead or dying quelea up to 20 km away from a spray target. Sick birds are likely to act as a magnet to raptors over a significant area (Fig. 21.6), but as the distance from the colony increases, the density of the birds will decrease. None the less, the acute oral toxicity of fenthion to Sparrow Hawks *Falco sparverius* and probably many other hawks is 1.3 mg/kg (Schafer 1972), and quelea are contaminated with 50–100 μg of fenthion following sprays of between 2.5 and 4.0 l/ha (Bruggers *et al.*, in press). Therefore, small raptors such as 600-g Pale Chanting Goshawks *Melierax poliopterus* and 60-g Pygmy Falcons *Poliohierax semitorquatus* would only need to eat between 6–12 and <1–3 contaminated quelea, respectively, to consume a lethal dose (Bruggers *et al.*, in press).

Despite the recorded adverse effects on non-target passerines and raptor species, it seems unlikely that quelea control is having much impact on non-

Fig. 21.6. Tawny Eagles *Aquila rapax* can scavenge on the quelea killed by aerial spraying and become secondarily contaminated (photo: R. Bruggers).

target species populations. Compared to agricultural spraying practices, the amount of quelea spraying being carried out is very small. For example, on average, Tanzania sprays about 1000 ha of colonies and roosts each year (C. Elliott, pers. obs.). In Kenya, the figure is variable but probably is less than 300 ha, and in Botswana is in the range of 70 ha (E. Bashir, pers. obs.). If raptor populations are adversely affected in local areas, it would be expected that in time the gap would be filled by immigration from neighbouring populations. Thomsett (1987) claimed that raptor populations in his study area had not recovered after 2 years, so in some circumstances it may take several years. However, the population decrease he observed probably was unusual due to the fact that several control methods were being used in the area.

Fenthion does not accumulate in the body, and it is possible that large birds of prey accustomed to doing without food for some days may be able to recover from fenthion poisoning. In the study by Bruggers *et al.* (in press), eagles and other raptors would have received severe exposure only on the first day after spraying. Thereafter, fenthion residues on dead and dying quelea rapidly decrease. Some studies indicated that fenthion has many physiological effects on animals (Grue *et al.* 1983), for example, on reducing parental activity in nestling care (Grue *et al.* 1982).

Environmental contamination also is a concern in inhabited areas, as control operations can pose a direct health hazard to humans. In situations where roosts are close to villages, the spray must drift away from the homes or the birds must be displaced before spraying. In many parts of Africa, sprayed birds are collected by villagers for food (Chapter 23).

According to Cisse (1981), the half-life of fenthion in soil is 25 days and in water is 3 days. Bruggers *et al.* (in press) found that residues on grasses decreased from 38 to 17 p.p.m. on the first day after spraying, and by the fourth day they had decreased to 1.1 p.p.m. As such, these residues would not pose a long-term contamination problem for forage eaten by wildlife or livestock. However, we must stress that further studies are needed to investigate more thoroughly all aspects of the environmental contamination of low-dosage spraying.

Conclusion

The aim of this chapter has been to describe the methods available for lethal control of quelea. The context under which it is carried out is discussed in the chapter on control strategies (Chapter 22). There is no doubt that current methods, having evolved over nearly 60 years, are now highly efficient. There are very few circumstances in which successful control of more than 90 per cent of a quelea roost or colony cannot be achieved, provided that the necessary equipment, including aircraft, is available when required. Given the near guarantee of success through properly conducted control operations, there is a clear temptation for control officers to respond to every anxiety of farmers and once again contemplate exterminating the quelea. We believe that there is a continual necessity to keep the quelea's effects on agriculture in perspective (Chapter 3) and to use lethal control only when a direct or imminent threat to agriculture occurs.

22

Quelea control strategies in practice

CLIVE C.H. ELLIOTT and
RICHARD G. ALLAN

Introduction

The objective of this chapter is to examine the practical problems of applying quelea control strategies to the realities of African agriculture. We start with two premises that complement one another. In Chapter 3 it was concluded that quelea, like any other bird pest elsewhere in the world causes only local damage to crops, albeit sometimes severe local damage. Quelea destroy an insignificant proportion of national cereal production in any one African country. The first premise is that the presence of quelea often causes severe local damage to cereals, and this together with farmers' apprehension affect the expansion of cereal production in major cereal-growing areas. We contend, therefore, that quelea control has to prevent local damage and, thus, alleviate the anxieties of farmers so that cereal production will be increased in the continent's general drive toward food self-sufficiency.

The second premise is derived from Ward's (1979) 'immediate crop protection' strategy, involving the 'destruction of only those concentrations (communal roosts or breeding colonies) existing in, or close to, an important cereal-producing area and confined in time to periods of the year when there are crops at a vulnerable stage in the fields'. Ward's concern was to get away from the idea that the quelea problem could be solved by overall population reduction or even by eradication of the species. Such a solution of 'long-term decimation of the bird population' had long been promoted by the West German GTZ team in north-eastern Nigeria (GTZ 1987, p. 236). We agree that any attempt at wide-scale eradication, apart from probably being impossible to achieve, is misguided both from economic and ecological points of view. Hence, our second premise is simply that impending crop damage should be combated. However, Ward's control strategy was essentially an academic proposal that he never translated into practical terms. We believe that there are a number of problems involved in trying to apply this strategy, and we shall now examine these problems in detail.

Management options

Non-lethal control

Dyer and Ward (1977) suggested that a number of management options exist when choosing a particular bird pest strategy. They ranked the options according to the degree of ecological disturbance, with the simplest and safest first (bird exclusion strategy) and the most complex and hazardous last (total population reduction strategy). However, if we consider the options presently available for quelea management, the main objective of which is the prevention of severe local damage to cereal crops either in large farms or agglomerations of small holdings, it appears that the choice is usually limited.

A variety of non-lethal techniques are available for preventing quelea damage. These techniques have been developed both traditionally and by using modern materials, chemicals, technology, and agronomy (Chapters 6, 18, 20). The range of lethal control methods has also been described (Chapter 21). For protecting large expanses of cereal crops, only lethal control can provide unequivocal results. This certainty is because after effective lethal control the birds are dead; non-lethal control techniques may only drive birds from one patch of cereals to another.

Despite this fundamental drawback to non-lethal methods, when properly applied some of the techniques are effective. The effectiveness of exclusion netting, of chemical repellents applied intensively at high dosages panicle by panicle, and of scaring by highly motivated humans is not disputed. However, the economics of farming situations and applying some of these methods often relegate their use to high-value crops. Effectiveness against quelea still has to be demonstrated for other methods of non-lethal control including 'bird-resistant' cereals, the application of repellents extensively or by 'spot-spraying' to large areas of crop, and the large variety of mechanical, bird-scaring gadgets.

Dyer and Ward (1977) suggested that the environmentally most acceptable way to avoid bird damage is to use agronomic techniques. Methods include crop substitution, crop calendar changes, and the use of short- or long-cycle varieties to avoid peaks of pest bird abundance (see also Chapter 20).

Farmers choose to grow a particular crop for many reasons, with the prospect of bird damage being only one of the factors. The variability of bird damage intensity from year to year and place to place means that this factor is rarely consistent enough to force farmers to make crop substitutions or agronomic adjustments to avoid birds. Apparently, Wood-pigeon *Columba palumbus* damage did convince farmers in England to abandon growing oilseed rape (Lane 1984), but in the Lake Erie region of North America, farmers refused to adopt fast-maturing but lower-yielding maize that would completely avoid bird damage (Dyer and Ward 1977). It appears that farmers generally choose to plant high-yielding varieties preferring the unpredictable risk of bird damage to the certainty of lower yields from recommended 'bird-resistant' varieties. In the event of serious bird damage, their choice does not prevent vociferous complaint and demands for action from responsible agents.

Bird pressure is said to have been the main factor in determining the crops grown in some regions of semi-arid Africa. In parts of Kenya (for example Machakos) and Tanzania (Dodoma), and in Ethiopia (Brhane G., pers. comm.), bird-vulnerable sorghum and millet have been replaced by non-vulnerable maize to the detriment of overall cereal production because the maize is less tolerant of low rainfall. Groundnuts have replaced millet in local areas of Chad (C. Elliott, pers. obs.). It is difficult to judge the historical influence of birds on crop changes in Africa. In modern Kenya and Tanzania, the change to maize appears to have been more strongly influenced by market pressures and human food preferences than birds. Maize can be grown as a cash crop of higher value than either sorghum or millet. In the two countries, food preferences have changed, perhaps partly because of food aid, so that maize flour is now the most popular staple. The reality is that farmers are gamblers and will often opt for short-term profit. By planting maize, farmers in eastern Africa hope to avoid the unpredictable risk of drought, while remaining confident that if their gamble fails they will receive food aid from their governments.

Somalia reportedly is one of the few countries to have had some success in persuading farmers in the riverine areas to switch from sorghum to maize in order to avoid heavy quelea damage (Bruggers 1980). However, no data are available on the extent of the change and whether or not it has been maintained. The policy in Tanzania is the reverse, encouraging farmers in semi-arid areas to switch from maize to traditional sorghum and millet to avoid crop failures in dry years. Farmer response has been poor, and only intense political pressure is now producing some change. The policy has had to be accompanied by increased bird control activities to prevent bird attacks being used by farmers as an excuse to resist the policy.

Crop calendar changes to avoid birds are equally difficult to apply. Farmers in northern Botswana would suffer little damage if they planted early (Dyer and Ward 1977), but in 1985 there was no evidence that this strategy was being followed (C. Elliott, pers. obs.). Irrigated rice on certain farm schemes in Chad/Cameroon would avoid much quelea damage if planted early (Elliott 1979), but the farm management has been unable to make the adjustment because other factors override the bird problem. Traditional small farmers in semi-arid Africa cannot afford to plant all of their crop at once because of unpredictable rainfall. They also cannot physically work all their land simultaneously and, by spreading planting time, at least part of the crop may receive the most favourable rainfall. Early harvesting by stooking sorghum in Botswana (COPR 1976) and by swathing wheat in Tanzania (C. Elliott, pers. obs.) reduce quelea damage or deflect it to unharvested areas. Both techniques require extra work, and farmers are reluctant to adopt the methods unless bird attacks are severe, by which time it is often too late to avoid damage.

At present, the outlook for using agronomic techniques for solving the quelea problem is not encouraging. Crop protection research in Africa has tended to evaluate one technique at a time (Bruggers *et al.* 1981*b*; Bullard 1979; Elliott 1979). It is possible that if a number of different non-lethal techniques were used in combination or sequence, better protection might be achieved. Chemical repellents have been tried in combination with human bird-scarers but, apart from that, a combination approach has not been tried on quelea. Even if a successful combination of techniques was found, there would be extension problems in inducing farmers to apply them correctly. Little new work is going ahead and African farmers must rely on traditional bird-scaring or lethal control. It is conceivable that in the future public opposition to lethal control may produce the pressure and the funds for more research into non-lethal methods.

Lethal control

Lethal control by aerial spraying of avicides (mainly the organophosphate fenthion) is the established control technique for quelea. Ground spraying is under development and may prove a viable, much more economic alternative if fears that it causes serious hazards to non-target species prove groundless. Explosives or fire-bombs are still occasionally used in West Africa and Kenya. Their use is likely to remain restricted due to the security implications.

All these methods are described in detail in Chapter 21 together with the multiplicity of other lethal techniques that have been tried and rejected. The possibility of avoiding toxic chemicals by developing a mass-capture technique for control is discussed in Chapter 23. We consider that such a technique would be ideal if it achieved both control and the harvesting of quelea as a food source.

Successful aerial spraying causes a massive reduction in the local quelea population and the cessation of significant damage or the prevention of an outbreak of damage to cereal crops in the area affected by that population. Common practice is to evaluate success in terms of the number of birds found dead at the target site (Elliott 1981*b*). Measurement of the result by estimating the drop in numbers of birds attacking crops has seldom been carried out. R. Luder (pers. comm.) recorded a fall from 90 000 to less than 5000 in the number of birds leaving wheat fields in Tanzania following aerial spraying of associated roosts. More often, success is indicated when quelea control teams are overwhelmed by the gratitude of local farmers lining the road to applaud the departing team (Jijiga, Ethiopia; M. Jaeger, pers. comm.), presenting the team with goats (Shinyanga, Tanzania; C. Elliott, pers. obs.), or the sudden silence in the fields when traditional bird-scaring is no longer necessary (Chifukulu, Tanzania, and Mwea-Tebere, Kenya; C. Elliott and R. Allan, pers. obs.).

The duration of calm varies according to the circumstances. The control of a large concentration of millions of quelea may prevent damage to thousands of hectares during a complete growing season. Sometimes there is a high population turnover, and the target site will have to be resprayed again after a few weeks if nearby crops are still vulnerable.

The main disadvantages of aerial spraying are the high cost and the use of environmentally hazardous toxic chemicals. Because aerial spraying is relatively easy for a ground team to direct, there is also a danger that more spraying may be carried out than is strictly necessary for agricultural purposes. The latter factor may apply even more to ground-spraying if efficient easy-to-handle machines are made available.

In the western world, the morality of killing living things is increasingly being questioned. We believe that the perilous agricultural situation in Africa, where countries are struggling to achieve self-sufficiency in food production and in some cases to ward off local famines, justifies lethal control of quelea where there is an obvious threat to crops. However, our level of scientific knowledge at present makes it sometimes difficult to be sure if a genuine threat exists. Detailed multidisciplinary research is needed to find out why quelea attack crops in some circumstances and not in others. Such research might help to reduce lethal control operations to a minimum.

When should quelea be sprayed?

Given that the main option presently available for preventing quelea damage to large areas of cereals is lethal control and that aerial spraying is the principal technique to be used, the next practical problem is to decide when to spray. Ward's (1979) strategy proposed three criteria to guide this decision: the 'closeness' of the birds to the crop, the 'importance' of the crop, and its 'vulnerability'.

Of these criteria, vulnerability is the easiest to assess. The principal crops attacked, sorghum, millet, wheat, and rice, are vulnerable from the milky stage through harvest. Once the harvest is in, the necessity for spraying ceases. This may seem obvious but in the early days of quelea control 30 or more years ago, the accepted control strategy was to reduce the total population or preferably eliminate the species, and spraying dry-season roosts in the absence of crops was widely practised, especially in Sudan and South Africa. One of the few points of unanimity in quelea control all over Africa today, is that once the harvest is in, control can stop.

The 'closeness' of quelea to a crop and the degree to which this poses a threat, depends on whether a roost or a breeding colony is involved. Most people who have worked on quelea would probably agree that the maximum distance quelea usually fly from a roost to a feeding ground is about 30 km. Exceptionally it may be as much as 50 km (Ward 1978). The minimum may be as little as 500 m. A roost further than 30 km from vulnerable cereals is not normally, therefore, a justifiable spray target.

The feeding range of adult birds in a breeding colony varies, but observations suggest it is between 2 and about 10 km (Bruggers *et al.* 1983; GTZ 1982; Jarvis and Vernon, in press-*a*). Cultivation further than 10 km from a colony is, therefore, unlikely to be attacked immediately. Yet, it is well known that the main threat posed by a colony does not come from the breeding adults but from the juveniles which are liable to move on to the most easily available food source adjacent to the colony, i.e. cereal cultivation (Jones 1972; Ward 1973*a*). Such juveniles are sometimes so hungry that they are almost impossible to frighten out of the fields.

When a quelea colony breaks up, most of the adults depart usually leaving the juveniles behind (Ward 1971). Little detailed study has been made of the subsequent behaviour of juveniles, but observations suggest that it is variable. Sometimes local crops are severely attacked (M. Jaeger, pers. comm.), and sometimes the juveniles move out of the colony area at the same time as the adults (R. Luder, pers. comm.). This variability relates to the difficult question of establishing whether crops further away than 10 km are at risk from a particular breeding area. Ringing has shown that quelea can move up to 500 km in East Africa (Disney 1960) and up to 1500 km in southern Africa (Ward 1971). Fluorescent-marked birds moved 700 km in Ethiopia (Jaeger *et al.* 1986). Theoretically, therefore, crops hundreds of kilometres away could be at risk from breeding quelea. However, given the high cost of aerial spraying, we consider that some direct evidence of a link between a breeding colony and cereal crops is important before control is undertaken.

Geographical features suggested to Jaeger and Erickson (1980) that quelea breeding in the Awash Valley, Ethiopia, might be channelled towards sorghum cultivation more than 100 km away. Control operations in the suspected source breeding areas produced a dramatic decline from 16 to 2 per cent in sorghum damage levels. Strong evidence was, thus, obtained for considering the Awash colonies sufficiently close to cereal crops to justify spraying. Unfortunately, such evidence is generally not available for deciding if colonies further than 10 km from crops pose a threat. Further research is required both to understand the variability of the behaviour of juveniles when breeding is completed and to understand post-breeding dispersal of all age classes. Seasonal movements may be situation-specific requiring each major breeding area to be investigated separately. Fluorescent marking (Chapter 5) would seem to offer the best hope for unravelling these movement patterns.

Ward's (1979) third criterion was the 'importance' of the crop. Earlier, Ward (1973*a*) clearly meant crop 'importance' to be a large expanse of cereals, the protection of which would be economical. If such an approach were strictly applied, only large, probably mechanized farms would be economical to protect because of their relatively high yield, high inputs, and high level of synchrony in planting. Planting synchrony is particularly significant in bird control because often one operation is sufficient to cover the ripening period. However, cereal production in Africa is mainly in the hands of smallholders and subsistence farmers. It is they who suffer most if bird damage is severe. If Africa is to break out of its dependence on food imports, it will probably be achieved through improved smallholder production. Given the anxiety that quelea cause at all levels of farming, we believe that to concentrate bird control on important major farms is neither in the interest of improving crop production nor is it acceptable politically.

'Importance' has to be assessed in broader terms. Firstly, local importance must be considered. A village may depend on 100 ha of cereals to provide its staple food for the next year. If so, that crop is exceedingly important. We suggest that any quelea colonies close to that crop must be controlled even if it may not be economic to do so. The economics of smallholder crop protection are difficult to evaluate. Data on hectarage, expected yield, etc., may be lacking. If serious losses due to birds occur, food aid may be required, costing much more than the grain originally lost.

Secondly, the national importance of a crop-growing area must be considered. The authorities may designate certain parts of the country for growing specific cereal crops because of soil suitability and rainfall pattern. For example, the policy may be to replace drought-prone maize with sorghum in semi-arid areas or to promote rice cultivation instead of root crops in swampy regions. Both policies encourage growing of crops at risk to bird attack and if they are to succeed, farmers must feel confident that their crops will be protected. Any concentration of quelea in such areas during the ripening season probably should be controlled.

Thirdly, the political importance of a crop has to be taken into account. If the crop's owner has powerful political connections allowing him direct or indirect access to a high authority in government, the quelea control officer risks serious criticism if he does not organize a spray when the owner demands one. It is, of course, undesirable that the officer should be subjected to such pressures, but we consider that political influence has to be accepted as part of the reality of decision-making in Africa. Undoubtedly, political pressure has sometimes resulted in unnecessary sprays being carried out. This can be minimized by having spray operations directed only by scientifically well-qualified staff, who have a better chance of explaining the evidence against spraying and obtaining the co-operation of a farmer.

Although we suggested above that quelea control could not be exclusively reserved for major farms, we are not suggesting that economics should be totally ignored. We consider that cost-effectiveness is an important aspect of spraying, but it is one that has not received much attention. The major problem is to estimate how much damage would have been caused if control had not been carried out. Given the vagaries of quelea feeding patterns, of their post-breeding movements, and our lack of understanding of the factors affecting both, predicting damage resulting from post-breeding dispersal is usually guesswork. A further complication is that the best guess would be based on the damage levels occurring in a number of years when no control takes place. With control operations firmly established in most countries in which the quelea is a serious pest, this information is normally not available.

Jaeger and Erickson (1980) were able to measure damage levels in lowland sorghum areas in Ethiopia for 2 years prior to introducing control. Losses varied from about $0.6 million to about $3.8 million (at $150/t). In the following 2 years in which control of source breeding colonies took place, losses dropped to about $0.25–0.4 million, saving between $0.2–3.75 million. Operation costs were estimated at $45 000, although no details are given as to how that figure was reached. Despite the variability in damage levels in the 2 years before control, the results suggest that even if damage had been low in the two control years, the control operations would have been cost-effective to the amount of at least $150 000.

The cost-effectiveness of control in other countries has to be assessed more indirectly than was done by Jaeger and Erickson (1980) because comprehensive data on quelea damage prior to control is not available. Very rough cost estimates can be made for aerial spraying in Tanzania. The annual budget for the National Bird Control Unit (excluding salaries) was about $310 000 in 1984 including externally derived aid of about $60 000. Since 1978, a helicopter has been contracted for about $90 000/year for seasonal survey and spray work. The Government also obtains the services of spray aircraft from the Desert Locust Control Organization of Eastern Africa at a basic running cost of about $200/h (1984 figure). Allowing 3 h flying for each spray sortie and about 50 sorties a year, $30 000 is added to the annual cost, bringing the total to $420 000/year. In 1982, 89 sorties were flown against quelea using fixed-wing aircraft and a helicopter, and 75 sorties were flown in 1983. The rough cost per sortie is, therefore, $4800–5700.

Using the figure of $5000 per sortie, each sortie in Tanzania would have to save about 10 t of wheat ($40 per 90-kg bag—1984 price) or about 6.25 t of rice ($72/90-kg bag—1985 price) to break even. If each quelea attacking the crop destroys about 10 g/bird per day (Chapter 3), this would be equivalent to one million bird-days for wheat and 625 000 for rice. Around Arusha, Tanzania roosts average about 100 000 birds, and if sampling revealed that 25 per cent were feeding exclusively on wheat, it would take 40 days to destroy the equivalent value of an aerial spraying operation. If less than 40 days remained before the crop was due to be harvested, it would not be economical to control the roost.

We suggest that an estimate of roost population, a sample to determine the percentage of birds eating the crop, and a knowledge of the cost of the spray sortie can allow a simple calculation of the cost-effectiveness for one particular roost spray operation. Such calculations should become a routine part of deciding whether or not to spray. If the break-even point is not reached and no other considerations outweigh the economic one, the decision to spray can be rejected or at least postponed pending an increase in the number of birds or in the percentage eating the crop. Obviously, the aim of any Control Unit would be to save crops worth many times more than the cost of bird control.

The break-even point will vary from country to country according to the costs of each sortie and the national value of a particular crop. In Kenya, a single spray sortie by a private contractor costs only about $160 (1983 price). The annual government budget allocated exclusively for bird control is about $69 000 (1984 figure), but the number of sorties rarely exceeds 30 (25 in 1983, and 20 in 1984). The total cost per sortie is estimated at $2750 against the price paid for wheat in Kenya of only $160/t, due to a markedly different official exchange rate. A factor of five times as many birds as in Tanzania would, therefore, be needed to justify a control operation cost-effectively.

Ground sprayers have proved highly efficient in controlling quelea and Golden Sparrows *Passer luteus* in Niger (Heckel 1983). In 15 roost sprays, an average of 588 666 birds was killed and an estimated $15 400 worth of crops saved. The annual cost to the Niger Crop Protection Unit of using the sprayer was about $106 000 or $7076 per spray. The high cost per spray is due in part to government salaries being included in these estimates.

It is debatable whether staff costs should be included in spraying and if so, which staff should be considered. In some countries, bird control staff have other duties which makes apportionment of the costs difficult. For Tanzania, the cost per sortie would go up by about $500. No doubt our estimates for the cost of bird control can be improved with up-dated figures. We suggest that simple charts could be prepared before each control season with the latest grain prices, government budgets, and aerial spray contract prices. These could then be used as guidelines in the decision-making process.

Discussion

Although our knowledge of the quelea's biology is probably more comprehensive than that of any other bird species in Africa, we are still a long way from being able to apply a scientifically sophisticated control strategy to prevent damage. There has been significant development from the early days of quelea control when every concentration was destroyed regardless of whether or not there were crops nearby to damage. We would also suggest that our expansion of Ward's (1979) control strategy represents a further improvement in that we have attempted to quantify his ideas and put them into practical terms.

The control strategy we have ended up with is the spraying of quelea colonies within 30 km of vulnerable crops, or further away if there is evidence to link the breeding area with subsequent damage, and the destruction of roosts when their damage potential greatly exceeds the cost of control. This strategy operates within the social and political considerations. We accept that it is still simplistic and will remain so until we have a better understanding of the quelea's movements from breeding and roosting sources to crops. It is also probable that government-sponsored quelea control teams will tend to give the social and political factors priority over the agricultural and economic factors. The result will be that more spraying is done than is necessary and that governments will have to shoulder greater costs than necessary.

Given the stresses and strains on African agriculture, it is perhaps surprising to find, in our experience, that there is little official concern with cost-effectiveness in quelea control or with reducing costs by reducing the number of operations. The main aspect of funding bird control that seems to present problems is the foreign-currency component. Expenditure in local currency appears to be considered a worthwhile investment. The anxiety of farmers is alleviated, and the government often gains political benefit from the farmers' strong political lobby. In Africa, there is little or no counter-lobby from the indigenous population against lethal control even if the evidence for a serious threat to crops is sometimes lacking. We do not see this situation changing soon unless governments run out of the foreign currency needed to continue lethal control. Nevertheless, economic or environmental constraints may in time force greater consideration of cost-effectiveness in quelea control. Meanwhile, it can only be stated that costs could be cut both through more attention to cost-effectiveness and through more research into quelea foraging behaviour and dispersal from breeding areas to crops. There is also a need for minimizing the secondary effects involved in the use of toxic chemicals for lethal control and into seeking alternative, non-toxic methods of lethal control such as mass-capture techniques (Chapter 23).

23

Quelea as a resource

MARGARET E. JAEGER and
CLIVE C.H. ELLIOTT

Introduction

Food shortages in Africa have been world news in recent years. Thousands of people have died of starvation, and millions more suffer chronic undernourishment and malnutrition. These problems have been caused by lack of staple cereals and proteins and poor-quality food. Many of these famine-stricken areas are occupied, for at least part of the year, by large numbers of quelea that constitute a potential source of protein. At the same time, control operations against these birds to protect crops produce literally tonnes of dead birds. Jaeger and Erickson (1980) estimated that colonies and roosts in the middle Awash River Valley of Ethiopia in 1978 contained 7.5 million adult quelea, 70 per cent of which were killed during control. At a weight of 7 g per dried prepared bird (Uk and Monks 1984), 37 t of potentially edible birds would have become available. A similar rate of control conducted in colonies found throughout Ethiopia, Kenya, Somalia, and Tanzania would yield about 345 t. The extreme abundance of this crop pest in areas of chronic food shortage raises a number of questions about the quelea's potential as a food resource and the possible integration of quelea exploitation with quelea control.

Nutritional value of quelea

In some parts of Africa quelea are eaten. Traditional ways of preparing quelea for food retain the bones and sometimes the viscera; both enhance the overall nutritional content. In East Africa, quelea are dropped into hot water, removed, and plucked. They are then eviscerated, heads and legs are removed, and the carcasses are split open to dry in the sun for about 2 days. When eaten, the dried birds are cooked with vegetables in oil or water and used as a sauce to accompany the staple cereal, usually stiff maize porridge.

In Chad, the birds are plucked directly and grilled as a kebab. Grilling singes off any remaining feathers and cooks the meat, which helps to preserve it on the way to market. If no market is nearby, the birds are simply split open, dried in the sun, and may be pounded into a powder for later use in a sauce (L. Bortoli, pers. comm.). In Zimbabwe, the whole bird is usually grilled, and the burnt crust of feathers and skin is broken away before eating (M. LaGrange, pers. comm.).

Quelea can also be used in modern cooking and in the field has often been a welcome addition to the diet of quelea researchers. A simple recipe is to remove the breast (with bone), skewer it, add salt, pepper, and butter, and grill over the coals. These kebabs can be improved by wrapping them in bacon or by marinating before grilling in oil, soy sauce, and herbs. The meat is dark and comparable in taste to that of larger game birds.

To investigate the food value of quelea, Jaeger and Jaeger (1977) mist-netted 200 birds, slit them open ventrally, and spread them out to dry in the sun for 1 day. Drying was completed in a kitchen oven on low heat for about 2 h. The feathers were singed off over a fire and beaks and feet cut off. The dried bodies were ground in a hand-cranked meat grinder, producing a dark-brown material with the texture of coarse sawdust and a faint, but not unpleasant odour. The meal was kept at room temperature for several months without visible indication of spoilage.

A sample of this quelea meal was analysed by the Ethiopian Nutrition Institute in Addis Ababa, and the results compared with a variety of foods commonly eaten in Ethiopia (Table 23.1). Dried quelea compared favourably with the other foods. The fact that the entire bird (less the undigestible beak, feet, and feathers) including bones and internal organs was used, presumably explains the higher calorific value and contents of fat, carbohydrates, calcium, phosphorus, and iron as compared to dried meat. Quelea provide proportionally less protein than dried meat, but four to six times the protein found in staple cereals, as well as significant amounts of minerals and vitamins. Traditional preparation methods that retain the bones and sometimes the internal organs and the cooking liquid enhance the nutritional qualities of quelea. Therefore, quelea might be a useful dietary addition for general health, especially for people who face protein deficiencies.

Traditional exploitation of quelea

The easiest way for villagers to obtain large numbers of quelea for food is to collect nestlings at breeding colonies; this is the most frequently used method, especially in eastern and southern Africa. The chicks also are easier to eat than the adults because they have softer bones. Chicks are collected at almost any age, although it would seem most productive to take them at 10–14 days

Table 23.1. Nutrient content of dried quelea and some common foods of Ethiopia (source of comparative data: Agren and Gibson 1968).

Food substance (100 g)	Calories	Moisture (g)	Ash (g)	Protein (g)	Fat (g)	Fibre (g)	Carbo-hydrate (g)	Ca (mg)	P (mg)	Fe (mg)	Vitamin A (mg)	Thiamine (mg)	Riboflavin (mg)	Niacin (mg)	Vitamin C (mg)
Dried quelea	412	9.3	10.2	47.2	18.7	1.3	13.3	2145	1544	25.8	0	0.19	1.00	10.1	7.0
Dried meat	363	6.2	4.5	64.2	9.8	1.3	7.3	42	730	15.2	0	([a])	([a])	15.8	([a])
Injera[b] (dried)	367	9.2	3.3	12.8	1.4	2.0	74.0	347	398	34.0	0	0.62	0.13	1.7	1.0
Sorghum (mixed red and white)	348	9.7	1.7	7.6	3.2	2.5	77.8	35	278	17.3	Trace	0.28	0.10	1.9	([a])
Maize (whole kernel)	356	12.4	1.3	8.3	4.6	2.2	73.4	6	276	4.2	0	0.38	0.09	2.1	2.0

[a]No value given.
[b]Injera, the traditional bread of Ethiopia, is a large, fermented pancake made from teff *Eragrostis tef* flour.

old just before they leave the nest. In one colony in southern Tanzania, we saw villagers collecting piles of naked chicks about 6 days old. In Zimbabwe, villagers often waited until the chicks were almost ready to fledge before raiding the colony (Jarvis and Vernon, in press-*b*). After such a raid, a small colony of a few hectares may be devastated, with intact nests remaining only in the tallest trees and the densest bush. Once the chicks are out of the nest, we have seen small boys collecting them in colonies with sticks and catapults, but they are unable to catch more than a few birds per day.

Presumably, quelea eggs also are nutritious. Thirty 1.7-g eggs (Lourens 1963) would be equivalent to one medium-sized chicken egg; the taste of a quelea egg omelette has been declared delicious (L. Bortoli, pers. comm.). However, the difficulty of collecting quelea eggs in quantity without breaking them seems to have deterred local people from attempting it.

A number of traditional techniques are used to capture free-flying quelea. These vary from throwing sticks about 0.5 m long into dense feeding flocks in Cameroon to using baited drop traps to catch quelea alive for food as well as for the cage bird trade in Senegal. Bird lime, an entangling glue made from boiled fish bones, is also commonly used in Tanzania and Cameroon. Sticky twigs covered with bird lime are placed in resting or drinking sites, and a catch can be made every few minutes at the right time of day. Such methods are generally used only by small boys supplementing their personal diet while occupied in other activities like herding livestock, rather than being used to collect food for whole families or villages. Within colonies, it is virtually impossible to fell adult birds with sticks and stones, but it might be possible to snare them at the nest entrance although we have not seen this done.

In Chad, professional trappers (piegeurs) catch quelea in roosts using hand-held nets (Plate 24). The triangular nets are made of local fish netting slung between two slender poles. The trappers enter a roost at night and quietly work their way under a tree in which birds are roosting. They jerk the nets backwards into the tree, clapping the two poles together to form a cylinder of netting. The startled quelea fly into the netting and are shaken down into a sack. The trappers claim that during a dark night one man can catch 1000 birds. These trappers concentrate on dry-season roosts which probably provide the heaviest density of birds per tree. The Chadian trappers provide the only example we know of where quelea trapped by traditional means are sold commercially for food. In 1983, the price per bird in the N'Djamena market was reported to be 40–50 CFA francs (approximately US $0.09–0.11) (S. Manikowski, pers. comm.).

Exploitation of quelea after control

Quelea control operations, using avicide spraying, explosives, or fire-bombs produce a large amount of potential food. In many parts of Africa, collecting dead quelea after control operations is now commonplace, but seems to depend on the traditions of tribes in an area (Plate 25). If people are accustomed to eating small birds or even to raiding quelea colonies, they may take advantage of the opportunity to collect birds from control sites. Thus in Tanzania, the Wasungu in the south, the Wasukuma in the west, and the Wasafa in the north collect dead quelea; but the Wagogo in the centre do not. The practice has also been reported as widespread in Nigeria (GTZ 1979, 1982), Cameroon, Zimbabwe (M. LaGrange, pers. comm.), and in certain parts of Kenya (R. Allan, pers. comm.).

In one 80-ha colony in Tanzania, we saw 300 villagers collecting dead birds from a spray site for almost 2 days. One man said that he and his family had collected six 75-kg sacks and had used his oxcart to haul them home. In some places in Tanzania, local people have been so anxious to obtain birds for food, that their complaints about bird damage seemed as much directed at being able to collect birds as to alleviate a particular damage problem.

In Zimbabwe, birds collected after spraying are sold on the local market for up to US $0.07 each. One enterprising individual bought birds from collectors at a spray site, transported them to villages 60 km away, and resold them for the equivalent of about US $0.17 each (M. LaGrange, pers. comm.).

Control by explosives and fire-bombs produces many uncontaminated, edible quelea, because birds are killed by the force of the blast or by the fire (Fig. 23.1). Many hundreds of quelea were killed during tests on the

Fig. 23.1. Control using explosives in West Africa provides a much appreciated source of uncontaminated quelea for food (photo: R. Bruggers).

efficacy of explosives in reducing damage to rice in Cameroon by blowing-up quelea day roosts around the farm's perimeter. This activity was so popular with farm workers that 150 of them abandoned harvesting rice and gathered to collect dead birds. Consequently, farm management asked that control by explosives be stopped (C. Elliott, pers. obs.).

Health hazards of eating contaminated quelea

Undoubtedly, certain African tribes greatly appreciate the quelea generated by control operations as an opportunistic food source. On numerous occasions, local people showed no anxiety at the birds having died due to avicides and told us that they had been eating quelea killed by fenthion every season for years, without any ill effect on themselves or their children. However, eating pesticide-contaminated food is a serious matter; therefore, we will consider it in more detail.

Organophosphates are the avicides primarily used in lethal control of quelea. These chemicals are nerve poisons that cause disruption of nervous impulses through the depression of cholinesterase activity. Death in quelea usually occurs through respiratory failure. The most commonly used chemical throughout Africa is fenthion. Parathion is occasionally used in western Africa, and cyanophos has been used experimentally in some areas.

Considerable variation exists in the mammalian toxicity of these three chemicals as measured by the acute oral LD_{50} in rats: parathion—13 mg/kg in males and 3.6 mg/kg in females; fenthion—190–315 mg/kg in males and 245–615 mg/kg in females; cyanophos—610 mg/kg in males and females (Worthing 1979). This variation in toxicity is complicated by the fact that these chemicals may metabolize to more toxic forms; for fenthion, the acute oral LD_{50} of both the sulphoxide and sulphone metabolites is 125 mg/kg in rats. Some species are more sensitive than others to organophosphates; fenthion is more toxic to dogs than to rats (Worthing 1979). Apart from studies on general toxicity, tests with mice have shown that fenthion does not seem to be mutagenic, carcinogenic, or teratogenic and does not affect reproduction in rodents (FAO/WHO 1980).

Toxicity of these chemicals to humans is difficult to quantify. There are records of men recovering from ingesting 57 g of Entex[R], a fenthion formulation, with recovery taking up to 30 days, and from ingesting 30 ml of a fenthion formulation, with recovery taking 6 days (FAO/WHO 1980). In both situations, the men were initially very ill. No information on the actual concentration ingested seems to be available.

In assessing the human health risks of eating quelea that have been sprayed with an organophosphate, two factors must be considered. The first is the amount of chemical present in each bird after spraying and after cooking.

Residue levels would be expected to be influenced by the application method (aerial or ground spray, high volume or ultra-low volume) and the application rate. The second factor is the amount of chemical contamination that can be tolerated in human food.

Results of residue analyses of quelea collected after spraying are given in Table 23.2. Residue levels varied at all stages of food preparation, yet a relationship sometimes seemed to exist between application rate and residue levels. Although all data showed decreasing residue concentrations from whole birds to plucked and gutted birds, to cooked birds, even cooked birds contained fenthion residues. Another point these data brought out is the wide variation in the application rates used for fenthion; even if an acceptable maximum residue level in birds could be established, the potential consumer on the ground will not know if the rate used on the birds he is collecting will achieve that level or less.

FAO/WHO have established official Acceptable Daily Intake (ADI) rates for various pesticides in foods. From a statistical view, it is generally accepted that the toxicity of pesticides to humans is normally distributed. The peak of the curve represents the LD_{50}. Since direct measurement of the LD_{50} in humans is impossible, the procedure usually used by FAO/WHO to determine an ADI is to take the lowest dose known to cause an effect in a mammal and divide this dose by 100, an amount which for fenthion is 0–0.001 mg/kg body weight per day (FAO/WHO 1980). Based on ingestion of low doses of fenthion by human volunteers, FAO/WHO (*loc. cit.*) has determined that the level at which fenthion causes no toxicological effects (NTE) in humans is 0.02 mg/kg body weight per day. Based on a residue level of 6 mg/kg (Table 23.2) and a prepared bird weight of 7 g, a 50-kg person would be able to eat 1.25 cooked birds per day to stay below the ADI. (The ADI of fenthion for that 50-kg person would be 0.05 mg; a 7-g prepared bird would contain 0.042 mg of fenthion.) A meal of quelea consisting of about 20 birds per person would exceed the ADI but would be below the NTE level. (At 6 mg/kg of residue, 20 quelea would contain a total of 0.44 mg; the NTE level would be 1.0 mg for the 50-kg consumer.) With lower residue levels of 1 mg/kg per cooked bird and 20 birds eaten, a 50-kg person would ingest 0.14 mg of fenthion and/or its metabolites, exceeding the ADI (0.05 mg) but remaining well below the NTE dose. At this residue level, an individual could eat seven cooked birds without exceeding the ADI.

These extrapolations would seem to suggest that, under certain circumstances, a small number of quelea containing fenthion residues can be safely eaten. However, a number of potentially serious difficulties exist when eating sprayed birds—the consumer has no way of determining residue levels in the birds he eats. The metabolites of fenthion appear to be more toxic than the parent compound. If the peak of a normal curve represents the LD_{50}, there will be some individuals at one end of that curve with much greater

Table 23.2. Residue analysis results from quelea sprayed with parathion and fenthion.

Application rate (mg/kg)	Application method	Residues in birds (mg/kg) Whole	Plucked/ gutted	Cooked	Source
Parathion					
7.0	Air	22.6	12.3	—	Cisse (1981)
Fenthion					
3.0	Air	46.3	19.4	5.65	Cisse (1981)
3.0	Air	92.0	26.3	8.39	Cisse (1981)
1.5	Air	10.1	—	—	GTZ (1979)
2.4	Air	—	5.9	2.80	Uk and Munks (1984)
0.6	Air	—	2.2	1.30	Uk and Munks (1984)
12.0	Ground	45.1	—	5.40	Jeremiah and Parker (1985)
6.0	Air	8.9	—	0.75	Jeremiah and Parker (1985)

sensitivities to the toxin. The ADI has been calculated for a 50-kg adult of presumed good health and nutritional status, so that individuals debilitated by chronic disease and/or poor diet would have a lower ADI. There is a possibility of adverse effects on children of pregnant and nursing women. Many people in rural Africa are exposed to other agricultural chemicals so that the effects of other organophosphate cholinesterase inhibitors could be additive. The NTE level seems to be of little help as it does not incorporate any chronic effect or safety factor into its equation as does the ADI (E. Schafer, pers. comm.). A related problem is that to be eaten, quelea must be collected from a spray site shortly after the control operation. The recommended re-entry time into a sprayed area or time to harvest after pesticide application is usually days or weeks, not hours. Villagers collecting birds at a spray site may absorb chemical through their skin or inhale it, as they are unlikely to have proper protective clothing.

Therefore, even though humans in many areas have been eating fenthion-sprayed birds for years, we cannot recommend this practice as a means of exploiting quelea as a resource. It probably is not necessary to go so far as to patrol spray sites to prevent people from collecting birds: this was unsuccessful when tried in Zimbabwe (M. LaGrange, pers. comm.). By educating local people and, more importantly, through developing alternative mass-capture control techniques, it may be possible to achieve successful control and to provide villagers with uncontaminated quelea for food.

Exploitation of quelea as cage birds

Presently, the only major commercial exploitation of quelea seems to be the cage bird trade largely in western Africa. Between 1974 and 1981, over 200 000 quelea were exported live from Senegal (Bruggers 1982; Ruelle and Bruggers 1983). Most of these went to West Germany, France, and the United States, although quelea are prohibited in the United States. It is difficult to determine the amount of foreign exchange earned by the trade in quelea, but the estimated overall value of the bird trade to Senegal is equivalent to approximately US $500 000 per annum, and quelea and other pest species (*Ploceus* spp., *Euplectes* spp., and others) constitute from 12 to 18 per cent of the annual total cage bird exports (Ruelle and Bruggers 1983).

Discussion

Individuals involved in quelea management have been concerned about the general waste produced by so many potentially nutritious quelea being left to rot. A desirable goal is that this large biomass of quelea should be managed in such a way that the birds can be used. A possible alternative to toxicants might be to develop an inexpensive, technologically simple, readily portable trap that poses minimal risks to non-target species. We believe that the most likely trap to fit these requirements would be a form of giant funnel trap, which is used in Tunisia and Morocco for trapping roosting Spanish Sparrows *Passer hispaniolensis* and European Starlings *Sturnus vulgaris*. Louis Bortoli (pers. comm.) has said that a team of seven trappers can catch several tens of thousands of birds in a single night. A modification of this trap, which involved a catching tent at the apex and floodlights as attractants, caught 672 000 blackbirds and starlings in 101 operations in the United States; the best night's catch was 120 000 birds (Mitchell 1963). Giant funnel traps have not been tried on a quelea roost, but the approach would have potential. Small traps have been used to catch quelea fledglings in colonies (Jackson 1974*b*). The roof of a simple wire cage, 1 m^3 in size, was fitted with sloping sheets of translucent plastic. Apparently, the fledglings were attracted by the shiny plastic in their search for water and slid into the trap through a slit in the roof. The trap caught 800 fledglings in 90 min, but its use would be limited to a very short period of the year. The hand-held net technique of the Chadian trappers could be exported to other areas as another possibility.

An idea is being developed in Zimbabwe to lure quelea to an artificially grown roost of Napier Grass *Pennisetum purpureum* where they can then be controlled (Jarvis and LaGrange, in press). In 1985, quelea were attracted to four of these trap roosts (M. LaGrange, pers. comm.). Possibly, a non-toxic surfactant could be used to kill these birds and permit their commercial exploitation (P. Mundy, pers. comm.).

Nylon mist-nets have been promoted in parts of Africa to protect crops; trapping results generally have been inconsistent. Because of their variable success, fragility, and non-selectivity of bird species, we concur with Bruggers and Ruelle (1982) that they should not be recommended.

Whether any of these trapping techniques can effectively control quelea is questionable. Probably some roosts, and perhaps a few colonies will be particularly suited for trapping, while others will not. Therefore, trapping is likely to be only an occasional alternative to lethal control. It is possible that trapping birds at roosts not necessarily involved in crop damage may provide a periodic source of uncontaminated birds. However, the migratory nature of quelea means that they will only be available seasonally in any particular roosting area.

There is little to recommend expansion of the cage bird trade for quelea. The market is limited. The wild bird trade is subject to abuse and contravention of such regulations as exist on the care and handling of birds, on traffic in endangered and pest species, and on quarantine restrictions (Nilsson 1981). If quelea should escape or be released in a suitable habitat such as the southern United States, there is a chance that they could become established as pests there. The introduced Black-headed Weaver *Ploceus cucullatus* has become a pest on the Caribbean island of Hispaniola (Fitzwater 1971, 1973).

The prospects for expanding local consumption of quelea exist, at least in areas where the birds already are appreciated as food. There are local commercial markets in Chad and Zimbabwe and they probably could be developed elsewhere. The introduction or expansion of quelea consumption among people who do not traditionally eat them is unlikely, owing to the cultural traditions and religious practices that play a strong role in determining dietary habits; such practices do not readily change in traditional societies.

The possibility exists of exporting quelea as a luxury food. There is a long tradition of eating small birds, either whole or as paté, in Mediterranean countries. Sparrows and starlings from Tunisia and Morocco are exported for consumption. Quelea were exported from Senegal to France in the 1950s under the brand name of 'Ortolan du Sénégal,' but the market collapsed when one consignment was contaminated with diesel oil (J. Roy, pers. comm.). A Zimbabwean company experimented with canning quelea as a potential export, but some problems, although not considered insurmountable, were encountered with the taste and texture of the product. A consignment of quelea was sent from Chad in 1978 to a processed food manufacturer in France, who found the taste of the birds acceptable and offered to purchase more birds at 18 centimes (US $0.05) per dressed bird. Asia may also be an export market for quelea; the Japanese have expressed interest in exporting quelea from Tanzania as a means of improving Tanzania's balance of trade, as well as exploiting a food resource. A similar interest has been expressed in Kenya.

Exporting quelea could earn vital foreign currency for some countries, but it would seem inappropriate to export birds for luxury food from areas where protein deficiency occurs alongside a tradition of eating birds. On the other hand, traditionally quelea would be eaten as an opportunistic bonus in the local diet rather than as a major dietary item, and on that basis it should be possible for local and export trades to coexist with benefit for locals and for a country as a whole.

It also is questionable whether eating small birds in developed countries should be promoted at all. Despite legislation and public education campaigns, indiscriminate hunting and trapping of many species of small birds, especially migrants, continues in southern Europe (Ryan 1981). It might be helpful to try to replace the eating of harmless songbirds with Africa's pest birds, but alternatively this might undermine efforts to put an end altogether to the practice of eating small birds. Although the size of the European luxury market is likely to be so limited as to make little impact on either of the above aspects, we suggest that any attempt to develop an export market for quelea as food needs to be examined closely.

Quelea could be used as a component in animal feed or as fertilizer. We have not heard of any instances of traditional or experimental use of quelea in either of these areas, but it surely must be feasible. Possibly, dried and ground quelea could be included in prepared livestock feed such as for pigs or poultry. In Kenya, the Turkana Fishermen Cooperative Society sells fish bones and dried offal to an animal feeds manufacturer (Kenya News Agency 1985). Perhaps quelea could be used in the same way. The Ethiopian Livestock Commission slaughterhouse in Addis Ababa sells blood and bonemeal for agricultural purposes. North American Indians of the Atlantic coast traditionally fertilized crops with fish, burying them with the seeds during planting. Possibly quelea, even those collected from control operations, could be used similarly.

Conclusion

We have shown that quelea are nutritious and are presently used as human food in some parts of Africa. Since we cannot recommend eating pesticide-contaminated birds, we encourage investigation into large-scale trapping techniques, both as a means of providing residue-free birds for consumption and as an alternative to toxic control in some situations. Contaminated birds are eaten in several African countries, therefore, we encourage further investigation of the toxicological problems. This might be directed at the local people involved, for instance by measuring any inhibition of cholinesterase activity, in individuals at various stages in the process of collecting, preparing, and consuming sprayed birds. Long-term effects should also be examined in areas where the practice has been going on for years. Quelea might be exported as a foodstuff, but only if local use were not adversely affected. We also recommend research on using quelea as an animal feed or as fertilizer to develop other ways of making this resource available to the human food chain.

24

Training and extension in quelea management and research

ALIOUNE N'DIAYE,
EL SADIG A. BASHIR, and
WILLIAM B. JACKSON

Introduction

Training of local staff in all aspects of quelea ecology and management has been an important component of the development projects concerned with overcoming the quelea problem since these began in the 1950s. In this chapter, we describe the size, scope, and nature of in-service and group training programmes conducted by either UNDP/FAO-assisted national and regional projects or other international aid projects (i.e. USAID/DWRC and GTZ) during the period 1979–1983. In addition, analysis of the patterns, trends, and problems in these training activities is attempted; important future improvements and policy issues arising are identified.

Methods of collecting data on training activities

Reliable quantitative data on training efforts are practically non-existent. However, we started by examining lists of trainees funded by OCLALAV, based on information obtained from the organization's records of per-diem payments during each training course. Questionnaires also were sent to appropriate officials in all countries concerned. Additional information was obtained from the UNDP/FAO eastern Africa regional quelea projects entrusted with group training for crop protection officers from Kenya, Somalia, Sudan, Tanzania, and Ethiopia. This training usually also involved experts from USAID/DWRC/GTZ and DLCO(EA). Other information was obtained from investigators who have been personally involved with quelea research and training actions throughout these regions during the period.

In-service apprenticeship-based programmes

In eastern Africa, in-service training usually is carried out by national bird control units in the Ministries of Agriculture in each country as a seasonal, routine operation. Pest management or crop protection officers and field scouts are given training on pest bird survey and control methods and organization and management of bird control campaigns. This training is offered by national experts who have been trained at the regional group level or have obtained overseas training.

In western Africa, OCLALAV has in the past been exclusively responsible for undertaking bird control programmes and conducting training. The personnel recruited from the Member States are given on-the-job training (Plate 26); and, over time, many of them developed high technical proficiency. They also are trained in the field by supervisory personnel during control operations. However, with the establishment of numerous irrigated cereal schemes, following the acute drought experienced in many countries, it became evident that a single organization could not handle all the control and training aspects of crop protection management in western Africa. As a consequence, many countries have begun training their own personnel—generally young graduates of agricultural colleges or national agronomic institutes.

Participants of in-service training are involved in locating bird concentrations and taking necessary operational actions to deal with bird damage problems. However, owing to the size of pest bird flocks and the political or socio-economic reasons for reducing crop losses as quickly as possible, very little training in applied research or biology is actually conducted.

In-service training also is designed to assist in establishing an infrastructure for a national bird control unit and to provide field survey personnel, control officers, and scouts with knowledge based on the concept of knowing 'a little about everything'. In-service training is particularly important, because it is a continuous process aimed at improving the infrastructure and the quality of pest management programmes through the acquired skills of local personnel. In vast countries like Sudan, where problems of crop damage by quelea are widely distributed, the national bird control unit has adopted a decentralized organizational policy, whereby provincial outstations or subcontrol units are encouraged to deal with in-service training at their level.

The background of bird control trainees varies from illiterate to university graduates, while their trainers range from secondary school educated to holders of M.A. or M.Sc. degrees. Most of the training in western Africa is in French and in eastern and southern Africa in English. Local languages are sometimes used including Arabic (in Sudan), Swahili (Tanzania), Somali

(Somalia), and Amharic (Ethiopia). Without the continuous provision of in-service training for local personnel, bird pest research and management findings aimed at improving the cost-effectiveness and efficiency of control operations will likely not be applied, and such national units will fail to achieve objectives.

Quelea survey and control operations in remote, semi-arid or subtropical savannah areas are naturally hazardous and difficult to carry out. They can effectively be achieved by a dedicated, physically tough personality, usually recruited from within the Ministry of Agriculture; but often they are treated as employees of no exceptional status and, therefore, not entitled to special treatment, such as payment of hardship allowances or incentives. National bird control units that fail to recognize the importance of such special treatment for bird control officers and field scouts are witnessing the gradual disintegration of their units.

Obviously, the maintenance of strong, stable, and responsible national units entrusted with in-service training is a government responsibility. Without attracting highly interested and motivated field personnel to carry out quelea management and participate in continuous in-service training, national bird control units can hardly continue to function effectively, particularly after the termination of international assistance projects.

In-service training provides too little attention to farmers. Farmers report verbally on incidents of bird concentrations, location, and damage. Also, in some countries (e.g. Sudan) they voluntarily contribute to preparing bush airstrips for aircraft involved in aerial applications of avicides, guiding survey officers to quelea colonies and roosts, assisting in marking colonies and roosts for aerial spraying, and collecting crop loss data.

Approximately 500 individuals have been involved in various types of training between 1979 and 1984. The total number of local personnel in eastern Africa receiving seasonal, in-service training at a regional level is shown in Table 24.1. It seems that Sudan, due to its vast area, huge quelea populations, and decentralized training policy, has been able to recruit more local staff than other countries. Yet, countries like Sudan and Somalia have been seriously affected in recent years by the exodus of trained personnel due to the lack of incentives. The same phenomenon, while existing in other countries, may have been under-represented in Table 24.1, since the figures quoted do not represent new individuals trained every year; rather they are personnel (i.e. employees of the unit) obtaining training. The continuous discovery of new quelea colonies and the opening of new agricultural schemes in individual countries should result in an increase in the need for bird control activities and a gradual adoption of decentralized systems, leading to recruitment of new personnel and more emphasis on in-service training.

Table 24.1. Number of individuals trained in quelea research and management through various types of training in eastern Africa (1979–1983) and western Africa (1980–1984).

	Type of training			
Location	In-service	Group[a]	Counterpart visits[a]	Joint surveys[a]
Eastern Africa (1979–1983)[b]			NA[c]	NA[c,d]
Ethiopia	20	51[e] (1980, 1983)		
Kenya	8	24[f] (1981)		
Somalia	10	—		
Sudan	30	—		
Tanzania	10	30 (1979)		
DLCO(EA)	—	8		
Western Africa (1980–1984)[a]		([a])	16[g]	12[g]
Benin	3			
Cameroon	—	100 (1982, 1984)[g]		
Ghana	2			
Guinea Conak	1			
Ivory Coast	1			
Liberia	3	15 (1984)		
Mali	3			
Mauritania	30			
Senegal	10	74 (1980, 1981, 1983)		
OCLALAV/ WARDA	18	11		

[a]Sponsored by UNDP/FAO Regional Projects.
[b]Sponsored by NBCU/UNDP/FAO.
[c]NA = data not available.
[d]Included DWRC participation.
[e]Included two Ugandans.
[f]Included three Mozambicans.
[g]Also sponsored by USAID.

Group training

Group training, carried out at the regional level, covers both theoretical and practical aspects of bird biology and management (Fig. 24.1). It is designed to bring together high-calibre participants from various national bird control units to learn bird biology and ecology, damage assessment and protection techniques, crop survey and control methods, and recent developments in quelea research and management. Participants gather in a member

Fig. 24.1. Training workshops held under field conditions provide participants with practical experience in quelea monitoring and control (photo: M. E. Jaeger).

country at a time when the country is likely to suffer from bird pests. Classes and field activities are conducted in eastern Africa by UNDP/FAO project experts assisted by invited international experts from organizations such as USAID/DWRC, GTZ, and DLCO(EA).

The purpose of inviting experienced individuals for training by international experts is to ensure the dissemination of basic scientific knowledge and to exchange information of recent developments in quelea research and management techniques. A group training course usually lasts for 10–15 days, with 4–6 days spent in classes and the rest in practical field activities. Methods are demonstrated for sampling and collecting birds from the wild and dissecting them for breeding condition, sex, age, moult, and quantity and type of food eaten. Lectures are given on identification, taxonomy and classification of bird pests, population dynamics, abundance, distribution, movements, feeding habits, damage patterns, and migratory movements. Damage assessment techniques also are practised. Aerial and ground spray techniques for applying avicides to bird concentrations and various other lethal and non-lethal crop protection methods also are explained, demonstrated, and evaluated. Group training usually is held near quelea breeding colonies or in areas where quelea problems exist.

Group training is designed to achieve three objectives: (1) to train leaders of the national bird control units in basic and new quelea research and management techniques so that they can provide comprehensive coverage of bird-pest/crop-damage situations during in-service training and crop protection practices in their own countries; (2) to provide a forum whereby participants and experts from various nations and organizations with different experiences can freely exchange ideas and formulate training plans; and (3) to facilitate dissemination of quelea information through establishing and maintaining personal contacts. This is especially important where socio-political differences obstruct the regular flow and exchange of information and movement of personnel and joint survey teams.

Group training is as important as in-service training, because it qualifies specialists expected to carry out in-service training in their own countries. The quality of a trainee, rather than the quantity of trainees, usually is the criterion emphasized. Training is based on the 'training of the trainer' strategy to produce trainers who are very knowledgeable about many things as opposed to in-service trainees knowing a little about many different aspects of pest bird management.

From 1979 to 1983, four group training courses were held in eastern Africa, with a total of 105 individuals trained (Table 24.1). The training, originally designed to take place annually with its location circulating among the five eastern African countries, was not fully consummated because of political and logistic considerations. The only eastern African training course attended by participants from all member countries was held in Kenya in 1981. The level of trainees varied from high secondary school graduates to holders of B.Sc. and M.Sc. degrees. The English language was used for this training.

The national bird control unit of the host country (assisted by its FAO national project) and the DLCO(EA) play important roles in group training in terms of field preparations and provision of spraying equipment and survey transport facilities. While emphasis has been put on classes rather than on practical application, the course is constrained by a fixed schedule; extra time taken on any topic risks encroaching on remaining lectures. Furthermore, to overrun the established schedule means additional per diem, which budgets simply will not permit.

Some member governments have tended to send the same personnel for repeat training. This has contributed to a decrease in the number of beneficiaries. It is worth pointing out that, with the exception of Somalia and, recently, Kenya, no country in the region has sent female participants for group training.

One drawback of group training is the tendency for high-calibre trainees to quit their national units once they think they have acquired sufficient knowledge on quelea research and management techniques and take highly paid posts with other institutions. Others return to their Ministries of Agriculture and assume more administrative responsibilities, cutting themselves off from fieldwork. This may be beneficial, if these individuals are involved in administering plant protection activities.

Vertebrate pest lectures in agronomic training schools and centres

A variety of more formalized training in vertebrate pest management on quelea control is available in universities and agronomic centres in both western and eastern Africa. In western Africa, UNDP/FAO project personnel have participated in courses at some of the following institutes:

BENIN	Department of Agronomic and Agrotechnical Studies of the National University (Agricultural School in Sekou and the Rural Training Centre in Cotonou)
LIBERIA	Department of Agronomic Studies of the University of Liberia
MALI	Institute of Agronomic Studies in Kahbougou and the Agricultural Training Centre in Molodo
NIGER	Institute of Rural Training in Kolo
MAURITANIA	Kaedi Agronomic Training Centre
SENEGAL	National Institute for Rural Development in Thies and the Technical College of the University of Dakar
TOGO	National Agronomic Training College of the University of Benin

Similar courses are offered at the University of Ife, Institute of Agriculture, Research and Training, Ibadan, Nigeria. In addition, several other crop protection training programmes are offered, including several lectures on the control of grain-eating birds. These include a 6-month course annually organized by WARDA units for supervisory personnel working in rice schemes in the Member States; courses at the Crop Protection Training Centres in Dakar and Yaounde; courses to upgrade personnel from national crop protection services in preparation for future agricultural campaigns; and sessions organized by various international universities at the request of different countries in the region.

Attendance at a university within a student's country is also desirable because of lower costs and adaptation of the curriculum to local needs. In eastern Africa, courses on particular aspects of Vertebrate Pest Management have been or are being offered at the University of Nairobi, Kenya, University of Khartoum, Sudan, and University of Gezira, Sudan. Graduate research specifically on birds also has been conducted at Makerere University, Uganda. In 1987, Kenyatta University, Kenya, has initiated an M.Sc. course in Vertebrate Pest Management. Hopefully, this can become an important academic training and research site.

Exchange visits

Exchange visits of crop protection personnel between eastern and western African countries have benefitted their respective organizations considerably. During these visits, the trainee (who usually has been sponsored by the UNDP/FAO regional project) benefits from direct involvement in all laboratory and field activities of the host bird control unit. The objectives are to acquaint a visiting trainee with quelea ecology and biology and quelea research and management techniques practised by the host unit, and to strengthen the co-ordination and exchange of information, particularly regarding joint survey methods. It was difficult to trace the total number of trainees involved in exchange visits between 1979 and 1983 due to the irregularity of such training and difficulties in cross-border movements of personnel. However, exchange visits occurred between personnel from Tanzania and Kenya; Somalia and Kenya, Senegal, Ivory Coast, and Nigeria; and Kenya and Senegal. One frustration of this training arises when a trainee returns home with useful ideas that cannot be implemented. In spite of the importance of training, governments have shown little interest in implementing 'new' ideas, particularly after the termination of external funding.

Joint regional surveys

Where quelea are known to migrate across a national frontier, training has sometimes been achieved through organizing joint surveys by teams from the two countries concerned. During these surveys, information on local and cross-border movements is collected, exchanged, and evaluated; and future co-operative plans are formulated. The objectives of joint regional surveys are to clarify the situation of potential quelea cross-border movements that may subsequently affect crop damage patterns and to strengthen bilateral and multilateral links between national units. Several joint surveys have been carried out between Kenya and Ethiopia and Tanzania and Kenya. These surveys began receiving emphasis in 1982–83, when data on cross-border movements of quelea versus mass-migratory and long-distance movements were re-examined. Joint survey and training efforts also have been carried out in western Africa (Table 24.1).

Local or international experts from participating countries usually are sponsored by national units of UNDP/FAO-assisted country projects. In addition, other bilateral organizations such as DWRC, have been actively involved in providing expertise for determining movements of quelea using techniques of mass-marking and radio-telemetry. Many local staff bird survey officers and field scouts from neighbouring countries assist in these operations and ultimately benefit from this training. Political priorities can limit the value of this type of training when governments restrict the initial activities to only one side of the border; however, follow-up work often can be conducted simultaneously by bird control units of both countries.

International and university degree training

An intensive effort was made to secure university degree training in the US for potential leaders of quelea projects in eastern Africa. This was accomplished largely with FAO and USAID funding of fellowships. Ten Eastern Africans have completed graduate courses at Bowling Green State University, Bowling Green, Ohio (Table 24.2); one of these students had previously completed his B.Sc. at California State University, San Louis Obispo, and another student currently is completing his Ph.D. at the University of California, Davis. Most of these graduates have returned to Africa and provided leadership to their respective bird control units. Several have provided instructional input into college training of agricultural specialists in Sudan. In Ethiopia, regular training sessions for extension workers include instruction by such individuals on quelea management.

Non-degree training at universities in the US and Europe also has been available and valuable. Each summer, Purdue University offers a course in Integrated Pest Management for international agricultural students, and lectures are given on vertebrate pests. Several individuals enrolled in regular academic programmes elsewhere in the US have participated in this course. Similarly, the Bird Control Seminars at Bowling Green State University and Vertebrate Pest Conferences in California are attended by students and leaders of African bird control units. Colorado State University, Fort Collins, Colorado, and the University of Wyoming, Laramie, Wyoming, now offer a graduate course in Vertebrate Pest Management. In England, non-degree and degree course work in vertebrate pests has been available at Reading University and in pesticide application at the International Centre for the Application of Pesticides at Cranfield Institute for Technology, Bedford. The Cranfield staff have participated frequently in the conferences and in-service training programmes in Africa. In West Germany, GTZ once provided a 1-year training programme in pesticide application techniques. Plant Protection personnel from Kenya, Ethiopia, and Somalia participated.

Table 24.2. Master of Science degrees in Vertebrate Pest Management received at Bowling Green State University, Bowling Green, Ohio; 1978–1985.

Name of student	Country	Year	Thesis title
Awash Teklehaimanot	Ethiopia	1978	Evaluation of methiocarb efficacy in reducing bird damage to blueberries in southwestern Michigan.
Ali Babiker	Sudan	1978	Mesurol as a bird repellent on ripening highbush blueberries in southwestern Michigan.
Mohamed Zeinelabdin	Sudan	1980	The potential of vegetable tannin as a bird repellent.
Joseph Ndege	Tanzania	1981	Evaluation of methiocarb in reducing bird damage to ripening wheat in Arusha-Tanzania.
Elsafi Elmahdi	Sudan	1982	Sensory cue enhancement of methiocarb repellency to the African Weaver-finch (*Quelea quelea*).
Francis Miano	Kenya	1982	House sparrow movements, food habits, and responses to 4-aminopyridine treatments in a suburban area.
Ephantus Njeru	Kenya	1982	The pest status of Red-collared (*Euplectes ardens*) and Jackson's (*Euplectes jacksoni*) Widow-birds in Nanyuki, Kenya.
William Erickson	USA	1984	Diets of five weaverbird species (Ploceidae) in the Awash River Valley of Ethiopia.
Damena Assefa	Ethiopia	1984	Food consumption and contamination by House Sparrows (*Passer domesticus*) and Feral Pigeons (*Columba livia*) in simulated bulk-grain store.
Hawa Muse Yusuf	Somalia	1985	Efficacy of Ornitrol as pigeon chemosterilant.
Thadeus Tarimo	Tanzania	1985	Assessment of blackbird damage to sunflower.

Funding

UNDP has been the major funding source for FAO activities. Experience has shown that even in-service training can hardly be executed effectively without UNDP/FAO funding assistance. All UNDP/FAO national and regional projects have had training components in their operational plans. With the exception of UNDP, which offers its contributions within the framework of defined projects, there is very little co-ordination among other sources of funding for pest management and training activities. Financing is erratic, usually in reaction to current headlines and always after the crisis. For example, following rat outbreaks, media elements often sensationalize the situation, and funds suddenly become obtainable. Likewise, in consideration of public opinion, organizations may agree to fund bird management training programmes but not crop protection activities against grain-eating birds.

Although initially involved as executing agencies of the projects financed by the UNDP, certain organizations such as the Office for Special Relief Operations (OSRO) and FAO, upon the request of Member States, have financed short-term projects. For example, during the period of acute drought, OSRO provided 2- and 3-year funding for a large number of training programmes for locust and bird control.

The future for these types of training is rather bleak. The problem of future funding of training activities has to be seriously considered within the context that training is fundamentally a national and regional responsibility. A comprehensive review of necessary training needs is an important prerequisite for seeking ways and means of funding such activities in the future.

Role of regional and national units

Food and Agriculture Organization of the United Nations Pest bird management requires sustained action if it is to be effective. The initial UNDP/FAO programme established in 1968 in western Africa was designed to train trainers. However, of the 10 individuals receiving training, only 2 are still at work in the field of bird control—an undeniably low ratio. Later projects were designed to provide individually tailored training programmes. The fact that the certificates awarded upon completion of training could not guarantee career advancement limited the appeal.

US Agency for International Development In western Africa, USAID has established important crop protection training centres in Dakar, Senegal, and Yaounde, Cameroon, and has contributed to certain bird control training actions. Between 1978–81, USAID financed and DWRC executed technical assistance projects in vertebrate pest management to Tanzania and Sudan. Since 1981, this aid has been in the form of technical consultancies designed in research and management techniques and participation in training activities. The development and use of techniques of radio-telemetry, mass-marking, trace-element analysis, and chemical repellents were introduced by USAID/DWRC. This fieldwork has provided considerable opportunities for in-service training by bird control unit personnel in these techniques.

Deutsche Gesellschaft für Technische Zusammenarbeit GTZ has been supporting national plant protection services with an emphasis on bird control for many years—in the 1960s in Sudan, the 1970s in Nigeria, and the early 1980s in Niger, and it has now embarked on a major project in Somalia. Each of these projects has had an important training component organized in-country. Occasionally, exchange visits have been arranged, for example, from Niger to Tanzania and from Somalia to Kenya. Internationally, GTZ has concentrated on participating in technical meetings and group training sessions, particularly with respect to evaluating and demonstrating ground-spraying techniques.

Desert Locust Control Organization for Eastern Africa DLCO(EA) has been entrusted with aerial application of avicides on quelea concentrations in its member countries (Kenya, Ethiopia, Sudan, Somalia, Tanzania, Uganda, and Djibouti). With the termination of UNDP/FAO regional quelea activities, DLCO(EA) has assumed the role of co-ordinating regional bird management activities and has played an important role in group training and joint surveys through the participation of its experts, pilots, and field staff and provision of spray aircraft and flying time, vehicles, and camping equipment. Their direct contribution in technical meetings, indirect contribution with in-service training, and their valuable service of regional radio network, which significantly improved information transfer, has been important to quelea research, management, and training programmes.

Organisation Commune de Lutte Antiacridienne et de Lutte Antiaviaire OCLALAV has provided active support to all the training actions undertaken in western Africa. However, in July 1982, the OCLALAV Administrative Council decided that OCLALAV would progressively reduce its involvement in bird management operations and would, instead, assist the Member States to establish or strengthen their national bird control units, under which training activities will be conducted.

Future training needs

Our surveys on national and international training and extension programmes in quelea management and research have brought up several important issues. UNDP/FAO-assisted regional and national projects, directed at the bird problem alone, terminated in 1986. The projects now going ahead are broader based, aimed at developing crop protection as a whole (in Kenya and in southern Africa) or at vertebrate pests in general (Uganda). As a result, appropriate policy and management decisions for improving the future quality and quantity of training activities will need to be made. These decisions should be considered at several levels. At least three objectives might be included to improve quelea training activities: better planning and increased emphasis on control strategies and methodologies; systematic and in-depth evaluation of locally oriented training; and concentration on both the quantity and quality of training, particularly in-service, of field scouts at the grassroots level.

It will be necessary to proceed by objectives. For each technical position, the competencies actually needed and the conditions under which they will be exercised will have to be determined. This will require establishing a comprehensive list of the duties; determining the frequency with which the principal duties are to be performed; establishing the level of responsibility of each assigned task; and providing career advancement possibilities for each post. These points may appear obvious to effective job performance, but such considerations have not often been specifically identified during the initial establishment period of bird control units.

Training efforts in eastern and western Africa have been similar in many respects, but language, economic, logistical, and political problems have reduced and restricted these efforts. It should, however, be borne in mind that results are more important than counting cost:benefit ratios. To ensure any regional organization of effective control units in its member states, a training division composed of a co-ordinator and two bird control training specialists is needed. It is important not to lose sight of traditional, rural dwellers whose interests are, in fact, the reason behind and justification for all our actions. The scheme presented below indicates the links ideally developed in training actions. Such a scheme can ensure that integrated management programmes can be further developed.

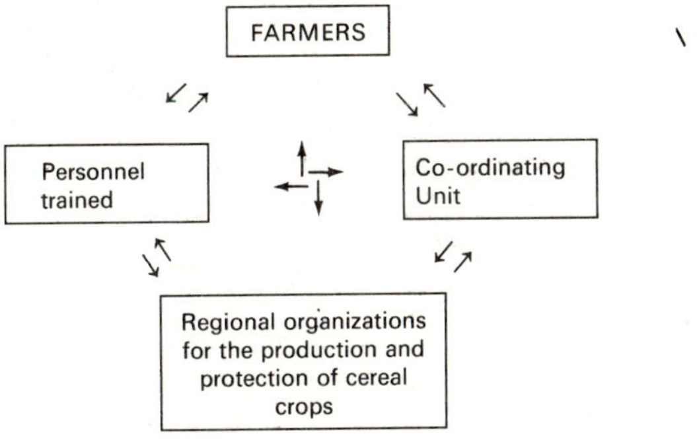

25

Conclusions and future perspectives

CLIVE C.H. ELLIOTT and
RICHARD L. BRUGGERS

A great deal of information has been collected on the quelea in the past 20 years. Enormous amounts of time and money have been spent on quelea research because of the species' notoriety as a crop pest. This major international effort has resulted in quelea being perhaps the most thoroughly investigated bird species in Africa. Results of these efforts include the following:

(1) an increased understanding of the taxonomy, biology, and ecology of the species in some parts of its range;
(2) the development of techniques such as miniature radio transmitters, fluorescent markers, and trace element and 'fingerprint' profiles to monitor local and migratory movements of birds relative to cropping and damage patterns;
(3) the development of techniques systematically to assess damage in selected important agricultural areas or nation-wide;
(4) the improvement of lethal control techniques, implementation of a selective control strategy, and investigation of non-lethal crop protection methods;
(5) the initiation of environmental assessments of control operations; and
(6) the establishment in several countries of effective control units, staffed with trained individuals to conduct proper control operations.

If quelea problems are to be better defined and control strategies to become more refined, we feel that the use of the new techniques described here by our colleagues will most likely lead to these improvements. These new techniques have not yet been utilized to their fullest extent because they have arrived at a time when the pendulum of opinion has been swinging against further research on quelea. However, to develop a truly effective, environmentally sound approach to protecting crops from quelea, using both lethal and non-lethal techniques, we believe future activities should be concentrated in the following five general categories:

(1) **Determination of susceptible crop areas and damage**. Identify important

crop areas and smallholdings of traditional dryland crops within a region, and implement the use of aerial mapping techniques and statistically sound ground damage assessment methods to quantify losses.

(2) **Identification of quelea populations within a region**. Using previously developed monitoring techniques, determine characteristics, range, and movement patterns of populations threatening cereal concentrations. Using new remote sensing satellites, attempt to detect potential quelea breeding areas or at least to detect those areas where poor rainfall and vegetative growth mean that quelea will not breed in that area that season. Based on this information, implement a selective lethal control strategy whereby only those birds causing damage or likely to cause damage are killed and evaluate effectiveness of the control strategy based on the amount of crop saved.

(3) **Assessment of potential environmental hazards of control techniques**. Determine the most appropriate aerial or terrestrial chemical application equipment and procedures so as to minimize hazards to non-target wildlife and the environment. Compare environmental hazards during different stages of breeding colony development, in roosts, and in different habitats. Screen chemicals to find an alternative to fenthion, and evaluate environmental impact of non-lethal chemicals used as repellents on ripening crops.

(4) **Implementation of integrated pest management**. Develop strategies for integrating lethal and non-lethal techniques for reducing damage. Investigate the feasibility of harvesting quelea as a food source.

(5) **Training and institutionalization activities**. Train individuals in statistics and information processing, in pesticide chemistry, and evaluation techniques. Develop systems within regions to collect, process, store, and analyse bird pest data.

There are several implications of future research efforts in some of these areas. The ecological factors that turn a local population of quelea from a minor nuisance to farmers into a major threat to their crops may prove to be so complex that collecting the data accurately to forecast such threats is beyond the resources of most bird control teams in Africa. However, if damage predictions can be linked to simple factors like rainfall and the distribution of breeding, it would allow key populations to be controlled to prevent subsequent damage. But if further research on quelea damage is not carried out, governments will be permanently committed to annual, increasingly expensive control activities against quelea concentrations, which may be more expensive than is really justified.

Our suggestion that the new technologies should be fully exploited, if improvements are to be made, is tempered by the realization that none of them can be fully utilized without considerable investments of resources and expertise. For example, widescale statistically satisfactory surveys of bird damage have not yet been done in any African country despite several worthy attempts. Such surveys are expensive and not worth doing unless they are repeated over several years. One survey in one year may not be representative. The surveys also require specialist teams that will not be side-tracked by other bird control activities. Similarly, if fluorescent-marking is to be used to determine the movements of quelea, it will require extensive inputs not only by the host country in which it is carried out, but also by neighbouring countries if adequate distribution of sampling points for recoveries is to be established. Experience has shown that the easiest part of the technology is to apply the fluorescent marker. The hard work involving a heavy investment of manpower comes in organizing the spray operation, making wing collections, and examining them for the marks. The work may need to be carried out by a team for which movement studies is the main occupation. Such a team could also collect the systematic samples needed to develop the potential of trace element analysis. Probably the most fruitful direction would be to obtain samples of juveniles from colonies in several different potential source areas and match them against juvenile feathers of birds causing damage.

Radio-telemetry has been one of the major contributors to advances in ornithology in the last 10 years. The technique has been used twice on quelea, with, as has so often been the case with this technology, surprising results on the movements of pre-breeding flocks of birds and local movements of roosting birds in agricultural areas. If more studies were conducted at the critical periods when breeding is completed and when roosting birds are causing crop damage, a better understanding of post-breeding dispersal and feeding ecology might be possible. The most severe damage often seems to occur when juvenile quelea disperse on to adjacent cultivation, and yet such dispersal is not an automatic event. Radio-telemetry might help to develop a system for predicting when post-breeding dispersal will constitute a serious threat to crops. Additional studies at roosts might also show how roosts change from nuisance status into threats. But, as with the other new technologies, such studies would be costly and require suitable expertise and aircraft so that radio-marked birds can be located quickly.

Traditionally, Africans have in parts of the quelea's range regarded the birds both as a resource as well as a pest. Nestlings from breeding colonies have systematically been harvested for food. In Chad, professional trappers collect at roosts and sell the birds on the local market. Yet the possibility of developing a quelea industry has never been investigated. It is clear that markets exist. Mass-capture techniques might allow toxic control to be reserved for the most urgent damage-threatening concentrations of quelea, and the rest could be used to support a food industry. If individual farmers received a bounty for every quelea caught on their land, as the basis for a food industry, perhaps farmers' attitudes to the birds might change.

One of the most remarkable of modern technologies is the remote sensing satellite. Two attempts have been made to utilize satellite-generated data to advance quelea studies. Voss (1986) used Landsat 5 spectral imagery to inventory certain aspects of quelea habitat. More recently, studies have been undertaken by D. Wallin of the University of Virginia, USA. D. Wallin has investigated the quelea habitat and vegetation development in Kenya and Tanzania through changes in spectral reflectance as detected by the National Ocean Agency Headquarters (NOAH) satellite. The work is continuing but preliminary results suggest that there is a difference between the normalized difference vegetation index (NDVI) for breeding habitats that support quelea colonies in one year but not another. It may be possible to predict when conditions in known breeding areas can support collonies and when the NDVI fails to reach a minimum threshold (D. Wallin, pers. comm.). If the technology is to be fully utilized, ground surveys based on computer predictions will need to be carried out to test the validity of the method.

The costs of controlling quelea remain high, relying as they do on aerial spraying. We believe that governments in Africa should be encouraged to look much harder at the cost-effectiveness of bird control. They should not be satisfied with awarding the quelea the status of an agro-political pest, whereby cost-effectiveness is not important because of the political gain achieved by control operations. We suggest that African countries cannot afford to spend large sums of money, often with a high foreign currency component, on quelea control without being sure that the birds being killed genuinely threaten local cereal production.

Another area of important concern is environmental contamination by chemical sprays used in quelea control. Minimizing non-target hazards should be an important consideration when developing spraying techniques. It has been assumed that lower spray rates would reduce hazards to non-target species. This assumption may be invalid, as low application techniques may actually increase environmental hazards because of drift and sublethal effects on quelea that permit contaminated birds to disperse over wider areas than if higher spray dosages are used. For this reason, efforts to continue to reduce spray quantities and further develop low-volume ground sprayers, which emit even smaller droplets, should be thoroughly evaluated before being recommended for general use. However, assuming lethal control will continue for the present time, it would seem appropriate to initiate screening studies into finding better chemicals and to be more selective in the choice of spray sites.

A last area with great potential for productive work is in the field of training and education. In all the training exercises on quelea that have been carried out over the past years, one target group has been almost completely ignored—the farmers themselves. Much of the quelea control presently carried out is in response to the calls from farmers for pest birds 'to be dealt with'. It follows that if farmers could better distinguish bird pests likely to seriously affect production, they would be more inclined to request bird control operations only when essential. Similarly, if they were able to see bird damage in the perspective of all the other constraints on cereal production, they would not so readily single out birds for special blame.

Our conclusion is that although control techniques have advanced to the point that any quelea concentration can be eliminated, the quelea problem itself is far from being resolved to the best advantage of the affected countries. Much remains to be done to understand the depredations of the quelea so that the most cost-effective crop protection strategies can be employed. We earnestly hope that this compilation of the efforts of some of the many investigators of this species will provide direction toward future research that may eventually lead to a more thorough understanding of the ecology and pest status of quelea and to appropriate, effective management.

References

Abifarin, A. D. (1984). The importance of rice awns in the reduction of bird damage. *West Afr. Rice Dev. Assoc. Tech. Newsl.*, **5,** 27–8.

Abramson, M. (1979). Vigilance as a factor influencing flock formation among curlews *Numenius arquata. Ibis*, **121,** 213–6.

Adesiyun, A. A. (1973). Bird damage to cereals grown in the dry season in some parts of northern Nigeria. *Samaru Agric. Newsl.*, **15,** 34–7.

Agren, G. and Gibson, R. (1968). *Food composition table for use in Ethiopia.* Swedish Int. Dev. Auth./Ethiopian Ministry of Health, Addis Ababa.

Alcock, J. (1973). Cues used in searching for food by red-winged blackbirds (*Agelaius phoeniceus*). *Behaviour*, **46,** 174–87.

Ali, S. and Ripley, S. D. (1969). *Handbook of the birds of India and Pakistan*, Vol. 3. Oxford University Press, Bombay.

Allan, R. (1983). The strategy for protecting crops from the depredations of quelea birds in Kenya. *Proc. 9th Bird Control Semin.*, Bowling Green, Ohio, **9,** 307–16.

Allan, R. G. (1975). Assessment of bird damage to irrigated wheat in Sudan. Unpubl. Internal Rep., FAO/UNDP Quelea Project RAF/73/055, FAO, Rome.

Allan, R. G. (1980). Quantitative and qualitative assessment of bird damage. *Proc. 2nd Annu. Tech. Meet.*, FAO/UNDP Regional Quelea Project RAF/77/042.

Alsager, D. E. (1976). The role of private consultants in vertebrate pest problems in Canada. *Proc. 7th Vertebr. Pest Conf.*, Monterey, California, **7,** 26–34.

Ankney, C. D. and Scott, D. M. (1980). Changes in nutrient reserves and diet of breeding brown-headed cowbirds. *Auk*, **97,** 684–96.

Anonymous (Undated). Report of a preliminary survey of vertebrate pest damage in the Northwest Frontier Province, Pakistan. Unpubl. Rep.

Anonymous (1954). La lutte contre les oiseaux granivores au Sénégal et en Mauritanie. *Protection des Végétaux*, Dakar.

Anonymous (1979). FAO crop protection manual, African grain-eating birds. Unpubl. Internal Rep., FAO/UNDP, Rome, Italy.

Armstrong, W. D., Rogler, J. C., and Featherston, W. R. (1974). Effect of tannin extraction on the performance of chicks fed bird resistant sorghum grain diets. *Poult. Sci.*, **53,** 714–20.

Ash, J. S. (1981). Qualitative and quantitative assessment of bird pests in Eastern Africa: Somalia. *Proc. 3rd Annu. Tech. Meet.*, FAO/UNDP Regional Quelea Project RAF/77/042.

Ash, J. S. and Miskell, J. E. (1983). Birds of Somalia their habitat status and distribution. *Scopus Special Suppl.*, No. 1.

Ashton, H. (1957). Sixth ringing report. *Ostrich*, **28,** 98–115.

Asquith, T. N., Izuno, C. C., and Butler, L. G. (1983). Characterization of the condensed tannin (proanthocyanidin) from a group II sorghum. *J. Agric. Food Chem.*, **31,** 1299–303.

Attwell, R.I.G. (1954). Crocodiles feeding on weaver birds. *Ibis*, **96,** 485–6.

Avery, M. L. (1979). Food preferences and damage levels of some avian rice field pests in Malaysia. *Proc. 8th Bird Control Semin.*, Bowling Green, Ohio, **8,** 161–6.

Barnard, C. J. (1980). Flock feeding and time budgets in the house sparrow (*Passer domesticus* L.) *Anim. Behav.*, **28,** 295–309.

Barré, M. H. (1983). Quelea movement patterns in Somalia (April 1982– April 1983). *Proc. 4th Annu. Tech. Meet.*, FAO/UNDP Regional Quelea Project RAF/81/023, pp. 40–4.

Barré, N. (1973). Incidence de la suppression d'un des parents sur le devenir de la couvée chez *Quelea quelea* (L.). Efficacité de la lutte chimique. Unpubl. Internal Rep., FAO/UNDP Quelea Project RAF/73/055, FAO, Rome.

Bashir, E. A. (1978). Review of parakeet damage in Pakistan and suggested control methods. *Proc. Semin. Bird Pest Problems in Agric.*, July 5–6, 1978, Karachi, Pakistan, pp. 22–7.

Bashir, E. A. (1979). A new 'parotrap' adapted from the MAC trap for capturing live parakeets in the field. *Proc. 8th Bird Control Semin.*, Bowling Green, Ohio, **8,** 167–71.

Bashir, E. A. (1983). An assessment of bird pest problems to rice in Liberia. Unpubl. Internal Rep., FAO/UNDP Project RAF/81/022, Dakar, Senegal.

Bashir, E. A. (1984). The ecology of birds and their damage to rice in Liberia. *West Afr. Rice Dev. Assoc. Tech. Newsl.*, **5,** 9–12.

Bayer, R. D. (1982). How important are bird colonies as information centers? *Auk*, **99,** 31–40.

Beesley, J.S.S. (1978). Extension of Botswana bird pest research project 1976–1978. Ministry of Overseas Dev./Govt. of Botswana ODM Research Scheme R.2664.

Benson, C. W. and Benson, F. M. (1977). *The birds of Malawi.* Montfort Press, Limbe.

Benson, C. W., Brooke, R. K., Dowsett, R. J., and Irwin, M.P.S. (1973). *The birds of Zambia*, 2nd edn. Collins, London.

Beri, Y. P., Jotwani, M. G., Misra, S. S., and Chander, D. (1969). Studies on relative bird damage to different experimental hybrids of bajara. *Indian J. Entomol.*, **30,** 69–71.

Bertram, B. C. (1978). Living in groups: predators and prey. In *Behavioural ecology* (eds. J. R. Krebs and N. B. Davies). Blackwell Scientific Publishers, Oxford.

Besser, J. (1971). Syllabus. Unpubl. Rep., Denver Wildlife Research Center, Denver, Colorado.

Besser, J. (1973). Protecting seeded rice from blackbirds with methiocarb. *Int. Rice Comm. Newsl.*, **22,** 9–14.

Besser, J. (1978). Improvements in the use of 4-aminopyridine for protecting agricultural crops from birds. *Proc. 8th Vertebr. Pest Conf.*, Sacramento, California, **8,** 51–3.

Besser, J. F., Berg, W. J., and Knittle, C. E. (1979). Late-summer feeding patterns of red-winged blackbirds in a sunflower-growing area of North Dakota. *Proc. 8th Bird Control Semin.*, Bowling Green, Ohio, **8,** 209–14.

Bille, J.-C. (1976). Etude de la production primaire nette d'un écosystème sahélien. Travaux et documents de l'ORSTOM, Nr. 65, ORSTOM, Paris.

Blem, A. R., Ames, R. B., Liew, C. S., and Pryzbylek, J. M. (1983). Effect of preharvest application of Dimethipin on grain moisture, milling quality and yield of rice. *Proc. 10th Annu. Meet. Plant Growth Regul. Soc. America*, pp. 241–7.

Blessin, C. W., Anderson, R. A., Deatherage, W. L., and Inglett, G. E. (1971). Effect of alkali dehulling on composition and wet-milling characteristics of sorghum grain. *Cereal Chem.*, **40,** 528–32.

Bocquet, C. and Roy, J. (1953). Lutte antiaviaire rapport de mission. *Protection des Végétaux*, Dakar.

Bortoli, L. (1970). Rapport de campagne-nidification 1970. Unpubl. Internal Rep., FAO/UNDP Quelea Project RAF/73/055, FAO, Rome.

Bortoli, L. (1974*a*). Mission en Haute Volta—du 4 au 11 juin 1974. Unpubl. Internal Rep., FAO/UNDP Quelea Project RAF/73/055, FAO, Rome.

Bortoli, L. (1974*b*). Nidification des principales espèces d'oiseaux granivores au Mali en 1974. Unpubl. Internal Rep., FAO/UNDP Quelea Project RAF/73/055, FAO, Rome.

Bortoli, L. (1975). Rapport sur la nidification de *Quelea quelea* dans le delta interieur du Niger et les regions adjacentes en 1975. Unpubl. Internal Rep., FAO/UNDP Quelea Project RAF/73/055, FAO, Rome.

Bortoli, L. (1978). Traditional crop protection methods. Unpubl. Internal Rep., FAO/UNDP Quelea Project RAF/73/055, FAO, Rome.

Bortoli, L. and Bruggers, R. L. (1976). Degats d'oiseaux sur sorgho de decrue dans La Vallée du Sénégal. Unpubl. Internal Rep., FAO/UNDP Quelea Project RAF/73/055, FAO, Rome.

Bortoli, L. and Jackson, J. (1972). The distribution of races of *Quelea quelea* in the project area. Unpubl. Internal Rep., FAO/UNDP Quelea Project RAF/73/055, FAO, Rome.

Bouchardeau, A. and Lefevre, L. (1965). Monographie du Lac Tchad. ORSTOM, Paris.

Boudet, G. (1975). Manuel sur les paturages tropicaux et les cultures fourragères. Ministère de la Coopération, Paris, IEMVT, Maisons-Alfort.

Boudreau, G. W. (1975). *How to win the war with pest birds.* Wildlife Technology, Hollister, California.

Bray, O. E. (1973). Radiotelemetry for studying problem birds. *Proc. 6th Bird Control Semin.*, Bowling Green, Ohio, **6,** 198–200.

Bray, O. E., Knittle, C. E., Jack, J. R., and Bowman, R. L. (1978). Locating and identifying blackbird-starling roosts by multispectral remote sensing. *Sci. Tech. Ser. Natn. Wildl. Fed.* **3**, 194–6.

Bray, O. E., Larsen, K. H., and Mott, D. F. (1975). Winter movements and activities of radio-equipped starlings. *J. Wildl. Manage.*, **39,** 795–801.

Breman, H., Cisse, A. M., Djiteye, M. A., and Elberse, W. Th. (1982). Le potentiel botanique des paturages. In *La productivité des paturages saheliens* (eds. F.W.T. Penning De Vries and M. A. Djiteye). Centre for Agricultural Publishing and Documentation, Wageningen, pp. 98–132.

Brooke, C. (1967). The heritage of famine in central Tanzania. *Tanzania Notes Rec.*, **67,** 15–22.

Brown, L. H. and Britton, P. L. (1980). *The breeding seasons of East African birds.* East African Natural History Society, Nairobi.

Brown, L. H., Urban, E. K., and Newman, K. (1982). *The birds of Africa*, Vol. 1. Academic Press, London, UK.

Bruggers, R., Ellis, J., Sedgwick, J. and Bourassa, J. (1981*a*). A radio transmitter for monitoring the movements of small passerine birds. *Proc. 3rd Int. Conf. Wildl. Biotelem.*, Laramie, Wyoming, **3,** 69–79.

Bruggers, R., Matee, J., Miskell, J., Erickson, W., Jaeger, M., Jackson, W. B., and Juimale, Y. (1981*b*). Reduction of bird damage to field crops in eastern Africa with methiocarb. *Trop. Pest Manage.*, **27,** 230–41.

Bruggers, R. L. (1979*a*). Evaluating Curb as a crop repellent to West African bird pests. In *Vertebrate pest control and management materials, ASTM STP 680* (ed. J. R. Beck). Am. Soc. for Testing and Materials, pp. 188–97.

Bruggers, R. L. (1979*b*). Summary of methiocarb trials against pest birds in Senegal. *Proc. 8th Bird Control Semin.*, Bowling Green, Ohio, **8,** 172–84.

Bruggers, R. L. (1980). The situation of grain-eating birds in Somalia. *Proc. 9th Vertebr. Pest Conf.*, Fresno, California, **9,** 5–16.

Bruggers, R. L. (1982). The exportation of cage birds from Senegal. *Traffic Bull.* **IV,** 12–22. Wildlife Trade Monitoring Unit, IUCN Conservation Monitoring Centre, Cambridge, UK.

Bruggers, R. L., Bohl, W. H., Bashir, S. El, Hamza, M., Ali, B., Besser, J. F., De Grazio, J. W., and Jackson, J. J. (1984*a*). Bird damage to agriculture and crop protection efforts in the Sudan. *FAO Plant Protect. Bull.*, **32,** 2–16.

Bruggers, R. L. and Bortoli, L. (1979). Laboratory trials using fluorescent dyes and paints as marking agents for quelea studies. In *Vertebrate pest control and management materials, ASTM STP 680* (ed. J. R. Beck). Am. Soc. for Testing and Materials, pp. 231–6.

Bruggers, R. L. and Jackson, W. B. (1981). Suggested methods for determining the efficacy of vertebrate control agents in developing countries. In *Vertebrate pest control and management materials, ASTM STP 752* (eds. E. W. Schafer, Jr. and C. R. Walker). Am. Soc. for Testing and Materials, pp. 15–28.

Bruggers, R. L., Jaeger, M. E., and Jaeger, M. M. (1985). Tisserins gendarmes (*Ploceus cucullatus abyssinicus*) et tisserins masqués (*Ploceus intermedius intermedius*) munis d'émetteurs radio et de rubans dans une colonie de nidification du sud de l'Ethiopie. *Oiseau Rev. Fr. Ornithol.*, **55,** 81-92.

Bruggers, R. L. and Jaeger, M. M. (1982). Bird pests and crop protection strategies for cereals of the semi-arid African tropics. In *Sorghum in the Eighties: Proc. Int. Symp. on Sorghum* (ed. J. Mertin). ICRISAT, Patancheru, A. P., India, pp. 303–12.

Bruggers, R. L., Jaeger, M. M., and Bourassa, J. B. (1983). The application of radiotelemetry for locating and controlling concentrations of red-billed quelea in Africa. *Trop. Pest Manage.*, **29,** 27–32.

Bruggers, R. L., Jaeger, M. M., Keith, J. O., Hegdal, P. L., Bourassa, J. B., Latigo, A. A., and Gillis, J. N. (In press). Impact of fenthion sprays on nontarget birds during quelea control in Kenya. *Wildl. Soc. Bull.*,

Bruggers, R. L., Murshid, A. A., and Miskell, J. (1981*c*). Accidental death of red-billed queleas roosting in lemon trees in Somalia. *Ostrich* **52,** 60–2.

Bruggers, R. L. and Ruelle, P. (1977). Bird losses in Senegal rice significantly cut. *Rice J.*, Nov/Dec, pp. 10–4.

Bruggers, R. L. and Ruelle, P. (1981). Economic impact of pest birds on ripening cereals in Senegal. *Protect. Ecol.*, **3,** 7–16.

Bruggers, R. L. and Ruelle, P. (1982). Efficacy of nets and fibres for protecting crops from grain-eating birds in Africa. *Crop Protect.*, **1,** 55–65.

Bruggers, R. L., Sultana, P., Brooks, J. E., Fiedler, L. A., Rimpel, M., Manikowski, S., Shivanarayan, N., Santhaiah, N., and Okuno, I. (1984*b*). Preliminary investigations of the effectiveness of trimethacarb as a bird repellent in developing countries. *Proc. 11th Vertebr. Pest Conf.*, Sacramento, California, **11,** 192–203.

Brunel, J. and Thiollay, J. M. (1969). Liste préliminaire des oiseaux de Côte-d'Ivoire. *Alauda*, **37,** 230–54.

Bullard, R. W. (1979). New developments in bird resistant sorghums. *Proc. 8th Bird Control Semin.*, Bowling Green, Ohio, **8,** 229–34.

Bullard, R. W., Bruggers, R. L., Kilburn, S. R., and Fiedler, L. A. (1983*c*.) Sensory-cue enhancement of the bird repellency of methiocarb. *Crop Protect.*, **2,** 387–9.

Bullard, R. W. and Elias, D. J. (1980). Sorghum polyphenols and bird resistance. In *Polyphenols in cereals and legumes, Proc. 36th Annu. Meet. Inst. Food Technol.* (ed. J. H. Hulse). Ottawa, Canada, Int. Dev. Res. Centre Publ. IDRC-145e, pp. 43–9.

Bullard, R. W., Garrison, M. V., Kilburn, S. R., and York, J. O. (1980). Laboratory comparisons of polyphenols and their repellent characteristics in bird-resistant sorghum grains. *J. Agric. Food Chem.* **28,** 1006–11.

Bullard, R. W., Schafer, E. W., Jr., and Bruggers, R. L. (1983*a*). Tests of the enhancement of avian repellent chemicals with sensory cues. In *Vertebrate pest control and management materials, ASTM STP 817* (ed. D. E. Kaukeinen). Am. Soc. for Testing and Materials, pp. 66–75.

Bullard, R. W. and Shumake, S. A. (1979). Two-choice preference testing of taste repellency in *Quelea quelea*. In *Vertebrate pest control and management materials, ASTM STP 680* (ed. J. R. Beck). Am. Soc. for Testing and Materials, pp. 178–87.

Bullard, R. W. and York, J. O. (1985). Breeding for bird resistance in sorghum and maize. In *Plant breeding progress reviews* (ed. G. E. Russell). Butterworths, Surrey, England, pp. 193–222.

Bullard, R. W., York, J. O., and Kilburn, S. R. (1981). Polyphenolic changes in ripening bird-resistant sorghums. *J. Agric. Food Chem.*, **29,** 973–81.

Bullard, R. W., Zeinelabdin, M. H., and Jackson, W. B. (1983b). Repellent potential of vegetable tannins on *Quelea quelea. Proc. 9th Bird Control Semin.*, Bowling Green, Ohio, **9,** 233–9.

Busnel, R. G. and Grosmaire, P. (1958). Enquête auprès des populations du fleuve Sénégal sur leur méthode acoustique de lutte traditionnelle contre le Quelea. *Bull. I.F.A.N.*, **20,** 623–33.

Calvi, C., Besser, J. F., De Grazio, J. W., and Mott, D. F. (1976). Protecting Uruguayan crops from bird damage with methiocarb and 4-aminopyridine. *Proc. 7th Bird Control Semin.*, Bowling Green, Ohio, **7,** 255–8.

Campbell, B. and Lack, E. (eds.) (1985). *A dictionary of birds.* BOU/Poyser, Calton, England.

Capretta, P. J. (1961). An experimental modification of food preferences in chicks. *J. Comp. Physiol. Psychol.*, **54,** 238–42.

Caraco, T. (1979). Time budgeting and group size: a theory. *Ecology*, **60,** 611–7.

Caraco, T., Martindale, S., and Pulliam, H. R. (1980). Avian flocking in the presence of a predator. *Nature*, **285,** 400–1.

Caughley, G. (1977). *Analysis of vertebrate populations.* Wiley, Chichester.

Cheke, R. A. and Walsh, J. F. (1980). Bird records from the Republic of Togo. *Malimbus*, **2,** 112–20.

Chen, P. Y. and Li, Y. (1980). The effect of wheat awns on grain weight and their physiological function. *Acta Agron.*, **7,** 279–82.

Chibber, B. A. K., Mertz, E. T., and Axtell, J. D. (1978). Effects of dehulling on tannin content, protein distribution, and quality of high and low tannin sorghum. *J. Agric. Food Chem.*, **26,** 679–83.

Church, B. M. (1971). The place of sample survey in crop loss estimation. In *Crop loss assessment methods, FAO manual on the evaluation and prevention of losses by pests, disease and weeds* (ed. L. Chiarappa), pp. 2.2/1–2.2/8.

Cisse, A. M. and Breman, H. (1982). La phytoécologie de Sahel et du terrain d'etude. In *La productivité des pâturages Sahéliens* (eds. F.W.T. Penning De Vries and M. A. Djiteye). Centre for Agricultural Publishing and Documentation, Wageningen, pp. 71–83.

Cisse, B. (1981). Lutte chimique contre le quelea (mange-mil) en Afrique de l'Ouest. Unpubl. Ph.D. Thesis, Univ. of Dakar, Faculty of Medicine and Pharmacy, Dakar, Senegal.

Clancey, P. A. (1960). A new race of red-billed quelea from southeastern Africa. *Bull. Br. Ornithol. Club*, **80,** 67–8.

Clancey, P. A. (1968). Subspeciation in some birds from Rhodesia II. *Durban Mus. Novit.*, **8,** 153–82.

Clancey, P. A. (1973). The subspecies of the *lathamii*-group of *Quelea quelea* (Linnaeus). *Durban Mus. Novit.*, **10,** 13–22.

Cochran, W. G. (1977). *Sampling techniques*, 3rd edn. Wiley, New York.

Collias, N. E. and Collias, E. C. (1970). The behaviour of the west African village weaverbird. *Ibis*, **112,** 457–80.

Collias, N. E. and Collias, E. C. (1971). Ecology and behaviour of the spotted-backed weaverbird in the Kruger National Park. *Koedoe*, **14,** 1–27.

Collar, N. J. and Stuart, S. N. (1985). *Threatened birds of Africa and related islands.* Int. Council for Bird Preservation and Int. Union for Conservation of Nature and Natural Resources, Cambridge, UK.

COPR. (1975). The problem of damage to sorghum by doves in Botswana, 1972–1974 Report. Unpubl. Int. Rep., Centre for Overseas Pest Research, London, UK.

COPR. (1976). Bird pest research project, Botswana. Final Rep. 1972-1975. Centre for Overseas Pest Research, London, UK.

COPR. (1977). Quelea investigations project, Nigeria. Final Rep. 1972–1975. Ministry of Overseas Dev./Fed. Military Govt., Nigeria.

Crase, F. T. and DeHaven, R. W. (1976). Methiocarb: its current status as a bird repellent. *Proc. 7th Vertebr. Pest Conf.*, Monterey, California, **7,** 46–50.

Crase, F. T. and DeHaven, R. W. (1978). Food selection by five sympatric California blackbird species. *Calif. Fish Game*, **64,** 255–67.

Crocker, J. (1984). How to build a better scarecrow. *New Scientist*, **1403,** 10–2.

Crook, J. H. (1960). Studies on the social behaviour of *Quelea q. quelea* (Linn.) in French West Africa. *Behaviour,* **16,** 1–55.

Crook, J. H. (1962). The adaptive significance of pair formation types in weaver birds. *Symp. Zool. Soc. Lond.*, **8,** 57–70.

Crook, J. H. (1964). The evolution of social organization and visual communication in the weaver birds (Ploceinae). *Behaviour [Suppl.]*, **10,** 1–178.

Crook, J. H. and Butterfield, P. A. (1970). Gender role in the social system of quelea. In *Social behaviour in birds and mammals* (ed. J. H. Crook). Academic Press, London, pp. 211–48.

Crook, J. H. and Ward, P. (1968). The quelea problem in Africa. In *The problems of birds as pests* (eds. R. K. Murton and E. N. Wright). Academic Press, London, pp. 211–29.

Curtis, D. L. (1965). Sorghum in West Africa. *Samaru Res. Bull.*, **59.** Nigeria.

Curtis, D. L. (1968). The relation between the date of heading of Nigerian sorghums and the duration of the growing season. *J. Appl. Ecol.*, **5,** 215–26.

Czaplicki, J. A., Borrebach, D. E., and Wilcoxin, H. C. (1976). Stimulus generalization of an illness-induced aversion to different intensities of colored water in Japanese Quail. *Anim. Learn. Behav.*, **4,** 45–8.

Da Camara-Smeets, M. (1977). Les dégâts d'oiseaux au berbéré au Tchad et au Nord-Cameroun. *Agron. Trop.*, **XXXII 3,** 262–78.

Da Camara-Smeets, M. and Affoyon, D. (1980). Mission de reconnaissance des oiseaux granivores depredateurs au sud-Cameroun II. Unpubl. Internal Rep., FAO/UNDP Project RAF/77/047, FAO, Rome.

Da Camara-Smeets, M. and Manikowski, S. (1979). Repères visuels utilisés par *Quelea quelea* et *Ploceus cucullatus* dans leurs choix alimentaires. *Malimbus*, **1,** 127–34.

Dar, C. (1974). *Summary of trials with CURB on cultivated vegetables and fruit from sowing to harvest.* Assiamaabarot Ltd., Israel.

Davies, N. B. (1977). Prey selection and the search strategy of the spotted flycatcher (*Muscicapa striata*): a field study on optimal foraging. *Anim. Behav.*, **25,** 1016–33.

Dawson, D. G. (1970). Estimation of grain loss to sparrows (*Passer domesticus*) in New Zealand. *N.Z. J. Agric. Res.*, **13,** 681–8.

De Grazio, J. W. (1978). World bird damage problems. *Proc. 8th Vertebr. Pest Conf.*, Sacramento, California, **8,** 9–24.

De Grazio, J. W. (Compiler) (1984). Progress of vertebrate pest management in agriculture, 1966–1982. Unpubl. USAID/DWRC Rep., Denver Wildlife Research Center, Denver, Colorado.

De Grazio, J. W. and Besser, J. F. (1970). Bird damage problems in Latin America. *Proc. 4th Vertebr. Pest Conf.*, Davis, California, **4,** 162–7.

De Grazio, J. W., Besser, J., and Schafer, E., Jr. (1971). Unpubl. Annu. Rep., Denver Wildlife Research Center, Denver, Colorado.

De Grazio, J. W. and Shumake, S. A. (1982). Controlling quelea damage to small grains in Africa with methiocarb. In *Alternative strategies for desert development and management*, Vol. **2.** Proc. UNITAR Int. Conf., Sacramento, California, 1977. Pergamon Press, New York, pp. 452–6.

De Groot, P. (1980). Information transfer in a socially roosting weaver bird (*Quelea quelea*; Ploceinae): an experimental study. *Anim. Behav.*, **28,** 1249–54.

Dekeyser, P. L. (1958). Recherches sur la biologie du travailleur a bec rouge (*Quelea quelea* Latham). In *Reunion de specialistes sur les Quelea*, Dakar, 31 October–6 November 1955. CCTA/CSA Joint Secretariat, London, pp. 1–8.

Devine, T. and Peterle, T. J. (1968). Possible differentiation of natal areas of North American waterfowl by neutron activation analysis. *J. Wildl. Manage.*, **32,** 274–9.

Dhindsa, M. S. and Toor, H. S. (1980). Extent of bird damage to rice nurseries in the Punjab and its control. *India J. Agric. Sci.*, **50,** 715–9.

Disney, H. J. de S. (1957). *Quelea quelea* in Tanganyika. Cage experiments. CCTA/CSA Africa (57)QB13. *CSA Symposium on Quelea*, Livingston, 1957. CCTA/CSA Joint Secretariat, Bukavu.

Disney, H. J. de S. (1960). Ringing and marking of quelea in Tanganyika, CCTA/CSA Quelea (60)9. *CCTA/FAO Symposia on Quelea*, Bamako, CCTA/CSA Publ. **58,** 143–9.

Disney, H. J. de S. (1964). Quelea control. In *A new dictionary of birds* (ed. A. Landsborough Thomson). Nelson, London and Edinburgh, pp. 673–4.

Disney, H. J. de S. and Haylock, J. W. (1956). The distribution and breeding behaviour of the Sudan dioch (*Quelea q. aethiopica*) in Tanganyika. *East Afr. Agric. J.*, **21,** 141–7.

Disney, H. J. de S., Lofts, B., and Marshall, A. J. (1959). Duration of the regeneration period of the internal reproductive rhythm in a xerophilous equatorial bird *Quelea quelea. Nature (Lond.)*, **184,** 1659–60.

Disney, H. J. de S., Lofts, B., and Marshall, A. J. (1961). An experimental study of the internal rhythm of reproduction in the red-billed dioch *Quelea quelea* by means of photo-stimulation, with a note on melanism induced in captivity. *Proc. Zool. Soc. Lond.*, **136,** 123–9.

Disney, H. J. de S. and Marshall, A. J. (1956). A contribution to the breeding biology of the weaver-finch *Quelea quelea* (Linnaeus) in East Africa. *Proc. Zool. Soc. Lond.*, **127,** 379–87.

Doggett, H. (1957). Bird-resistance in sorghum and the quelea problem. *Field Crop Abstracts*, **10,** 153–6.

Doggett, H. (1970). *Sorghum.* Longmans Green and Co. Ltd., London, UK.

Doggett, H. (1982). Factors reducing sorghum yields *Striga* and birds. In *Sorghum in the Eighties: Proc. Int. Symp. on Sorghum* (ed. J. Mertin). ICRISAT, Patancheru, A. P., India, pp. 313–16.

Doggett, H., Curtis, D. L., Laubscher, F. X., and Webster, O. J. (1970). Sorghum in Africa. In *Sorghum production and utilization* (eds. J. S. Wall and W. M. Ross). AVI Publishing Company, Inc., Westport, Connecticut, pp. 288–326.

Dolbeer, R. A. (1980). Blackbirds and corn in Ohio. *U.S. Fish Wildl. Serv. Resour. Publ.*, **136.**

Dolbeer, R. A., Stickley, A. R., and Woronecki, P. P. (1978). Starling *Sturnus vulgaris* damage to sprouting wheat in Tennessee and Kentucky, U.S.A. *Protect. Ecol.*, **1,** 159–69.

Dolbeer, R. A., Woronecki, P. P., and Stehn, R. A. (1984). Blackbird (*Agelaius phoeniceus*) damage to maize: crop phenology and hybrid resistance. *Protect. Ecol.*, **7,** 43–63.

Dorow, E. (1973). Grunddaten und Ueberlegungen zum Einsatz einer wirkungsvollen Spruehausruestung für die Altvogelbekaempfung im Nistplatz. Unpubl. Internal Rep., GAWI, Frankfurt, West Germany.

Drees, E. M. (1980). Bird pests in agriculture in West Africa and their control. Unpubl. Internal Rep., Wageningen Agric. Univ. Natuurbeheer.

Dunbar, R.I.M. and Crook, J. H. (1975). Aggression and dominance in the weaver bird, *Quelea quelea*. *Anim. Behav.*, **23,** 450–9.

Duncan, R. R. (1980). Methiocarb as a bird repellent on ripening grain sorghum. *Can. J. Plant Sci.*, **60,** 1129–33.

Dunnet, G. M. and Patterson, I. J. (1968). The rook problem in North-east Scotland. In *The problems of birds as pests* (eds. R. K. Murton and E. N. Wright). Academic Press, London, pp. 119–39.

DWRC. (1978). Vertebrate damage control research in agriculture. Unpubl. Annu. Rep., Denver Wildlife Research Center.

Dyer, M. I. and Ward, P. (1977). Management of pest situations. In *Granivorous birds in ecosystems* (eds. J. Pinowski and S.C. Kendeigh). Cambridge University Press, Cambridge, pp. 267–300.

Eastman, P. (1980). An end to pounding: a new mechanical flour milling system in use in Africa. Monograph, IDRC-152e, Int. Dev. Res. Centre, Ottawa, Canada.

Edwards, W. R. and Smith, K. E. (1984) Exploratory experiments on the stability of mineral profiles of feathers. *J. Wildl. Manage.*, **48**, 853–66.

Elgood, J. H., Fry, C. H., and Dowsett, R. J. (1973). African migrants in Nigeria. *Ibis*, **115,** 375–411.

Elias, D. (1977). Vertebrate pests in Latin American agriculture. Unpubl. Internal Rep., Denver Wildlife Research Center, Denver, Colorado.

Elliott, C.C.H. (1979). The harvest time method as a means of avoiding quelea damage to irrigated rice in Chad/Cameroun. *J. Appl. Ecol.*, **16,** 23–35.

Elliott, C.C.H. (1980*a*). Monitoring and research in *Quelea quelea intermedia*. *Min. 2nd Annu. Tech. Meet.*, FAO/UNDP Regional Quelea Project RAF/77/042.

Elliott, C.C.H. (1980*b*). Sex ratio in two ploceids. *Acta 17th Congr. Int. Ornithol., Berlin*, (ed. R. Nöhring), pp. 1359–60.

Elliott, C.C.H. (1980*c*). A regional quelea survey/spray helicopter service for Eastern Africa. *Min. 2nd Annu. Tech. Meet.*, FAO/UNDP Regional Quelea Project RAF/77/042, pp. 184–93.

Elliott, C.C.H. (1981*a*). Monitoring of *Quelea quelea* in eastern Africa - Part II. The relationship between quelea breeding and rainfall. Quelea moult studies. *Proc. 3rd Annu. Tech. Meet.*, FAO/UNDP Regional Quelea Project RAF/77/042, pp. 35–42.

Elliott, C.C.H. (1981*b*). Methods for assessing the efficiency of aerial spraying control operations on quelea colonies and roosts. In *Vertebrate pest control and management materials, ASTM STP 752* (eds. E. W. Schafer, Jr. and C. R. Walker). Am. Soc. for Testing and Materials, pp. 62–73.

Elliott, C.C.H. (1981*c*). Overview of bird pest problems in Eastern Africa. *Proc. 3rd Annu. Tech. Meet.*, FAO/UNDP Regional Quelea Project RAF/77/042.

Elliott, C.C.H. (1983*a*). The quelea bird as a pest of wheat in eastern and southern

Africa. *Proc. Regional Wheat Workshop East, Central and Southern Africa*, Arusha, Tanzania. Nakuru Press, Kenya, pp. 140–6.

Elliott, C.C.H. (1983*b*). Quelea movement patterns at the national level—Tanzania. *Proc. 4th Annu. Tech. Meet.*, FAO/UNDP Regional Quelea Project RAF/81/023.

Elliott, C.C.H. (In press). The quelea as a major problem in a food-deficient continent. In *The quelea problem in southern Africa* (eds. P. J. Mundy and M.J.F. Jarvis). Baobab Books, Zimbabwe.

Elliott, C.C.H. and Beesley, J.S.S. (1980). Bird damage to cereal crops— Tanzania 1980. *Proc. 2nd Annu. Tech. Meet.*, FAO/UNDP Regional Quelea Project RAF/77/042.

Elliott, C.C.H. and Jarvis, M.J.F. (1970). Fourteenth ringing report. *Ostrich*, **41,** 1–117.

Elliott, C.C.H. and Jarvis, M.J.F. (1972–1973). Fifteenth ringing report. *Ostrich*, **43,** 236–95; **44,** 34–78.

Elliott, C.C.H. and Manikowski, S. (1976). A review of scouting methods used during the 1976 bird-control campaign in Chad/Cameroun and proposals for their improvement. Unpubl. Internal Rep., FAO/UNDP Quelea Project RAF/73/055, FAO, Rome.

Elmahdi, E. M. (1982). Sensory cue enhancement of methiocarb repellency to the African weaver-finch (*Quelea quelea*). Unpubl. M.S. thesis, Bowling Green State University, Bowling Green, Ohio.

Elmahdi, E. M., Bullard, R. W., and Jackson, W. B. (1985). Calcium carbonate enhancement of methiocarb repellency for quelea. *Trop. Pest Manage.*, **31,** 67–72.

Emlen, S. T. and Demong, N. J. (1975). Adaptive significance of synchronized breeding in a colonial bird. *Science*, **188,** 1029–31.

Endler, J. A. (1977). *Geographic variation, speciation, and clines.* Princeton University Press, Princeton.

Erickson, W. A. (1979). Diets of the red-billed quelea (*Quelea quelea*) in the Awash River Basin of Ethiopia. *Proc. 8th Bird Control Semin.*, Bowling Green, Ohio, **8,** 185–200.

Erickson, W. A. (1984). Diets of five weaverbird species (Ploceidae) in the Awash River Valley of Ethiopia. Unpubl. M.S. thesis, Bowling Green State University, Bowling Green, Ohio.

Erickson, W. A. and Damena, A. (1982). Breeding of red-billed queleas (*Quelea quelea*) in relation to rainfall patterns in Ethiopia. Unpubl. Internal Rep., FAO/UNDP Quelea Project ETH/77/022, FAO, Rome.

Erickson, W. A., Jaeger, M. M., and Bruggers, R. L. (1980). The development of methiocarb for protecting sorghum from birds in Ethiopia. *Ethiop. J. Agric. Sci.*, **2,** 91–100.

Evans, J. and Griffith, R. E., Jr. (1973). A fluorescent tracer and marker for animal studies. *J. Wildl. Manage.*, **37,**73–81.

Ewing, K., Crabb, A. C., Martin, L. R., and Moitoso, R. (1976). Preliminary laboratory and field trials of Curb, a possible avian repellent. *Proc. 7th Bird Control Semin.*, Bowling Green, Ohio, **7,** 239–41.

Fahlund, L. A. (1965). Report of the United States observer to the Food and Agriculture Organization of the United Nations. *Conf. on Quelea, Bird and Water Hyacinth Control in Africa*, Duala, Cameroon, **V,** 1–118.

FAO. (1978). *Bird scout's handbook*. United Nations Dev. Prog./Food and Agricultural Organization.

FAO. (1979*a*). *Min. 1st Annu. Tech. Meet.*, FAO/UNDP Regional Quelea Project RAF/77/042.

FAO. (1979*b*). Crop protection manual—African grain-eating birds. FAO/UNDP Publ. AGOA, RAF/73/055.

FAO. (1980*a*). Cereal crop pests in Africa, with particular reference to birds. Unpubl. Internal Rep., FAO/UNDP, Rome, Italy.

FAO. (1980*b*). *Min. 2nd Annu. Tech. Meet.*, FAO/UNDP Regional Quelea Project RAF/77/042.

FAO. (1980*c*). Coordination of cooperative action to reduce bird damage to crops in eastern Africa. *Min. 2nd Annu. Tech. Meet.*, FAO/UNDP Regional Quelea Project RAF/77/042.

FAO. (1981*a*). An assessment of the bird pest problem in Sudan, Ethiopia, Somalia, Kenya, Tanzania. Unpubl. Internal Rep., FAO/UNDP, Rome, Italy.

FAO. (1981*b*). The infrastructure for monitoring quelea in eastern Africa. Unpubl. Internal Rep., FAO/UNDP, Rome, Italy.

FAO. (1982*a*). FAO Month. Bull. Stat., **5,** 1–68, Food and Agriculture Organization of the United Nations, Rome, Italy.

FAO. (1982*b*). Regional technical assistance to OCLALAV for crop protection against grain-eating birds: conclusions and recommendations of the project. Final Report AG:DP/RAF/77/047. FAO/UNDP, Rome.

FAO. (1984*a*). Agroclimatological data for Africa. Unpubl. Internal Rep., FAO/UNDP, Rome, Italy.

FAO. (1984b). *Proc. 5th Annu. Tech. Meet.*, FAO/UNDP Regional Quelea Project RAF/81/023.

FAO/WHO. (1980). Pesticide residues in food—1980; evaluations 1980. Food and Agriculture Organization/World Health Organization, *FAO Plant Prod. Protect.*, **26,** 218–34.

Farris, M. A. E. (1975). The general bird problem in grain sorghum. *Proc. Int. Sorghum Workshop* (ed. Publication Staff). U.S. Agency for International Development, Washington, D.C., pp. 289–304.

Feare, C. (1984). *The starling*. Oxford University Press, Oxford and New York.

Feare, C. J. (1974). Ecological studies of the rook (*Corvus frigilegus* L.) in North-East Scotland. Damage and its control. *J. Appl. Ecol.*, **11,** 897–914.

Federer, W. T. (1955). *Experimental design: Theory and application*. Oxford and IBH Publ. Co., Calcutta.

Fitzwater, W. D. (1971). The weaver finch of Hispaniola. *Pest Control*, **39,** 19–20; 56–9.

Fitzwater, W. D. (1973). Madam Saga—an approach to an animal damage problem. *Proc. 6th Bird Control Semin.*, Bowling Green, Ohio, **6,** 47–52.

Fleming, T. H. (1981). Winter roosting and feeding behaviour of pied wagtails *Motacilla alba* near Oxford, England. *Ibis*, **123,** 463–76.

Fogden, M.P.L. (1972). The seasonality and population dynamics of equatorial forest birds in Sarawak. *Ibis*, **114,** 307–43.

Fogden, M.P.L. and Fogden, P. M. (1979). The role of fat and protein reserves in the annual cycle of the Grey-backed camaroptera in Uganda (Aves: Sylviidae). *J. Zool. (Lond.)*, **189,** 233–58.

Froman, B. and Persson, S. (1974). *An illustrated guide to the grasses of Ethiopia.* Chilalo Agricultural Development Unit, Asella, Ethiopia.

Funmilayo, O. and Akande, M. (1977). Vertebrate pests of rice in southwestern Nigeria. *PANS*, **23,** 38–48.

Fuggles-Couchman, N. R. (1952). The destruction of rice-eating birds. *East Afr. Agric. J.*, **19,** 77–8.

Gadgil, M. (1972). The function of communal roosts: relevance of mixed roosts. *Ibis*, **114,** 531–3.

Garrison, M. V. and Libay, J. L. (1982). Potential of methiocarb seed treatment for protection of sprouting rice from Philippine bird pests, *Lonchura* spp. *Philipp. Agric.*, **65,** 363–6.

Gaston, A. (1973). Equisse de reconnaissance des groupements vegetaux de la zone de recherches écologiques intensives du projet *Quelea quelea* (Region de N'Djamena). Unpubl. Internal Rep., FAO/UNDP Quelea Project RAF/67/087, FAO, Rome.

Gaston, A. and Lamarque, G. (1976). Travaux phytoécologiques en relation avec la lutte contre *Quelea quelea*—Bilan de quatre années. Rapport final, FAO/UNDP Regional Quelea Project RAF/67/087. Inst. Elev. Med. Vet. Pays Trop. Maisons Alfort.

Gaudchau, M. D. (1967). Report on control of the red-billed weaver bird (*Quelea quelea aethiopica*) in the Republic of the Sudan during 1964/65/66/67. Unpubl. Rep., Khartoum, Ministry of Agriculture, Plant Protection Division.

Ghosh, B. (1945). Efficiency of rectangular plots of different shapes and sizes in field experiments or sample surveys. *Proc. 32nd Indian Sci. Congr.*, Sec. XII, No. 48.

Gillet, H. (1974). Tapis végétal et paturages du Sahel. In *UNESCO Le Sahel: bases écologiques de l'aménagement.* Notes techniques MAB UNESCO, Paris, pp. 21–7.

Gillette, K., Irwin, J. D., Thomas, D. K., and Bellingham, W. P. (1980). Transfer of coloured food and water aversions in domestic chicks. *Bird Behav.*, **2,** 37–47.

Ginn, H. B. and Melville, D. S. (1983). *Moult in birds.* British Trust for Ornithology, Guide No. 19.

Goldstein, J. L. and Swain, T. (1963). Changes in tannins in ripening fruit. *Phytochemistry*, **2,** 371–83.

Goss-Custard, J. D. (1977). Optimal foraging and the size selection of worms by redshank, *Tringa totanus*, in the field. *Anim. Behav.*, **25,** 10–29.

Gramet, Ph. (1974). Rapport de mission en République du Mali et du Sénégal du 24/9 au 14/10/74. Unpubl. Internal Rep., FAO/UNDP Quelea Project RAF/73/055, FAO, Rome.

Grant, C. L. (1953). Spectrographic analysis of ashes of feathers and bones of ruffed grouse. Unpubl. Internal Rep., New Hampshire Fish and Game Department, Concord, New Hampshire.

Gras, G., Hasselman, C., Pellissier, C., and Bruggers, R. (1981). Residue analysis of methiocarb applied to ripening sorghum as a bird repellent in Senegal. *Bull. Environ. Contam. Toxicol.*, **26,** 393–400.

Grist, D. H. and Lever, R.J.A.W. (1969). *Pests of rice.* Longmans, London, UK.

Grosmaire, P. (1955). Essai sur l'evolution de la population de *Quelea* dans la vallée du fleuve Sénégal. Variation de cette population depuis Mai 1953 jusqu'au 15

Octobre 1955. Efficacité de la lutte entreprise par l'Organisme de Lutte Anti-aviaire (OLA) du Sénégal. CSA Reunion des Specialistes du Quelea, Dakar, 1955. Bukavu, Secretariat Conjoint CCTA/CSA.

Grue, C. E., Fleming, W. J., Busby, D. G., and Hill, E. F. (1983). Assessing hazards of organophosphate pesticides to wildlife. *Proc. North Am. Wildl. Conf.*, **48,** 200.

Grue, C. E., Powell, G.V.N., and McChesney, M. J. (1982). Care of nestlings by wild female starlings exposed to an organophosphate pesticide. *J. Appl. Ecol.*, **19,** 327–35.

GTZ. (1979). Pesticide residue problems in the Third World. Unpubl. Rep., Deutsche Gesellschaft für Technische Zusammenarbeit (GTZ), Eschborn, West Germany.

GTZ. (1982). Die Oekologie and Bekaempfung des Blutschnabelwebervogels [*Quelea quelea* (L.)] in Nordostnigeria. Deutsche Gesellschaft für Technische Zusammenarbeit (GTZ), Eschborn, West Germany.

GTZ. (1986). Rotations-Driftspruehanlage. Einsatz- und Bedienungshandbuch. Spec. Publ. No. 186, Deutsche Gesellschaft für Technische Zusammenarbeit (GTZ), Eschborn, West Germany.

GTZ. (1987). The ecology and control of the Red-billed Weaver Bird (*Quelea quelea* L.) in Northeast Nigeria. Spec. Publ. No. 199, Deutsche Gesellschaft für Technische Zusammenarbeit (GTZ), Eschborn, West Germany.

Guarino, J. L. (1972). Methiocarb, a chemical bird repellent: a review of its effectiveness on crops. *Proc. 5th Vertebr. Pest Conf.*, Fresno, California, **5,** 211–6.

Gupta, R. K. and Haslam, F. (1980). Vegetable tannins—Structure and biosynthesis. In *Polyphenols in cereals and legumes, Proc. 36th Annu. Meet. Inst. Food Technol.* (ed. J. H. Hulse). Ottawa, Canada, Int. Dev. Res. Centre Publ. **IDRC-145e,** pp. 15–24.

Hagerman, A. E. and Butler, L. G. (1980). Condensed tannin purification and characterization of tannin-associated proteins. *J. Agric. Food Chem.*, **28,** 947–52.

Hailu, K. (1984). Lethal control of red-billed quelea (*Quelea quelea*) in the southern and central Rift Valleys during 1983 and 1984 control seasons. *Proc. 5th Annu. Tech. Meet.*, FAO/UNDP Regional Quelea Project RAF/81/023, pp. 102–7.

Haldane, J.B.S. (1955). The calculation of mortality rates from ringing data. *Proc. Int. Ornithol. Congr.*, **11,** 454–8.

Hall, B. P. and Moreau, R. E. (1970). *An atlas of speciation in African Passerine birds.* British Museum (Natural History), London, UK.

Hamza, M., Ali, B., El Haig, I., Bohl, W., Besser, J., De Grazio, J., and Bruggers, R. L. (1982). Evaluating the bird repellency of methiocarb. *Malimbus*, **4,** 33–41.

Hanson, H. C. and Jones, R. L. (1968). Use of feather minerals as biological tracers to determine the breeding and molting grounds of wild geese. *Ill. Nat. Hist. Surv. Biol. Notes* **60.**

Hanson, H. C. and Jones, R. L. (1976). The biogeochemistry of blue, snow and Ross' geese. *Ill. Nat. Hist. Surv. Spec. Publ.*, **1.**

Harrel, C. G. and Dirks, B. M. (1955). Cereals and cereal products. In *Handbook of food and agriculture* (ed. F. C. Blanck). Reinhold, New York, N.Y., pp. 411–52.

Harris, H. B. (1969). Bird resistance in grain sorghum. *Proc. 24th Annu. Corn Sorghum Res. Conf.* (eds. J. I. Sutherland and R. J. Falasea). American Seed Trade Association, Washington, D.C., pp. 113–22.

Hartigan, R. (1979). Sorghum tannins: inheritance, seasonal development, and biological value. Unpubl. M.S. thesis, Purdue Univ., Lafayette, Indiana.

Haylock, J. W. (1955). *Quelea quelea*—movements. Unpubl. Rep.: Moshi 25.1.55.

Haylock, J. W. (1957). Preliminary notes on the Sudan dioch (*Quelea quelea aethiopica*) and its control by the Department of Agriculture in Kenya Colony. CCTA/CSA Africa (57) QB 12. *CSA Symp. Quelea*, Livingstone, 1957. CCTA/CSA Joint Secretariat, Bukavu.

Haylock, J. W. (1959). *Investigations on the habits of quelea birds and their control.* Nairobi, Government Printers.

Heckel, J.-U. (1983). GTZ—bird control activities in the Republic of Niger during 1981/82 and future goals. *Proc. 4th Annu. Tech. Meet.*, FAO/UNDP Regional Quelea Project RAF/81/023.

Heisterberg, J. F., Knittle, C. E., Bray, O. E., Mott, D. F., and Besser, J. F. (1984). Movements of radio-instrumented blackbirds and European starlings among winter roosts. *J. Wildl. Manage.*, **48,** 203–9.

Hermann, G. and Kolbe, W. (1971). L'enrobage de la semence avec le Mesurol pour la lutte contre les oiseaux dans les cultures de mais, compte tenu de la tolérance des variétés et des effets secondaires. *Pflanzenschutz-Nachrichten Bayer*, **24,** 290–331.

Holler, N. R., Naquin, H. P., Lefebvre, P. W., Otis, D. L., and Cunningham. D. J. (1982). Mesurol® for protecting sprouting rice from blackbird damage in Louisiana. *Wildl. Soc. Bull.*, **10,** 165–70.

Holyoak, D. T. (1970). Sex-differences in feeding behaviour and size in the carrion crow. *Ibis*, **112,** 397–400.

Hoogland, J. L. and Sherman, P. W. (1976). Advantages and disadvantages of bank swallows (*Riparia riparia*) coloniality. *Ecol. Monogr.*, **46,** 33–58.

Horn, H. S. (1968). The adaptive significance of colonial nesting in the Brewer's blackbird (*Euphagus cyanocephalus*). *Ecology*, **49,** 682–94.

Hoshino, T. and Duncan, R. R. (1981). Bird damage and tannin content in grain sorghum hybrids under different environments. *Jpn. J. Crop Sci.*, **50,** 332–7.

Howard, W. E., Park, J. S., Shin, Y. M., and Cho, W. S. (1975). Rodent control in Republic of Korea. Inst. Agric. Sci. Office of Rural Development.

Hudson, R. H., Tucker, R. K., and Haegele, M. A. (1984). Handbook of toxicity of pesticides to wildlife. *USFWS Resour. Publ.* **153.**

Huffnagel, H. P. (1961). *Agriculture in Ethiopia.* Food and Agriculture Organization of the United Nations, Rome, Italy.

Hulse, J. H., Laing, E. M., and Pearson, O. E. (1980). *Sorghum and the millets: their composition and nutritive value.* Academic Press, London, UK.

Humphries, D. A. and Driver, P. M. (1970). Protean defence by prey animals. *Oecologia*, **5,** 285–302.

Inglis, I. R. (1980). Visual bird scarers: an ethological approach. *Proc. Bird Problems in Agric. Symp.* (eds. E. N. Wright, I. R. Inglis, and C. J. Feare). University of London, BCPC Publ., pp. 121 43.

Irwin, M.P.S. (1981). *The birds of Zimbabwe*. Quest Publishing, Harare.

Jackson, J. (1973). Summary of data on distribution and migration of quelea in the Lake Tchad Basin and the Benoue Watershed in Tchad and Cameroon. Unpubl. Internal Rep., FAO/UNDP Quelea Project RAF/73/055, FAO, Rome.

Jackson, J. and Park, P. O. (1973). The toxic effects of fenthion on a nesting population of queleas during experimental control by aerial spraying. *Proc. 6th Bird Control Semin.*, Bowling Green, Ohio, **6,** 53–73.

Jackson, J. J. (1971). A bird resistant millet from South Chad. Unpubl. Internal Rep., FAO/UNDP Quelea Project RAF/73/055, FAO, Rome.

Jackson, J. J. (1974*a*). Nesting success of *Quelea quelea* with one parent removed and observations on roosting behavior, with implications for control. *Proc. 6th Vertebr. Pest Conf.*, Anaheim, California, **6,** 242–5.

Jackson, J. J. (1974*b*). A trap for fledgling *Quelea quelea*. Unpubl. Internal Rep., FAO/UNDP Quelea Project RAF/73/055, FAO, Rome.

Jackson, J. J. (1974*c*). The relationship of Quelea migrations to cereal crop damage in the Lake Chad basin. *Proc. 6th Vertebr. Pest Conf.*, Anaheim, California, **6**, 238–42.

Jackson, W. B. (1979). Subcommittee on the estimation of bird damage to grain crops. *Min. 1st Annu. Tech. Meet.*, FAO/UNDP Regional Quelea Project RAF/77/042.

Jackson, W. B. and Jackson, S. S. (1977). Estimates of bird depredations to agricultural crops and stored products. Plant Health Newsl.: Colloquium on crop protection against starlings, pigeons, and sparrows. *EPPO Publ. Ser. B*, **84,** 33–43.

Jaeger, M. E. and Jaeger, M. M. (1977). Quelea as a resource. Unpubl. Internal Rep., FAO/UNDP Quelea Project RAF/73/055, FAO, Rome.

Jaeger, M. M. (1984). Seasonal distribution and movement patterns of quelea in eastern Africa: A current perspective. *Proc. 5th Annu. Tech. Meet.*, FAO/UNDP Regional Quelea Project RAF/81/023.

Jaeger, M. M., Bruggers, R. L., Johns, B. E., and Erickson, W. A. (1986). Evidence of itinerant breeding of the red-billed quelea *Quelea quelea* in the Ethiopian Rift Valley. *Ibis*, **128,** 469–82.

Jaeger, M. M., Cunningham, D. J., Bruggers, R. L., and Scott, E. J. (1983). Assessment of methiocarb-impregnated sunflower achenes as bait to repel blackbirds from ripening sunflowers. *Proc. 9th Bird Control Semin.*, Bowling Green, Ohio, **9,** 207–24.

Jaeger, M. M., Elliott, C. C., Lenton, G. M., Allan, R. G., Baṣhir, S., and Ash, J. S. (1981). Monitoring of *Quelea quelea* in eastern Africa (July 1978–October 1981). Mask index and the distribution of quelea. *Proc. 3rd Annu. Tech. Meet.*, FAO/UNDP Regional Quelea Project RAF/77/042.

Jaeger, M. M. and Erickson, W. A. (1980). Levels of bird damage to sorghum in the Awash Basin of Ethiopia and the effects of the control of quelea nesting colonies (1976–1979). *Proc. 9th Vertebr. Pest Conf.*, Fresno, California, **9,** 21–8.

Jaeger, M. M. and Erickson, W. A. (1981). Lethal control of quelea nesting colonies in the Awash Valley during 1981. *Proc. 3rd Annu. Tech. Meet.*, FAO/UNDP Regional Quelea Project RAF/77/042, pp. 62–63.

Jaeger, M. M., Erickson, W. A., and Jaeger, M. E. (1979). Sexual segregation of red-billed queleas (*Quelea quelea*) in the Awash River Basin of Ethiopia. *Auk*, **96,** 516–24.

James, F. C. (1983). Environmental component of morphological differentiation in birds. *Science*, **221,** 184–6.

James, F. C., Engstrom, R. T., Nesmith, C., and Laybourne, R. (1984). Inferences about population movements of red-winged blackbirds from morphological data. *Am. Midl. Nat.*, **111,** 319–31.

James, H. W. (1928). The nesting of the southern pink-billed weaver (*Quelea quelea lathami*). *Oologists' Rec.*, **8,** 84–5.

Jarvis, M.J.F. and LaGrange, M. (In press). Conservation, quelea control, and the trap roost concept. In *The quelea problem in southern Africa* (eds. P. J. Mundy and M.J.F. Jarvis). Baobab Books, Zimbabwe.

Jarvis, M.J.F. and Vernon, C. J. (In press-*a*). Food and feeding habits of quelea in southern Africa. In *The quelea problem in southern Africa* (eds. P. J. Mundy and M.J.F. Jarvis). Baobab Books, Zimbabwe.

Jarvis, M.J.F. and Vernon, C. J. (In press-*b*). Notes on quelea ecology in southern Africa. In *The quelea problem in southern Africa* (eds. P. J. Mundy and M.J.F. Jarvis). Baobab Books, Zimbabwe.

Jensen, J. V., and Kirkeby, J. (1980). *The birds of The Gambia.* An annotated check-list and guide to localities in the Gambia. Aros Nature Guides, Denmark.

Jeremiah, H. E. and Parker, J. D. (1985). Health hazard aspects of fenthion residues in quelea birds. Int. Cent. for the Application of Pesticides, Cranfield Institute of Technology, UK.

Johnston, R. F. (1969). Character variation and adaptation in European sparrows. *Syst. Zool.*, **18,** 206–31.

Johnston, R. F. and Klitz, W. J. (1977). Variation and evolution in a granivorous bird: the house sparrow. In *Granivorous birds in ecosystems* (eds. J. Pinowski and S. C. Kendeigh). Cambridge University Press, Cambridge, England, pp. 15–51.

Jones, P. J. (1972). The status of *Quelea quelea* in Botswana and recommendations for its control. Unpubl. Rep., Centre for Overseas Pest Research, to Government of Botswana, London, UK.

Jones, P. J. (1976). The utilization of calcareous grit by laying *Quelea quelea. Ibis*, **118,** 575–6.

Jones, P. J. (1980). The annual mortality of *Quelea quelea* in South Africa from ringing recoveries during a period of intensive quelea control. *Proc. Pan-Afr. Ornithol. Congr.*, **4,** 423–7.

Jones, P. J. (1983). Haematocrit values of breeding red-billed queleas *Quelea quelea* (Aves: Ploceidae) in relation to body condition and thymus activity. *J. Zool. (Lond.)*, **201,** 217–22.

Jones, P. J. and Pope, G. (1977). Wheat damage by quelea in Zambia. Unpubl. Int. Rep., Centre for Overseas Pest Research, London, UK.

Jones, P. J. and Ward, P. (1976). The level of reserve protein as the proximate factor controlling the timing of breeding and clutch-size in the red-billed quelea *Quelea quelea. Ibis*, **118,** 547–74.

Jones, P. J. and Ward, P. (1979). A physiological basis for colony desertion by red-billed queleas (*Quelea quelea*). *J. Zool. (Lond.)*, **189,** 1–19.

Joslyn, M. A. and Goldstein, J. L. (1964). Astringency of fruits and fruit products in relation to phenolic content. *Adv. Food Res.*, **13,** 179–217.

Jowett, D. (1967). Breeding bird-resistant sorghum in East Africa. *Plant Breeding Abstracts*, **37,** 85.

Kalmbach, E. R. (1937). Blackbirds and the rice crop on the gulf coast. *Wildlife Resource Management Leaflet B5–96*, U.S. Bureau of Biological Survey, Washington, D. C.

Kaske, R. F. (1970). Trials to control weaver-birds by non aerial operations in the Sudan. Unpubl. Internal Rep., GAWI, Frankfurt, West Germany; PPD Khartoum.

Kelsall, J. P. and Burton, R. (1977). Identification of origins of lesser snow geese by X-ray spectrometry. *Can. J. Zool.*, **55,** 718–32.

Kelsall, J. P. and Burton, R. (1979). Some problems in identification of origins of lesser snow geese by chemical profiles. *Can. J. Zool.*, **57,** 2292–302.

Kelsall, J. P. and Calaprice, J. R. (1972). Chemical content of waterfowl plumage as a potential diagnostic tool. *J. Wildl. Manage.*, **36,** 1088–97.

Kelsall, J. P., Pannekoek, W. J. and Burton, R. (1975). Chemical variability in plumage of wild lesser snow geese. *Can. J. Zool.* **53,** 1369–75.

Kendall, M. D. (1980). Avian thymus glands: a review. *Dev. Comp. Immunol.*, **4,** 191–210.

Kendall, M. D. and Ward, P. (1974). Erythropoiesis in an avian thymus. *Nature (Lond.)*, **249,** 366–7.

Kendall, M. D., Ward, P., and Bacchus, S. (1973). A protein reserve in the Pectoralis major flight muscle of *Quelea quelea*. *Ibis*, **115,** 600–01.

Kendeigh, S. C. and West, G. C. (1965). Caloric values of plant seeds eaten by birds. *Ecology*, **46,** 553–5.

Kenya News Agency. (1985). Fish waste sold. *Kenya Nation*, 26 January 1985.

Kieser, J. A. and Kieser, G. A. (1978). Birds of the De Aaar district. *South. Birds* **5.**

King, J. R. (1973). Energetics of reproduction in birds. In *Breeding biology of birds* (ed. D. S. Farner). National Academy of Sciences, Washington, D.C., pp. 77–107.

Kitonyo, F. M. (1981). Indirect control achievements: Kenya (October 1980–October 1981). *Proc. 3rd Annu. Tech. Meet.*, FAO/UNDP Regional Quelea Project RAF/77/042, pp. 64–70.

Kitonyo, F. M. (1983). Control achievements of the bird control unit in Kenya 1982/83—KEN/82/003. *Proc. 4th Annu. Tech. Meet.*, FAO/UNDP Regional Quelea Project RAF/81/023, pp. 56–63.

Kitonyo, F. M. and Allan, R. G. (1979). Quantitative and qualitative assessment of bird damage in Kenya. *Proc. lst Annu. Tech. Meet.*, FAO/UNDP Regional Quelea Project RAF/77/042.

Klopfer, P. H. (1958). Influence of social interactions on learning rates in birds. *Behaviour*, **14,** 282–99.

Knittle, C. E. and Guarino, J. L. (1976). A 1974 questionnaire survey of bird damage to ripening grain sorghum in the United States. *Sorghum Newsl.*, **19,** 93–4.

Knittle, C. E., Linz, G. M., Johns, B. E., Cummings, J. L., Davis, J. E. Jr., and Jaeger, M. M. (1987). Dispersal of male red-winged blackbirds from two spring roosts in central North America. *J. Field Ornithol.*, **58,** 490–8.

Krebs, J. R. and McCleery, R. H. (1984). Optimization in behavioural ecology. In *Behavioural ecology* (eds. J. R. Krebs and N. B. Davies). Blackwell Scientific Publications, Oxford, England, pp. 91–121.

Krebs, J. R., Stephens, D. W., and Sutherland, W. J. (1983). Perspectives in optimal foraging. In *Perspectives in ornithology* (eds. A. H. Brush and G. A. Clark, Jr.). Cambridge University Press, Massachusetts, pp. 165–216.

Lack, D. (1954). *The natural regulation of animal numbers*. Oxford University Press, Oxford, UK.

Lack, D. (1966). *Population studies of birds*. Clarendon Press, Oxford, UK.

Lack, D. (1968). *Ecological adaptations for breeding in birds*. Methuen, London, UK.

LaGrange, M. (In press-*a*). The effect of rainfall on the numbers of quelea destroyed in Zimbabwe. In *The quelea problem in southern Africa* (eds. P. J. Mundy and M.J.F. Jarvis). Baobab Books, Zimbabwe.

LaGrange, M. (In press-*b*). Past and present control methods for quelea in Zimbabwe. In *The quelea problem in southern Africa* (eds. P. J. Mundy and M.J.F. Jarvis). Baobab Books, Zimbabwe.

Lamarche, B. (1981). Liste commentée des oiseaux du Mali, Part II. *Malimbus*, **3,** 73–102.

Lamm, D. W. (1955). Local migratory movements in southern Mozambique. *Ostrich*, **26,** 32–7.

Lane, A. B. (1984). An inquiry into the response of growers to attacks by insect pests in oilseed rape (*Brassica napus* L.), a relatively new crop in the United Kingdom. *Protect. Ecol.*, **7,** 73–8.

Latigo, A.A.R. and Meinzingen, W. (1986). Guided application dose (GAD) for aerial control of quelea (*Quelea quelea*). *Proc. 1st Quelea Tech. Meet.*, Desert Locust Control Organization for Eastern Africa, Nairobi, Kenya.

Lawlor, D. W., Day, W., and Legg, B. J. (1979). Metabolism of water- stressed barley. *Field Crop Abstracts*, **32,** 944.

Lazarus, J. (1979). The early warning function of flocking in birds: An experimental study with captive quelea. *Anim. Behav.*, **27,** 855–65.

LeClerg, E. L. (1971). Field experiments for assessment of crop losses. In *Crop loss assessment methods*. FAO manual on the valuation and prevention of losses by pests, disease and weeds (ed. L. Chiarappa), pp. 2.1/1–2.1/11.

Leinati, L. (1968). Contribution to the knowledge of repellents against game birds. *Proc. 22nd Congr. Italian Soc. Vet. Sci.*, Grado, Italy, 26–29 September 1968.

Lenton, G. (1981). Qualitative and quantitative assessment of bird pests in Eastern Africa: Sudan. *Proc. 3rd Annu. Tech. Meet.*, FAO/UNDP Regional Quelea Project RAF/77/042.

Lenton, G. M. (1980). Monitoring and research on *Quelea quelea aethiopica* in Sudan 1979–1980. *Min. 2nd Annu. Tech. Meet.*, FAO/UNDP Regional Quelea Project RAF/77/042.

Leuthold, D. and Leuthold, B. (1972). Blutschnabelweber *Quelea quelea* als Beute von Greif- und Stelzvoegeln. *Vogelwarte*, **26,** 352–4.

Linz, G. M. and Fox, G. (1983). Food habits and molt of red-winged blackbirds in relation to sunflower and corn depredation. *Proc. 9th Bird Control Semin.*, Bowling Green, Ohio, **9,** 167–80.

Lofts, B. (1962). Photoperiod and the refractory period of reproduction in an equatorial bird (*Quelea quelea*). *Ibis*, **104,** 407–14.

Lofts, B. (1964). Evidence of an autonomous reproductive rhythm in an equatorial bird (*Quelea quelea*). *Nature (Lond.)*, **201,** 523–4.

Lofts, B. and Murton, R. K. (1968). Photoperiodic and physiological adaptations regulating avian breeding cycles. *J. Zool. (Lond.)*, **155,** 327–94.

Loman, J. and Tamm, S. (1980). Do roosts serve as 'information centers' for crows and ravens? *Am. Nat.*, **115,** 284–9.

Lourens, D. C. (1957). Parathion versus Quelea. CSA Symposium on Quelea, Livingstone, 1957. CCTA/CSA Joint Secretariat, Bukavu.

Lourens, D. C. (1960). Contribution: Union of South Africa. CCTA/CSA Quelea (60) 6. *CCTA/FAO Symp. on Quelea*, Bamako, 1960. Lagos, Nairobi and London. CCTA/CSA Publ. **58,** 95–118.

Lourens, D. C. (1961). Comments on the new race of the red-billed quelea. *Ostrich*, **32,** 187.

Lourens, D. C. (1963). The red-billed quelea. Unpubl. Ph.D. thesis, Pretoria University, South Africa.

Luder, R. (1985*a*). Weeds influence red-billed quelea damage to ripening wheat in Tanzania. *J. Wildl. Manage.*, **49,** 646–7.

Luder, R. (1985*b*). Guidelines to estimate the first possible installation dates of red-billed quelea colonies from daily rainfall figures. Unpubl. Internal Rep., FAO/UNDP Quelea Project URT/81/013, FAO, Rome.

Luder, R. and Elliott, C.C.H. (1984). Monitoring quelea at the national level: Tanzania. *Proc. 5th Annu. Tech. Meet.*, FAO/UNDP Regional Quelea Project RAF/81/023.

Mabbayad, B. B. and Tipton, K. W. (1975). Tannin concentration an in vitro dry matter disappearance of seeds of bird-resistant sorghum hybrids. *Philipp. Agric.*, **59,** 1–6.

MacCuaig, R. G. (1984). Terminal report of avian toxicologist. *Proc. 5th Annu. Tech. Meet.*, FAO/UNDP Regional Quelea Project RAF/81/023, pp. 19–25.

MacCuaig, R. G. (1986). Avicide index. Monograph, Food and Agriculture Organization, Rome, Italy.

Mackworth-Praed, C. W. and Grant, C. H. B. (1973). *Birds of west central and western Africa.* Ser. III, Vol. II. Longmans, London, UK.

Maclean, G. L. (1957). A summary of the birds of Westminster, O.F.S. and surroundings. *Ostrich*, **28,** 217–32.

Magor, J. (1974). Quelling the quelea-bird plague of Africa. *Spectrum*, **118,** 8–11.

Magor, J. I. and Ward, P. (1972). Illustrated descriptions, distribution maps and bibliography of the species of Quelea (weaver-birds: Ploceidae). *Trop. Pest. Bull.*, **1**, 1–23. Centre for Overseas Pest Research, London, UK.

Mallamaire, A. (1959*a*). Control of weaverbirds in Africa. *FAO Plant Protect. Bull.*, **7,** 105–12.

Mallamaire, A. (1959*b*). La lutte contre le quelea en Afrique-Occidentale francaise. *Bull. Phytosanitaire FAO*, **7,** 109–16.

Mallamaire, A. (1961). La lutte contre les oiseaux granivores en Afrique Occidentale (Mauretanie, Senegal, Soudan, Niger). *J. Agric. Trop. Bot. Appl.* **8,**141–265.

Manikowski, S. (1975). The influence of vegetation and meteorological conditions in the Lake Chad Basin on the distribution of *Quelea quelea*. Part I. Dry season. Unpubl. Internal Rep., FAO/UNDP Quelea Project RAF/73/055, FAO, Rome.

Manikowski, S. (1980). The dynamics of the Chari-Logone population of *Quelea quelea* and its control. *Proc. 4th Pan-Afr. Ornithol. Congr.*, **4,** 411–21.

Manikowski, S. (1981). Les resultats d'études sur les *Quelea quelea* dans le delta central du Niger. Unpubl. Internal Rep., FAO/UNDP Quelea Project RAF/77/047, FAO, Rome.

Manikowski, S. (1984). Birds injurious to crops in West Africa. *Trop. Pest Manage.*, **30,** 349–87.

Manikowski, S. (1988). Aerial spraying of quelea. *Trop. Pest Manage.*, **34,** 133–40.

Manikowski, S. and Da Camara-Smeets, M. (1975*a*). Estimation de dégâts d'oiseaux sur la sorgho dans la region de N'Djamena. Unpubl. Internal Rep., FAO/UNDP Quelea Project RAF/73/055, FAO, Rome.

Manikowski, S. and Da Camara-Smeets, M. (1975*b*). Observations sur les dégâts d'oiseaux dans la zone de Maroua-Lere-Pala. Unpubl. Internal Rep., FAO/UNDP Quelea Project RAF/73/055, FAO, Rome.

Manikowski, S. and Da Camara-Smeets, M. (1979*a*). Estimating bird damage to sorghum and millet in Chad. *J. Wildl. Manage.*, **43,** 540–4.

Manikowski, S. and Da Camara-Smeets, M. (1979*b*). Preferences alimentaires chez *Quelea quelea quelea* (L.). *Terre Vie*, **33,** 6ll-22.

Marshall, A. J. and Disney, H. J. de S. (1956). Photostimulation of an equatorial bird (*Quelea quelea*, Linnaeus). *Nature (Lond.)*, **177,** 143–4.

Marshall, A. J. and Disney, H. J. de S. (1957). Experimental induction of the breeding season in a xerophilous bird. *Nature (Lond.)*, **180,** 647–9.

Martin, L. (1976). Tests of bird damage control measures in Sudan, 1975. *Proc. 7th Bird Control Semin.*, Bowling Green, Ohio, **7,** 259–66.

Martin, L. R. (1979). Effective use of sound to repel birds from industrial waste ponds. *Proc. 8th Bird Control Semin.*, Bowling Green, Ohio, **8,** 71–6.

Martin, L. R. and Jackson, J. J. (1977). Field testing a bird repellent chemical on cereal crops. In *Vertebrate pest control and management Materials, ASTM STP 680* (ed. R. E. Marsh). Am. Soc. for Testing and Materials, pp. 177–85.

Mason, J. R., Glahn, J. F., Dolbeer, R. A., and Reidinger, R. F., Jr. (1985). Field evaluation of dimethyl anthranilate as a bird repellent livestock feed additive. *J. Wildl. Manage.*, **49,** 636–42.

Mason, J. R. and Reidinger, R. F. (1982). Observational learning of food aversion in red-winged blackbirds (*Agelaius phoeniceus*). *Auk*, **99,** 548–54.

Mason, J. R. and Reidinger, R. F., Jr. (1981). Effects of social facilitation and observational learning on feeding behavior of the red-winged blackbird (*Agelaius phoeniceus*). *Auk*, **98,** 778–84.

Mathew, D. N. (1976). Ecology of the weaver birds. *J. Bombay Nat. Hist. Soc.*, **73,** 249–60.

Mayo, E. S. and Lesur, J.-C. (1985). The control of quelea and other weaverbird pests by direct treatment of wheat with the avicide fenthion. Unpubl. Internal Rep., FAO/UNDP Project URT/81/013, Tanzania.

Mayr, E. (1971). *Populations, species, and evolution.* Belknap Press, Cambridge, Massachusetts.

McCourtie, W. D. (1973). Traditional farming in Liberia. Unpubl. Rep., FAO/UNDP Project, College of Agriculture and Forestry, University of Liberia.

McCullough, R. A. (1953). Supplementary whole grouse study to evaluate laboratory analysis of ruffed grouse wing and tail study. Unpubl. Internal Rep., New Hampshire Fish and Game Department, Concord, New Hampshire.

McGrath, R. M., Kaluza, W. Z., Daiber, K. H., Van der Riet, W. B., and Glennie, C. W. (1982). Polyphenols of sorghum grain, their changes during malting, and their inhibitory nature. *J. Agric. Food Chem.*, **30,** 450–6.

McLachlan, G. R. (1961). Seventh ringing report. *Ostrich*, **32,** 36–47.

McLachlan, G. R. (1962). Eighth ringing report. *Ostrich*, **33,** 29–37.

McLachlan, G. R. (1963). Ninth ringing report. *Ostrich*, **34,** 102–9.

McLachlan, G. R. (1964). Tenth ringing report. *Ostrich*, **35,** 101–10.

McLachlan, G. R. (1965). Eleventh ringing report. *Ostrich*, **36,** 214–23.

McLachlan, G. R. (1966). The first ten years of ringing in South Africa. *Ostrich [Suppl.]*, **6,** 255–63.

McLachlan, G. R. (1967). Twelfth ringing report. *Ostrich*, **38,** 17–26.

McLachlan, G. R. (1969). Thirteenth ringing report. *Ostrich*, **40,** 37–50.

McLachlan, G. R. and Liversidge, R. (1971). *Roberts birds of South Africa.* John Voelcker Bird Book Fund, South Africa.

Mead, C. J. and Watmough, B. R. (1976). Suspended moult of Trans-Saharan migrants in Iberia. *Bird Study*, **23,** 187–96.

Meanley, B. (1971). Blackbirds and the southern rice crop. *U.S. Fish Wildl. Serv. Resour. Publ.*, **100.**

Meanley, B. and Royall, W. C. (1976). Nationwide estimates of blackbirds and starlings. *Proc. 7th Bird Control Semin.*, Bowling Green, Ohio, **7,** 39–40.

Means, J. W., Jr. (1981). X-ray microanalysis of Kirtland's warbler feathers for possible population discrimination. Unpubl. M.S. thesis, Ohio State University, Columbus, Ohio.

Meinzingen, W. (1980). Development of aerial application for the control of *Quelea quelea* in Africa. Unpubl. Internal Rep., FAO/UNDP Regional Quelea Project RAF/81/023.

Meinzingen, W. (1983). Comparison study of droplet behaviour with an application rate of 2 l and 4 l/ha. Unpubl. Internal Rep. FAO/DLCO-EA.

Meinzingen, W. (1984). Effect of different application rates in quelea control in Ethiopia 1984. *Proc. 5th Annu. Tech. Meet.*, FAO/UNDP Regional Quelea Project RAF/81/023, pp. 54–6.

Meinzingen, W. and Latigo, A.A.R. (1986). A new technique for mass-marking of quelea (*Quelea quelea*). *Proc. 1st Quelea Tech. Meet.*, Desert Locust Control Organization for Eastern Africa, Nairobi, Kenya.

Mierzejewski, K. (1981). The physics of aerial and groundbased spraying for quelea control. Unpubl. Internal Rep., FAO/UNDP Regional Quelea Project URT/78/022.

Mitaru, B. N., Reichert, R. D., and Blair, R. (1983). Improvement of the nutritive value of high tannin sorghums for broiler chickens by high moisture storage (reconstitution). *Poult. Sci.*, **62,** 2065–72.

Mitchell, R. T. (1963). The floodlight trap—a device for capturing large numbers of blackbirds and starlings at roosts. *U.S. Fish Wildl. Serv. Spec. Sci. Rep. Wildl.*, **77.**

Moreau, R. E. (1960). Conspectus and classification of the ploceine weaver-birds, Part I and Part II. *Ibis*, **102,** 298–321;443–71.

Morel, G. (1965). La riziculture et les oiseaux dans la vallée du Sénégal. *Congr. Protect. Cultures Trop.*, Marseille, pp. 640–2.

Morel, G. (1968). L'impact écologique de *Quelea quelea* (L.) sur les savanes sahéliennes raisons du pullulement de ce plocéide. *Terre Vie*, **1,** 69–98.

Morel, G. and Bourlière, F. (1955). Recherches écologiques sur *Quelea quelea quelea* (L.) de la basse vallée du Sénégal. I. Données quantitatives sur le cycle annuel. *Bull. Inst. Fr. Afr. Noire Ser. A*, **17,** 617–63.

Morel, G. and Bourlière, F. (1956). Recherches écologiques sur les *Quelea quelea quelea* (L.) de la basse vallée du Sénégal. II. La reproduction. *Alauda*, **24,** 97–122.

Morel, G., Morel, M.-Y., and Bourlière, F. (1957). The blackfaced weaver bird or dioch in West Africa: an ecological study. *J. Bombay Nat. Hist. Soc.*, **54,** 811–25.

Morel, G. J. and Morel, M.-Y. (1978). Recherches écologiques sur une savane sahélienne du Ferlo septentrional, Sénégal. Etude d'une communaute avienne. *Cah. ORSTOM Ser. Biol.*, **XIII,** 3–34.

Morel, J. G. (1980). Liste commentee des oiseaux du Sénégal et de la Gambie. Suppl. No. 1, ORSTOM, Dakar.

Morel, J. G. and Morel, M.-Y. (1982). Dates de reproduction des oiseaux de Senegambie. *Bonn. Zool. Beitr.*, **33,** 249–68.

Morse, D. H. (1980). *Behavioral mechanism in ecology*. Harvard University Press, Cambridge, Massachusetts.

Moseman, A. H. (1966). Pest control: its role in the United States economy and in the world. Scientific aspects of pest control. *Natl. Acad. Sci.*, **1402,** 26–38. Washington, D.C.

Mosha, A. S. and Munisi, E. N. (1983). Focus on research for rainfed wheat production in Tanzania. *Proc. Regional Wheat Workshop East, Cent. and Southern Africa*, Arusha, Tanzania. Nakuru Press, Kenya, pp. 20–3.

Mott, D. F., Guarino, J. L., Schafer, E. W., Jr., and Cunningham, D. C. (1976). Methiocarb for preventing blackbird damage to sprouting rice. *Proc. 7th Vertebr. Pest Conf.*, Monterey, California, **7,** 22–5.

Muhammed, A. and Khan, A. (1982). Perspective of edible oils research and production in Pakistan. *Pakistan Agric. Res. Council*, Islamabad, Pakistan, Unnumbered Rep.

Munck, L., Knudsen, K. E. B., and Axtell, J. D. (1982). Industrial milling of sorghum for the 1980s. In *Sorghum in the Eighties: Proc. Int. Symp. on Sorghum* (ed. J. Mertin). ICRISAT, Patancheru, A. P., India, pp. 565–70.

Murton, R. K. (1965). Natural and artificial population control in the woodpigeon. *Ann. Appl. Biol.*, **55,** 177–92.

Murton, R. K. and Westwood, N. J. (1977). *Avian breeding cycles*. Oxford University Press, Oxford, UK.

Nakamura, K. and Matsuoka, S. (1983). The food-searching and foraging behaviours of rufous turtle dove, *Streptopelia orientalis* (Lathem), in soybean fields. *Proc. 9th Bird Control Semin.*, Bowling Green, Ohio, **9,** 161–6.

Naude, T. J. (1955*a*). The quelea problem in the Union of South Africa CCTA/CSA Africa (55)120. CSA Réunion des Specialistes du Quelea, Dakar, 1955. Secretariat Conjoint CCTA/CSA, Bukavu.

Naude, T. J. (1955*b*). Quelea control South Africa. Foreign correspondence, Vol. **I,** 1952–56. Unpubl. Rep., Govt. of South Africa, Pretoria.

Ndege, J. O. (1982). Evaluation of methiocarb efficacy in reducing bird damage to ripening wheat in Arusha-Tanzania. Unpubl. M.S. Thesis, Bowling Green State University, Bowling Green, Ohio.

Ndege, J. O. and Elliott, C.C.H. (1984). Quelea control achievements of the Tanzanian Bird Control Unit, June 1983–October 1984. *Proc. 5th Annu. Tech. Meet.*, FAO/UNDP Regional Quelea Project RAF/81/023, pp. 135–57.

Ndiaye, A. (1974). Fluctuation des populations aviaires dans la vallée du fleuve Sénégal. Unpubl. Internal Rep., FAO/UNDP Quelea Project RAF/73/055, FAO, Rome.

Ndiaye, A. (1979). OCLALAV experience in the field of bird control in West Africa. *Min. 1st Annu. Tech. Meet.*, FAO/UNDP Regional Quelea Project RAF/77/042.

Neth, J. W. (1971). Identifying natal areas of Ohio-hatched Canada geese by neutron activation analyses. Unpubl. M.S. thesis, Ohio State University, Columbus, Ohio.

Newby, J. (1980). The birds of Ouadi Rime-Ouadi Achim Faunal Reserve. A contribution to the study of the Chadian avifauna, Part II. *Malimbus* **2,** 29–50.

Newton, I. (1967). The adaptive radiation and feeding ecology of some British finches. *Ibis*, **109,** 33–98.

Newton, I. (1968). Bullfinches and fruit buds. In *The problems of birds as pests* (eds. R. K. Murton and E. N. Wright). Academic Press, London, pp. 199–209.

Nice, M. M. (1953). The question of ten day incubation periods. *Wilson Bull.*, **65,** 81–93.

Nicolaus, L. K., Cassel, J. F., Carlson, R. B., and Gustavson, C. R. (1983). Taste aversion conditioning of crows to control predation on eggs. *Science*, **220,** 212–4.

Nikolaus, G. (1981). Wir und die Voegel. *Deutsches Buchverzeichnis* **13,** 16.

Nilsson, G. (1981). *The bird business—A study of the commercial cage bird trade.* Animal Welfare Institute, Washington, D.C.

Nur, N. (1984). The consequences of brood size for breeding blue tits. I. Adult survival, weight change and the costs of reproduction. *J. Anim. Ecol.*, **53,** 479–96.

Orians, G. H. (1961). The ecology of blackbird (*Agelaius*) social systems. *Ecol. Monogr.*, **31,** 285–312.

ORSTOM. (1970). Monographie hydrologique de Bassin du Niger. 2ème partie La Cuvette Lacustre. Unpubl. Rep., Office de la Recherche Scientifique et Technique Outre-Mer, Paris.

Oswalt, O. L. (1975). Estimating the biological effects of tannins in grain sorghum. *Proc. Int. Sorghum Workshop*, (ed. Publication Staff). U.S. Agency for International Development, Washington, D.C., pp. 530–54.

Otis, D. L. (1984). A method for estimating sorghum loss to birds over large areas of Eastern Africa. Unpubl. Consultancy Rep., RAF/81/023, to FAO/UNDP, Rome, Italy.

Otis, D. L., Holler, N. R., Lefebvre, P. W., and Mott, D. F. (1983). Estimating bird damage to sprouting rice. In *Vertebrate pest control and management materials, ASTM STP 817* (ed. D. E. Kaukeinen). Am. Soc. for Testing and Materials, pp. 76–89.

Otis, D. L., Knittle, C. E., and Linz, G. M. (1986). A method for estimating turnover in spring blackbird roosts. *J. Wildl. Manage.*, **50,** 567–71.

Park, P. O. (1973). Attacks by bird enemies of rice and their control. Plant protection for the rice crop. *Proc. Semin. Liberia.* Unpubl. Rep., FAO/UNDP, Rome, Italy.

Park, P. O. (1974). Granivorous bird pests in Africa; towards integrated control. *Span*, **17,** 126–8.

Park, P. O. (1975). The socio-economic effects of the control of grain-eating birds. Unpubl. Internal Rep., FAO/UNDP Quelea Project RAF/73/055, FAO, Rome.

Park, P. O., Adam, J., and Lubazo, R. (1975). Trials of repellency for the protection of sorghum at Deli. Unpubl. Internal Rep., FAO/UNDP Quelea Project RAF/73/055, FAO, Rome.

Park, P. O. and Adam, J. A. (1976). Trials of repellents for the protection of rice in the ear—Cameroons, 1975. Unpubl. Internal Rep., FAO/UNDP Quelea Project RAF/73/055, FAO, Rome.

Park, P. O. and Assegninou, W. (1973). Trials at Deli of chemical repellents to protect sorghum against grain-eating birds. Unpubl. Internal Rep., FAO/UNDP Quelea Project RAF/73/055, FAO, Rome.

Parker, J. D. (1986). A novel sprayer for the control of quelea birds. *Trop. Pest Manage.*, **32,** 243–5.

Parker, J. D. and Casci, F. M. (1983). Report of a consultancy carried out with URT/81/013-FAO/UNDP Quelea Bird Control Project, Arusha, Tanzania.

Parrish, J. R., Rogers, D. T., Jr., and Prescott Ward, F. (1983). Identification of natal locales of peregrine falcons (*Falco peregrinus*) by trace-element analysis of feathers. *Auk*, **100,** 560–7.

Pavlov, A. N. and Kolesnik, T. I. (1979). The attracting ability of caryopses as one of the factors determining level of protein accumulation in wheat grain. *Field Crop Abstracts*, **32,** 394.

Payne, R. B. (1972). Mechanics and control of molt. In *Avian biology*, Vol. 2 (eds. D. S. Farner and J. R. King). Academic Press, New York and London.

Payne, R. B. (1980). Seasonal incidence of breeding, moult and local dispersal of red-billed firefinches *Lagonosticta senegala* in Zambia. *Ibis*, **122,** 43–56.

Peña, M. (1977). Proposal for studying and establishing a control program for Madam Sarah (*Ploceus cucullatus*) in Hispaniola. National Zoological Park, Santo Domingo, Dominican Republic.

Pepper, S. R. (1973). Observations on bird damage and traditional bird pest control methods on ripening sorghum. Unpubl. Internal Rep., FAO/UNDP Quelea Project RAF/73/055, FAO, Rome.

Perrins, C. M. (1970). The timing of birds' breeding seasons. *Ibis*, **112,** 242–55.

Perumal, R. S. and Subramaniam, T. R. (1973). Studies on panicle characteristics associated with bird resistance in sorghum. *Madras Agric. J.* **60,** 256–8.

Pienaar, V. de V. (1969). Observations on the nesting habits and predators of breeding colonies of red-billed quelea, *Quelea quelea lathami*, in the Kruger National Park. *Bokmakierie*, **21,** [Suppl.],11–5.

Pinowski, J. (1973). The problem of protecting crops against harmful birds in Poland. *European and Mediterranean Plant Protect. Organization OEPP/EPPO Bull.*, **3,** 107–10.

Pinowski, J., Tomek, T., and Tomek, W. (1972). Food selection in the tree sparrow, *Passer m. montanus* (L.). Prelim. Rep. In *Productivity, population dynamics and systematics of granivorous birds* (eds. S. C. Kendeigh and J. Pinowski). Polish Scientific Publishers, Warszawa, Poland, pp. 263–73

Pitman, C.R.S. (1957). Further notes on aquatic predators of birds. *Bull. Br. Ornithol. Club*, **77,** 89–97,105–10,122–6.

Pitman, C.R.S. (1961). More aquatic predators of birds. *Bull. Br. Ornithol. Club*, **81,** 57–62,78–81,l05–6.

Plowes, D.C.H. (1950). The red-billed quelea—a problem for grain- sorghum growers. *Rhod. Agric. J.*, **47,** 98–101.

Plowes, D.C.H. (1953). Report on red-billed queleas, Oct. 1952–Feb. 1953. Unpubl. Rep., Nyamandhlovu.

Plowes, D.C.H. (1955). Queleas in Southern Rhodesia, CCTA/CSA Africa (55) 121. CSA Réunion des Specialistes du Quelea, Dakar, 1955. Secretariat Conjoint CCTA/CSA., Bukavu.

Poché, R. M., Karim, Md. A., and Haque, Md. E. (1980). Bird damage control in sprouting wheat. *Bangladesh J. Agric. Res.*, **5**,41–6.

Pope, G. G. and King, W. J. (1973). Spray trials against the Red-billed Quelea (*Quelea quelea*) in Tanzania. Misc. Rep. 12, Centre for Overseas Pest Research, London, UK.

Pope, G. G. and Ward, P. (1972). The effects of small applications of an organophosphorus poison, fenthion, on the weaver-bird *Quelea quelea. Pestic. Sci.*, **3,** 197–205.

Power, D. M. (1970). Geographic variation of red-winged blackbirds in central North America. *Univ. of Kansas Publ. of the Mus. Nat. Hist.* **19,** 1–83.

Prakash, I. (1982). Vertebrate pest problems in India. In *Proc. Conf. on The Organisation and Practice of Vertebrate Pest Control* (ed. A. C. Dubock). Imperial Chemical Industries PLC, Dramrite Printers Ltd., London, pp. 29–35.

Price, M. L., Butler, L. G., Rogler, J. C., and Featherston, W. R. (1979). Overcoming the nutritionally harmful effects of tannin in sorghum grain by treatment with inexpensive chemicals. *J. Agric. Food Chem.*, **27,** 441–5.

Price, M. L., Van Scoyoc, S., and Butler, L. G. (1978). A critical evaluation of the vanillin reaction as an assay for tannin in sorghum grain. *J. Agric. Food Chem.*, **26,** 1214–8.

Prozesky, O. P. (1964). Comprehensive bird concentration at Lake Ngami. *Afr. Wildl.*, **18,** 137–42.

Pulliam, H. R. (1973). On the advantages of flocking. *J. Theor. Biol.*, **38,** 419–22.

Pulliam, H. R. (1975). Diet optimization with nutrient constraints. *Am. Nat.*, **109,** 765–8.

Pyke, G. H., Pulliam, H. R., and Charnov, E. L. (1977). Optimal foraging: a selective review of theory and tests. *Q. Rev. Biol.*, **52,** 137–54.

Quesnel, V. C. (1968). Fractionation and properties of the polymeric leucocyanidin of the seeds of *Theobrama cacao. Phytochemistry*, **7,** 1583–92.

Raju, A. S. and Shivanarayan, N. (1980). Extent of damage in some early rice varieties due to bird pests at Maruteru. *Int. Rice Comm. Newsl.*, **29,** 44–5.

Ramachandra, G., Virupaksha, T. K., and Shadaksharaswamy, M. (1977). Relationship between tannin levels and in vitro protein digestibility in finger millet (*Eleusine coracana* Gaertn.). *J. Agric. Food Chem.*, **25,** 1101–4.

Rattray, J. M. (1960). Tapis gramineens d'Afrique. Etudes Agricoles No. **49.** FAO, Rome.

Reichert, R. D., Fleming, S. E., and Schwab, D. J. (1980). Tannin deactivation and nutritional improvement of sorghum by anaerobic storage of H_2O, HCl, or NaOH-treated grain. *J. Agric. Food Chem.*, **28,** 824–9.

Reichert, R. D. and Youngs, C. G. (1977*a*). Dehulling cereal grains and grain legumes for developing countries. I. Quantitative comparison between attrition- and abrasive-type mills. *Cereal Chem.*, **53,** 829–39.

Reichert, R. D. and Youngs, C. G. (1977*b*). Dehulling cereal grains and grain legumes for developing countries. II. Chemical composition of mechanically and traditionally dehulled sorghum and millet. *Cereal Chem.*, **54,** 174–8.

Ricklefs, R. E. (1973). Fecundity, mortality and avian demography. In *Breeding biology of birds* (ed. D. S. Farner). National Research Council, Washington, D.C.

Roberts, N. (1909). *Pyromelana oryx* and its nesting parasites. *J. South Afr. Ornithol. Union*, **5,** 22–4.

Roberts, T. J. (1974). Bird damage to farm crops in Pakistan with special reference to sunflower (*Helianthus annus*). Vertebrate Pest Control Centre, Karachi, Pakistan.

Roberts, T. J. (ed.) (1981). *Handbook of Vertebrate Pest Control in Pakistan.* Pakistan Agric. Res. Council and Food and Agriculture Organization of the United Nations, Vertebrate Pest Control Research Centre, Karachi, Pakistan.

Rogers, J. G., Jr. (1974). Responses of caged red-winged blackbirds to two types of repellents. *J. Wildl. Manage.*, **38,** 118–23.

Rogers, J. G., Jr. (1978*a*). Repellents to protect crops from vertebrate pests: some considerations for their use and development. In *Flavor chemistry of animal foods* (ed. R. W. Bullard). ACS Symp. Ser. No. 67, American Chemical Society, Washington, DC. pp. 150–84.

Rogers, J. G., Jr. (1978*b*.) Some characteristics of conditioned aversion in red-winged blackbirds. *Auk*, **95,** 362–9.

Rogers, J. G., Jr. (1980). Conditioned taste aversion: its role in bird damage control. In *Bird problems in agriculture* (eds. E. N. Wright, I. R. Inglis, and C. J. Feare). British Crop Protection Council (BCPC) Publications, Croydon, UK, pp. 173–9.

Rooke, I. J. (1983). Conditioned aversion by Silvereyes *Zosterops lateralis* to food treated with methiocarb. *Bird Behav.*, **4,** 86–9.

Rooney, L. W. and Murty, D. S. (1982). Color of sorghum food products. In *Proc. Int. Symp. Sorghum Grain Quality* (eds. L. W. Rooney and D. S. Murty). ICRISAT, Patancheru, A.P., India, pp. 323–7.

Rosa Pinto, A. A., da. (1960). O problema 'Quelea' e a agricultura em Angola. *Melhoramento*, **13,** 79–113.

Rosa Pinto, A. A., da and Lamm, D. W. (1960). Memorias do Museu Dr Alvaro de Castro, no. 5. Lourenco Marques.

Rowan, M. K. (1964). An analysis of the records of a South African ringing station. *Ostrich*, **35,** 160–87.

Royama, T. (1966). A re-interpretation of courtship feeding. *Bird Study*, **13,** 116–29.

Royama, T. (1970). Factors governing the hunting behaviour and selection of food by the great tit (*Parus major* L.). *J. Anim. Ecol.*, **39,** 619–68.

Ruelle, P. and Bruggers, R. L. (1979). Evaluating bird protection to mechanically sown rice seed treated with methiocarb at Nianga, Senegal, West Africa. In *Vertebrate pest control and management materials, ASTM STP 680* (ed. J. R. Beck). Am. Soc. for Testing and Materials, pp. 211–6.

Ruelle, P. and Bruggers, R. L. (1982). Traditional approaches for protecting cereal crops from birds in Africa. *Proc. 10th Vertebr. Pest Conf.*, Monterey, California, **10,** 80–6.

Ruelle, P. and Bruggers, R. L. (1983). Senegal's trade in cage birds 1979–81. *U.S. Fish Wildl. Serv. Wildl. Leafl.*, **515.**

Ruelle, P. J. (1983). Control of granivorous bird pests of rice using the partial crop treatment method in West Africa. *Trop. Pest Manage.*, **29,** 23–6.

Ryan, J. (1981). Songbird stew. *Int. Wildl.*, **11,** 44–8.

Salvan, J. (1967). Contribution a l'étude des oiseaux du Tschad. *Oiseau Rev. Fr. Ornithol.*, **37,** 255–84.

Salvan, J. (1969). Contribution a l'étude des oiseaux du Tschad. *Oiseau Rev. Fr. Ornithol.*, **39,** 38–69.

Schafer, E. W., Jr. (1972). The acute oral toxicity of 369 pesticidal, pharmaceutical and other chemicals to wild birds. *Toxicol. Appl. Pharmacol.*, **21,** 315–30.

Schafer, E. W., Jr. (1979). Registered bird damage chemical controls. *Pest Control*, **47,** June:36–9.

Schafer, E. W., Jr. (1981). Bird control chemicals—nature, modes of action, and toxicity. In *CRC handbook of pest management in agriculture* (ed. D. Pimentel) Vol. III. CRC Series in Agriculture, Boca Raton, Florida, pp. 129–39.

Schafer, E. W., Jr. and Brunton, R. B. (1971). Chemicals as bird repellents: two promising agents. *J. Wildl. Manage.*, **35,** 569–72.

Schafer, E. W., Jr., Brunton, R. B., Lockyer, N. F., and De Grazio, J. W. (1973). Comparative toxicity of seventeen pesticides to the *Quelea*, house sparrow and red-winged blackbird. *Toxicol. Appl. Pharmacol.*, **26,** 154–7.

Schildmacher, H. (1929). Über den Wärmehaushalt kleiner Körnerfresser. *Ornithol. Monatsber.*, **37,** 102–6.

Schmutterer, H. (1969). *Pests of crops in north-east and central Africa.* Gustav Fisher, Stuttgart, West Germany, and Portland, USA.

Schuler, W. (1980). Factors influencing learning to avoid unpalatable prey in birds relearning new alternative prey and similarity of appearance of alternative prey. *Z. Tierpsychol.*, **54,** 105–43.

Seber, G. A. (1970). Estimating time-specific survival and reporting rates for adult birds from band returns. *Biometrika*, **57,** 313–8.

Seber, G.A.F. (1973). *The estimation of animal abundance and related parameters.* Charles Griffin, London, UK.

Selander, R. K. and Johnston, R. F. (1967). Evolution in the house sparrow. I. Intrapopulation variation in North America. *Condor*, **69,** 217–58.

Sengupta, S. (1973). Significance of communal roosting in the common mynah, *Acridotheres tristis* (L.). *J. Bombay Nat. Hist. Soc.*, **70,** 204–6.

Serrurier, A. (1965). Ecologie du *Quelea quelea quelea. Congr. Protect. Cultures Trop.*, Marseille, pp. 643–5.

Serrurier, A. (1966). La lutte anti-aviaire en Afrique sahelienne. *Mach. Agric. Trop.*, **13,** 28–33.

Shannon, J. G. and Reid, D. A. (1976). Awned vs awnless isogenic winter barley grown at three environments. *Crop Sci.*, **16,** 347–9.

Shefte, N., Bruggers, R. L., and Schafer, E. W., Jr. (1982). Repellency and toxicity of three bird control chemicals to four species of African grain-eating birds. *J. Wildl. Manage.*, **46,** 453–7.

Shepherd, A. D. (1981). How a typical sorghum peels. *Cereal Chem.*, **58,** 303–6.

Shivanarayan, N. (1980). Role of birds in agriculture. *Souvenir: Int. Meet. on Wild. Resources in Rural Development.* July 7–11, 1980, Hyderabad, India, pp. 25–30.

Shumake, S. A., Gaddis, S. E., and Garrison, M. V. (1983). Development of a preferred bait for quelea control. In *Vertebrate pest control and management materials: 4th Symp., ASTM STP 817* (ed. D. E. Kaukeinen). Am. Soc. for Testing and Materials, pp. 118–26.

Shumake, S. A., Gaddis, S. E., and Schafer, E. W., Jr. (1976). Behavioral response of quelea to methiocarb (Mesurol®). *Proc. 7th Bird Control Semin.*, Bowling Green, Ohio, **7,** 250–4.

Sinclair, A.R.E. (1978). Factors affecting the food supply and breeding season of resident birds and movements of Palaearctic migrants in a tropical African savannah. *Ibis*, **120,** 480–97.

Slater, P.J.B. (1980). Bird behaviour and scaring by sounds. *Proc. Bird Problems in Agric. Symp.* (eds. E. N. Wright, I. R. Inglis, and C. J. Feare). University of London, BCPC Publ., pp. 105–20.

Smith, J.N.M. and Sweatman, H.P.A. (1974). Food-searching behavior of titmice in patchy environments. *Ecology*, **55,** 1216–32.

Sonnier, J. (1957). Report on the action taken during 1956, in Senegal and Mauritania by the Department for Bird Control. *CSA Symp. Quelea*, Livingstone, 1957. CCTA/CSA Joint Secretariat, Bukavu.

Stewart, D. R. (1959). The red-billed quelea in northern Rhodesia. *North Rhod. J.*, **4,** 55–62.

Stickley, A. R., Otis, D. L., Bray, O. E., Heisterberg, J. F., and Grandpre, T. F. (1979*a*). Bird and mammal damage to mature corn in Kentucky and Tennessee. *Proc. Annu. Conf. Southeast. Assoc. Fish. Wildl. Agencies*, **32,** 228–33.

Stickley, A. R., Jr., Otis, D. L., and Palmer, D. T. (1979*b*). Evaluation and results of a survey of blackbird and mammal damage to mature field corn over a large (three-state) area. In *Vertebrate pest control and management materials, ASTM STP 680* (ed. J. R. Beck). Am. Soc. for Testing and Materials, pp. 169–77.

Stone, C. P. and Mott, D. F. (1973*a*). Bird damage to sprouting corn in the United States. *U.S. Fish Wildl. Serv., Spec. Sci. Rep. Wildl.*, **173.**

Stone, C. P. and Mott, D. F. (1973*b*). Bird damage to ripening field corn in the United States, 1971. *U.S. Fish Wildl. Serv. Wildl. Leaf.*, **505,** 1–8.

Stone, R. J. (1976). Chemical repellents can save crops. *World Crops*, May/June, pp. 132–3.

Stresemann, E. (1965). Die Mauser der Huehnervoegel. *J. Ornithol.*, **106,** 58–64.

Stroosnijder, L. and van Hempst, H.D.J. (1982). La meteorologie du sahel et du terrain d'etude. In *La productivité des paturages sahéliens* (eds. F.W.T. Penning De Vries and M. A. Djiteye). Centre for Agricultural Publishing and Documentation, Wageningen, pp. 37–51.

Sultana, P., Brooks, J. E., and Bruggers, R. L. (1986). Repellency and toxicity of bird control chemicals to pest birds in Bangladesh. *Trop. Pest Manage.* **32,** 246–8.

Taber, R. D. and Cowan, I. McT. (1969). Capturing and marking wild animals. In *Wildlife management techniques* (ed. R. H. Giles). Wildlife Society, Washington, D.C., pp. 227–318.

Tarboton, W. (1987). Redbilled Quelea spraying in South Africa. *Gabar*, **2,** 38–9.

Taylor, L. E. (1906). The birds of Irene, near Pretoria, Transvaal. *J. S. Afr. Ornithol. Union*, **2,** 55–83.

Thiollay, J. M. (1975). Exemple de predation naturelle sur une population nicheuse de *Quelea qu. quelea* L. au Mali. *Terre Vie*, **29,** 31–54.

Thiollay, J. M. (1978*a*). Production et taux de mortalité dans les colonies de *Quelea quelea* (Aves: Ploceidae) en Afrique Centrale. *Trop. Ecol.*, **19,** 7–24.

Thiollay, J. M. (1978*b*). Les migrations des rapaces en Afrique occidentale; adaptations écologiques aux fluctuations de production des écosystèmes. *Terre Vie*, **32,** 89–133.

Thompson, B. W. (1965). *The climate of Africa.* Oxford University Press, London, UK.

Thompson, J. and Jaeger M. M. (1984). Regional mass-marking and fingerprinting analysis during 1984. *Proc. 5th Annu. Tech. Meet.*, FAO/UNDP Regional Quelea Project RAF/81/023.

Thomsett, S. (1987). Raptor deaths as a result of poisoning quelea in Kenya. *Gabar*, **2,** 33–8.

Tinbergen, J. M. and Drent, R. H. (1980). The starling as a successful forager. In *Bird problems in agriculture* (eds. E. N. Wright, I. R. Inglis, and C. J. Feare). BCPC Publications, Croydon, England, pp. 83–97.

Traylor, M. A. (1963). *Check-list of Angolan birds.* Museu do Dundo, Lisbon.

Treca, B. (1976). Les oiseaux d'eau et la riziculture dans le Delta du Sénégal. *Oiseau Rev. Fr. Ornithol.*, **45,** 259–65.

Tree, A. J. (1962). The birds of the Leopardshill area of the Zambesi escarpment. *Ostrich*, **33,** 3–23.

Uk, S. and Munks, S. (1984). Fenthion residues in quelea birds from experimental aerial spraying of Queletox at Makayuni, Tanzania, in June 1983. International Centre for the Application of Pesticides, Cranfield Institute of Technology, UK.

UNESCO. (1959). Carte de la vegetation de l'Afrique au sud du tropique du cancer. Unpubl. Rep., United Nations Educational, Scientific, and Cultural Organization (UNESCO).

Urban, E. K. and Brown, L. H. (1971). *A checklist of the birds of Ethiopia.* Haile Sellassie I University Press, Addis Ababa, Ethiopia.

Van Ee, C. A. (1973). Cattle egrets prey on breeding queleas. *Ostrich*, **44,** 136.

Van Someren, V.G.L. (1922). Notes on the birds of East Africa. *Novit. Zool.*, **29,** 1–246.

Vernon, C. J. (In press). The quelea in natural ecosystems. In *The quelea problem in southern Africa* (eds. P. J. Mundy and M.J.F. Jarvis). Baobab Books, Zimbabwe.

Vesey-FitzGerald, D. F. (1958). Notes on the breeding colonies of the red-billed quelea in S. W. Tanganyika. *Ibis*, **100,** 167–74.

Voss, F. (1986). *ATLAS: Quelea habitats in East Africa.* Food and Agriculture Organization, Rome

Walsberg, G. E. and King, J. R. (1980). The thermoregulatory significance of the winter roost-sites selected by robins in eastern Washington. *Wilson Bull.*, **92,** 33–9.

Ward, P. (1965*a*). Feeding ecology of the black-faced dioch *Quelea quelea* in Nigeria. *Ibis*, **107,** 173–214.

Ward, P. (1965*b*). The breeding biology of the black-faced dioch *Quelea quelea* in Nigeria. *Ibis*, **107,** 326–49.

Ward, P. (1965*c*). Biological implications of quelea control in West Africa. *Congres de la Protection des Cultures Tropicales* 661-6, Marseilles.

Ward, P. (1965*d*). Seasonal changes in the sex ratio of *Quelea quelea* (Ploceidae). *Ibis*, **107,** 397–9.

Ward, P. (1966). Distribution, systematics, and polymorphism of the African weaver-bird (*Quelea quelea*). *Ibis*, **108,** 34–40.

Ward, P. (1969). The annual cycle of the yellow-vented bulbul *Pycnonotus goiavier* in a humid equatorial environment. *J. Zool. (Lond.)*, **157,** 25–45.

Ward, P. (1971). The migration patterns of *Quelea quelea* in Africa. *Ibis*, **113,** 275–97.

Ward, P. (1972*a*). East Africa tropical bird-pest research project. Final Rep., Centre of Overseas Pest Research. ODA Res. Scheme R. 2092.

Ward, P. (1972*b*). Synchronisation of the annual cycle within populations of *Quelea quelea* in East Africa. *Proc. Int. Ornithol. Congr.*, **15,** 702–3.

Ward, P. (1973*a*). A new strategy for the control of damage by queleas. *PANS*, **19,** 97–106.

Ward, P. (1973*b*). *Manual of techniques used in research on quelea birds.* AGP:RAF/67/087 Working Paper (Manual), United Nations Development Programme/FAO, Rome.

Ward, P. (1978). The role of the crop among red-billed quelea *Quelea quelea. Ibis*, **120,** 333–7.

Ward, P. (1979). Rational strategies for the control of queleas and other migrant bird pests in Africa. *Philos. Trans. R. Soc. Lond. B Biol. Sci.*, **287,** 289–300.

Ward, P. and Jones, P. J. (1977). Pre-migratory fattening in three races of the red-billed quelea *Quelea quelea* (Aves: Ploceidae), an intra-tropical migrant. *J. Zool. (Lond.)*, **181,** 43–56.

Ward, P. and Kendall, M. D. (1975). Morphological changes in the thymus of young and adult red-billed queleas *Quelea quelea* (Aves). *Philos. Trans. R. Soc. Lond. B Biol. Sci.*, **273,** 55–64.

Ward, P. and Pope, G. G. (1972). Flight-tunnel experiments with red-billed queleas to determine the distribution of a solution sprayed onto birds in flight. *Pestic. Sci.*, **3,** 709–14.

Ward, P. and Zahavi, A. (1973). The importance of certain assemblages of birds as 'information centres' for food finding. *Ibis*, **115,** 517–34.

WARDA. (1983). Preliminary analysis of socio-economic baseline data. West Africa Rice Development Association WARDA/TAT/83/ARR-8A.

Weatherhead, P. J. (1983). Two principal strategies in avian communal roosts. *Am. Nat.*, **121,** 237–43.

Weatherhead, P. J., Tinker, S., and Greenwood, H. (1982). Indirect assessment of avian damage to agriculture. *J. Appl. Ecol.*, **19,** 773–82.

Weidner, T. (1983). Why do pesticides cost so much $$$? *Pest Control Technol.*, July, pp. 50–2,76.

Wiens, J. A. and Dyer, M. I. (1975). Simulation modeling of red-winged blackbird impact on grain crops. *J. Appl. Ecol.*, **12,** 63–82.

Wiens, J. A. and Dyer, M. I. (1977). Assessing the potential impact of granivorous birds in ecosystems. In *Granivorous birds in ecosystems* (eds. J. Pinowski and S. C. Kendeigh). Cambridge University Press, Cambridge, England, pp. 205–66.

Wiens, J. A. and Johnston, R. F. (1977). Adaptive correlates of granivory in birds. In *Granivorous birds in ecosystems* (eds. J. Pinowski and S. C. Kendeigh). Cambridge University Press, Cambridge, England, pp. 301–340.

Wilkinson, G. S. and English-Loeb, G. M. (1982). Predation and coloniality in cliff swallows (*Petrochelidon pyrrhonota*). *Auk*, **99**, 459–67.

Williams, J. G. (1954). The quelea threat to Africa's grain crops. *East Afr. Agric. J.*, **19,** 133–6.

Wilson, S. W. (1978). Food size, food type, and foraging sites of red-winged blackbirds. *Wilson Bull.*, **90,** 511–20.

Winstanley, D., Spencer, R., and Williamson, K. (1974). Where have all the whitethroats gone? *Bird Study*, **21,** 1–14.

Wolfson, A. and Winchester, D. P. (1959). Effect of photoperiod on the gonadal cycle in an equatorial bird *Quelea quelea. Nature (Lond.)*, **184,** 1658–9.

Woronecki, P. P. and Dolbeer, R. A. (1980). The influence of insects in bird damage control. *Proc. 9th Vertebr. Pest Conf.*, Fresno, California, **9,** 53–9.

Woronecki, P. P., Dolbeer, R. A., and Stehn, R. A. (1981). Response of blackbirds to Mesurol and Sevin applications on sweet corn. *J. Wildl. Manage.*, **45,** 693–701.

Woronecki, P. P., Stehn, R. A., and Dolbeer, R. A. (1980). Compensatory response of maturing corn kernels following simulated damage by birds. *J. Appl. Ecol.*, **17,** 737–46.

Worthing, C. R. (Ed.) (1979). *The pesticide manual—a world compendium*, 6th edn. British Crop Protection Council, UK.

Wright, E. N. (1981). Chemical repellents—a review. In *Bird problems in agriculture, Proc. Conf. 'Understanding agricultural bird problems'* (eds. E. N. Wright, I. R. Inglis and C. J. Feare). Royal Holloway College, University of London, April 4–5, 1979, pp. 164–72.

Yahia, G. (1957). A note on the occurrence and control of the Red-billed Weaver (*Quelea quelea aethiopica*) in the Sudan. *CSA Symp. Quelea*, Livingstone, 1957. CCTA/CSA Joint Secretariat, Bukavu.

Yates, F. and Zacopanay, B. A. (1935). The estimation of the efficacy of sampling, with special reference to sampling for yield in cereal experiments. *J. Agric. Sci.*, **25,** 545–77.

Yom-Tov, Y., Imber, A., and Otterman, J. (1977). The microclimate of winter roosts of the starling *Sturnus vulgaris. Ibis*, **119,** 366–8.

York, J. O., Howe, D. F., Bullard, R. W., Nelson, T. S., and Stallcup, O. T. (1981). The purple testa in sorghum, *Sorghum bicolor* (L.) Moench. *Proc. 12th Biennial Grain Sorghum Research Utilization Conf.* (ed. D. E. Weibel). Grain Sorghum Producers Association and Texas Grain Sorghum Producers Board, Lubbock, Texas, p. 113.

York, J. O., Bullard, R. W., Nelson, T. S., and Stallcup, O. T. (1983). Dry matter digestibility in purple testa sorghums. *Proc. 37th Annu. Corn Sorghum Research Conf.*, Chicago, Illinois (eds. H. T. Loden and D. Wilkinson). American Seed Trade Association, Washington, D.C., pp. 1–9.

Zahavi, A. (1971). The function of pre-roost gatherings and communal roosts. *Ibis*, **113**, 106–9.

Zaske, J. (1973). Tropfengroessenanalyse unter besonderer Beruecksichtigung der Zerstaeubung im chemischen Pflanzenschutz. Dissert. Tech. Univ. Berlin.

Zeinelabdin, M. H. (1980). The potential of vegetable tannin as a bird repellent. Unpubl. M.A. thesis, Bowling Green State University, Bowling Green, Ohio.

Zeinelabdin, M. H., Bullard, R. W., and Jackson, W. B. (1983). Mode of repellent activity of condensed tannin to quelea. *Proc. 9th Bird Control Semin.*, Bowling Green, Ohio, **9**, 241–6.

Index